# Regulatory Toxicology

# Regulatory Toxicology
## Second edition

Shayne C. Gad

London and New York

First edition published 1995 by Raven Press, New York.

Second edition published 2001 by Taylor & Francis Ltd.
11 New Fetter Lane, London EC4P 4EE

Simultaneously published in the USA and Canada
by Taylor & Francis
29 West 35th Street, New York, NY 10001

*Taylor & Francis is an imprint of the Taylor & Francis Group*



Typeset in Garamond by Deerpark Publishing Services, Shannon
Printed and bound in Great Britain by TJ International Ltd, Padstow, Cornwall



Every effort has been made to ensure that the advice and information in this book is true and accurate at the time of going to press. However, neither the publisher nor the authors can accept any legal responsibility or liability for any errors or omissions that may be made. In the case of drug administration, any medical procedure or the use of technical equipment mentioned within this book you are strongly advised to consult the manufacturer's guidelines.

*British Library Cataloguing in Publication Data*
A catalogue record for this book has been requested

*Library of Congress Cataloging in Publication Data*
Regulatory toxicology/[edited by] Shayne C. Gad.--2nd ed.
p. ; cm.
Includes bibliographical references and index.
(hbk : alk. Paper)
1. Poisons--Law and legislation--United States. 2. Chemicals--Law and legislation--United States. 3. Toxicity testing--Law and legislation--United States. I. Gad, Shayne C., 1948-

KF3958.R44 2001
363.17'91'0973--dc21 00–066799

ISBN 0-415-23919-2

To Samantha, Katina and Jake: I am so proud of how you have grown. Though in Dad's heart you will always be loved as the children you were, you have become adults so soon.

And to the love of my life, Joyce.

# Contents

# Contributors

**Charles B. Breckenridge**, Novartis Crop Protection AG, 410 Swing Road, Greensboro, NC 27409, USA.

**Christopher P. Chengelis**, WIL Research Laboratories, Inc., 1487 George Rd, Ashland, OH 44805-9281, USA.

**Joseph C. DiNardo**, Pharma Cosmetix Research, LLC., Colonial Heights, VA 23834, USA.

**Patricia Frank**, Patricia Frank & Associates, Inc., 1310 Brummel Road, Evanston, IL 60202, USA.

**Vasilios H. Frankos**, ENVIRON International Corporation, Arlington, VA, USA.

**Shayne C. Gad**, Gad Consulting Services, 1818 White Oak Road, Raleigh, NC 27608, USA.

**Judith L. Garfield**, Stephen B. Harris Group, 6109 Madra Avenue, San Diego, CA 92120-3905, USA.

**Stephen B. Harris**, Stephen B. Harris Group, 6109 Madra Avenue, San Diego, CA 92120-3905, USA.

**Debora L. Hooper**, The Lubrizol Corporation, 29400 Lakeland Blvd., Wickliffe, OH 44092, USA.

**Richard C. Kraska**, The Lubrizol Corporation, 29400 Lakeland Blvd., Wickliffe, OH 44092, USA.

**Joseph V. Rodricks**, The Life Sciences Consultancy, 750 17th Street, N.W. Suite 1000, Washington, DC 20006, USA.

**James H. Schafer**, Schafer Veterinary Consultants, 800 Helena Court, Fort Collins, CO 80524, USA.

**Charles B. Spainhour**, The Summit Group, P.O. Box 756, Waverly, PA 18471, USA.

**James T. Stevens**, Novartis Crop Protection AG, 410 Swing Road, Greensboro, NC 27409, USA.

# Preface

Toxicology has made tremendous strides in the sophistication of the models used to identify and understand the mechanisms of agents that can harm or kill humans and other higher organisms. Early on, other people were used as surrogates for monarchs or others. Other animals then came to be used, and, until recently, this use, while becoming increasingly refined, also came to serve as the "gold standard" against which truth (at least in regulatory, legal, and economic senses) was judged.

Nonanimals or *in vitro* models timely started to gain significant use in the 1960s. For reasons of concern about animal welfare, economics, and the need for greater sensitivity and understanding of mechanisms, interest in *in vitro* models has increased.

As the contents of this volume demonstrate, there now exists a broad range of *in vitro* models for use in either identifying or understanding most forms of toxicity. The availability of *in vitro* models spans both the full range of endpoints (irritation, sensitization, lethality, mutagenicity, and developmental toxicity) and the full spectrum of target organ systems (skin, eye, heart, liver, kidney, nervous system, etc.). This volume devotes chapters to each of these specialty areas from a perspective of presenting the principal models and their uses and limitations.

Chapters that overview the principles involved in the general selection and use of models, and that address the issues of safety concerns and regulatory acceptance of these methods are also included.

By the time this book sees print, as in any such volume, portions will be dated but not obsolete. The authors and I hope this will provide a sound basis for broad understanding and utilization of these models.

Shayne C. Gad

# Chapter 1

# Introduction

*Shayne C. Gad and Christopher P. Chengelis*

## Historical perspective

Safety and toxicity testing is a necessary and vital part of bringing any chemical-containing product to market. Stripped to its technical essence, toxicological testing could be described as the process of giving large amounts of chemicals to large numbers of experimental animals, which are then exhaustively studied and measured for evidence of chemical toxicity. This, of course, oversimplifies the process and understates the purpose of such testing, which brings us to the question, What exactly is the purpose of such testing? Very simply, a great deal of toxicological testing and research is performed to comply with governmental regulations. While there are moral and ethical reasons for testing products prior to human exposure, testing requirements are codified by the law. Ours, after all, is a civilized society. Many books and texts have been published in the last decade on the science of toxicology. It is a burgeoning field. The first edition of *Casarett and Doull's Toxicology* had only 482 pages in 1975, while the fifth edition in 1996 had 1111 pages.

Other primary texts on the science of toxicology are listed in Table 1.1. In most of these texts the emphasis is, rightfully, on the science and technology of toxicology. Most include a chapter on regulatory toxicology, yet most students receive at best a perfunctory introduction to regulatory concerns. For most doctoral students in toxicology, the Good Laboratory Practices (GLPs) are only a distant drone that hampers and confuses real science with meaningless detail. As a result, most industrial toxicologists begin their careers with little basic cognizance of the regulations that will govern the context, if not the content, of the job or product. It is the object of this book to address this regulatory gap. This is a scientist's guide to the regulations. We presume that the reader has a toxicological background and is familiar with basic study designs and terminology. This is not exhaustive or encyclopedic, but provides a toxicologist, health scientist, or other professional who has little legal interest or training with a guide to regulations. The central focus of this book is on the use of toxicology in a regulatory and legal arena. The science of toxicology is a secondary concern and, in fact, is discussed in only a cursory fashion.

Other than as an instrument of torture and execution, the use of toxicology by governing bodies is a relatively new phenomenon. While the appropriate use of chemicals has been a central part of the industrial revolution that has resulted in a high standard of living, the unrestricted sale and use of chemicals has resulted in more than a few problems. For a variety of factors involved in industrialization (with centralization

*Table 1.1* Primary texts in toxicology

| *Text* | *Author, year* |
|---|---|
| *Casarett and Doull's Toxicology* | Klaassen, 1996[a] |
| *General and Applied Toxicology* | Ballantyne *et al.*, 1999[b] |
| *Handbook of In Vivo Toxicity Testing* | Arnold *et al.*, 1990[c] |
| *Principles and Methods of Toxicology* | Hayes, 2001[d] |
| *Product Safety Evaluation Handbook* | Gad, 1999[e] |
| *Safety Assessment for Pharmaceuticals* | Gad, 1994[f] |

[a] Klaassen CD, Amdur MO, Doull J. *Casarett and Doull's Toxicology*. New York Pergamon Press, 1996.
[b] Ballantyne B, Marrs T, Turner P. *General Applied Toxicology*. New York: Stockton Press, 1999.
[c] Arnold DL, Derice HC, Krewski DB. *Handbook of In Vivo Toxicology Testing*. New York: Academic Press, 1990.
[d] Hayes AW. *Principles and Methods of Toxicology*, 4th ed. New York: Taylor & Francis, 2001.
[e] Gad SC. *Product Safety Evaluation Handbook*, 2nd ed. New York: Marcel Dekker, 1999.
[f] Gad SC. *Safety Assessment for Pharmaceuticals*. New York: Van Nostrand Reinhold, 1994.

of the food supply) and urbanization in the nineteenth century, large numbers of people were increasingly coming into contact with toxic materials and had little understanding of the consequences and no control on such exposures. Episodes of mass poisoning have often resulted. Such incidents in the US have spawned the growth of various consumer activist and environmental protectionist organizations (some of which, despite noble intention, verge on hysterical chemophobia). It was perhaps inevitable that those in government would find it politically wise to enact laws to regulate the preparation and distribution of food products, drugs, and chemicals in commerce. For example, in 1901 a diphtheria epidemic broke out in St. Louis that was eventually linked to improperly manufactured diphtheria toxin. In response to the public outcry, the US Congress passed the Virus Act of 1902. Among other things, this dictated that only licensed establishments could introduce vaccines, serums, or antitoxins into interstate commerce. Thus, the modern era of government regulations in the US was born. The process followed similar paths in other countries, although the time lines were different.

## Regulations and agencies

Toxicological data are required to meet a vast array of legal and regulatory purposes, particularly in the areas of product development, registration, and regulation. In addition, not only are toxicological data regulatory required, but often the specific procedures for recording of data (e.g. GLPs) are also dictated by regulations.

Bureaucracies exist that regulate the production, testing, and distribution of just about all chemical products. The regulations and regulatory bodies are the focus of this book. The primary laws and responsible regulatory agencies are summarized in Table 1.2. In the US, the two main regulatory bodies covering most chemicals in commerce of any nature are the Food and Drug Administration (FDA) and the Environmental Protection Agency (EPA).

The role of the FDA in the regulation of human pharmaceuticals and medical devices, veterinary drugs, food additives and (to the extent that they are regulated)

*Table 1.2* Summary of US Federal Regulations

| *Category* | *Agency* | *Act* | *Law* | *Regulations* |
|---|---|---|---|---|
| General chemical | EPA | Toxic Substance Control Act, 1976 | 15 USC 2601 | 40 CFR 700-700 |
| Pesticides | EPA | Federal Insecticide, Fungicide Rodenticide Act, 1972 | 7 USC 136 | 40 CFR 162-180 |
| Human pharmaceuticals | FDA | Federal Food, Drug and Cosmetic Act, 1938; numerous amendments | 21 USC 301 | 21 CRF 200-499 600-680 |
| Human medical devices | FDA | Federal Food, Drug and Cosmetic Act, 1938; amendment in 1976 | 21 USC 307 | 21 CFR 800-895 |
| Veterinary medicines | FDA | Federal Food, Drug and Cosmetic Act, 1938; numerous amendments | 21 USC 301 | 21 CFR 500-589 |
| Food additives | FDA | Federal Food, Drug and Cosmetics Act, 1938; numerous amendments | 21 USC 301 | 21 CFR 170-189 |
| Human over-the-counter and cosmetics | FDA | Federal Food, Drug and Cosmetics Act, 1938; numerous amendments | 21 USC 301 | 21 CFR 300-391 700-790 |
| Consumer products | CPSC | Federal Hazardous Substances Act, 1960; numerous amendments | 15 USC 1261 | 16 CFR 1500-1512 |
| | CPSC | Consumer Product Safety Act, 1972; numerous amendments | 15 USC 2051 | 16 CFR 1000-1406 |
| Worker safety | OSHA | Occupational Safety and Health Act, 1970 | 29 USC 651 | 29 CFR 1910-1926 |
| Animal care and use | USDA | Animal Welfare Act, 1966; amended 1970, 1976, and 1985. | | 9 CFR 3 |

cosmetics are discussed in Chapters 2, 3, 4, and 5, respectively. The role of the EPA in regulating pesticides (as defined by the Federal Insecticide, Fungicide and Rodenticide Act (FIFRA)) and industrial (non-pesticide) chemicals (as defined by Toxic Substances Control Act (TSCA)) are covered in Chapters 7 and 8. The consumer Product Safety Commission (CPSC) is responsible for chemicals containing products not covered by FDA or EPA. This can include items such as soaps, household cleansers, fire extinguishments and retardants, and paint or artist's supplies. The CPSC is primarily covered in Chapter 6. Toxicological information is required by manufacturing employers in order to comply with Occupational Health and Safety Regulations covered in Chapter 9. Scientists interested in laboratory management must realize that worker safety is not simply a manufacturing concern. The Occupational Safety and Health Administration (OSHA) has recently promulgated safe laboratory working standards that dictate certain laboratory work practices.

Each chapter has been authored by a toxicologist with experience in the representative area. While style and substance may vary, each author focuses on the following:

*Table 1.3* Regulatory newsletters

| *Newsletter* | *Publisher* | *Coverage* |
|---|---|---|
| Federal Register | US Government | All proposed and final rules and guidelines |
| FDA Medical Bulletin | FDA | New indications, changes in claims, toxicity warnings |
| Pesticide & Toxic Chemical News | Food Chemical News, Inc. | FIFRA, TSCA issues, regulatory actions, guideline changes |
| BELLE Newsletter | University of Massachusetts Public Health | Biological effects of low-level exposure of chemicals in the environment |
| Animal Pharmaceuticals | PJB Publications | Animal health and nutrition |
| Washington Drug Letter | Washington Business Information, Inc. | Policy, problems, etc. at the FDA |
| Warning Letter | Washington Information Source | FDA enforcement activities, inspections, and compliance |
| Covance Regulatory Review | Covance | Changes in FIFRA, TSCA, FDA regulations and operations, trials, scientific news |
| F-D-C Reports ("pink sheets") applicable to the pharmaceutical industry | F-D-C Reports, Inc. | Regulatory actions, scientific news, business information |
| Bio World Week | Bioworld Publishing | Regulatory, scientific and financial happenings in the biotechnology world |

- History that led to original legislation. What were the major social and historical events that led to the development of original legislation and major subsequent changes?
- Administrative divisions and responsibilities for each agency.
- Description and discussion of the basic documents each agency/division requires to be filed.
- The toxicity data needs.
- The use of these data in some form of risk analysis.

While the emphasis of this book is on the regulations involved in causing testing to occur for product safety and environmental and manufacturing practices, there are other regulations that govern the manner in which this information is generated or gathered. These are the GLP Regulations and the Animal Welfare Laws, covered in Chapter 10. A new chapter also covers Proposition 65 in California.

## Non-governmental organizations

A significant part of the regulatory process is actually not performed by government agencies, but rather by non-governmental organizations (NGOs). Most NGOs have a historical basis or origin in practice from years past, and tend to have been created by professional societies, regulated industries, or both in coordination with government.

NGOs tend to regulate by developing standards for various things or operations. Examples are animal care (AALAS), marker exposure standards for chemicals

(ACGIH), test methods for assessing toxicity or suitability of non-health care products (ASTM), determinations of relative hazard of human carcinogenicity (IARC), harmonizing international standards for assessing safety of drugs and biologies (ICH), international standards for devices and for quality assurance (ISO), and specifications for drug potency and biocompatibility (USP and other pharmacopoeias). The standards established are incorporated officially or unofficially into government-enforced legal requirements.

## International harmonization

With the development of a true global economy and the moves towards removing trade barriers and free market worldwide, there has been significant progress towards standardizing regulatory requirements for establishing the safety of both drugs and medical devices. Throughout this volume the chapters present not just requirements in the US, but also point out the differences (and agencies) associated with key other countries.

The process of resolving differences in regulatory requirements between countries is called harmonization, and it has been led by two NGOs, International Conference on Harmonization (ICH) for drugs and International Standards Organization (ISO) for medical devices. The three largest markets are the European Union (EU, which encompasses most of Western Europe), Japan, and the US. The latter two are governed by single national governments, while the EU has a double layer of the Union's government (headquartered in Brussels) and separate national governments. The process of achieving harmonized standards has taken years, going through a series of steps. But it is now almost complete.

## Regulation on the web

Another change of striking proportions is that one no longer needs to wait for (and order) paper copies of regulations and guidelines. These are widely (for the US, universally) available online from the appropriate web sites. For example, one can go to www.fda.gov/cder for the text of recent or new regulations and guidelines. A summary of available sites is presented in Appendix 2.

## Good laboratory practices

The GLPs were first issued by the FDA in 1978 in response to a variety of instances that led the agency to conclude that some of the data it had obtained were not trustworthy. A copy of the current FDA GLP regulations are in Appendix B. The EPA has two sets of GLPs: one for Office of Pesticides (FIFRA) and one for Office of Toxic Substances (TSCA). They are each similar to each other and to the FDA's regulations. The inclusion of the FDA GLPs should be sufficient to give the reader a taste of the regulation. The GLPs require that all pivotal preclinical safety studies (those used to make direct decisions on human exposure) be conducted under a well-defined protocol according to written standard operating procedures by (documented) trained personnel under the direction of a study director. All work must be reviewed by an independent Quality Assurance Unit (QAU).

Record-keeping requirements are rigorous. For example, no notebook entry can be changed without a footnoted explanation, and the change must be made in such a way that the original entry is still legible.

Both the EPA and the FDA have offices of compliance. There are field auditors who inspect toxicity research facilities. All laboratories that conduct toxicological assessments that are submitted to either the EPA or the FDA must undergo periodic capability audits (approximately every 2 years). One should be prepared for a team of auditors to show up without advance warning and to review facilities and procedures in excruciatingly minute detail. At the end of an FDA inspection, the auditors will issue an Enforcement Inspection Report (EIR) to the agency and a report on inspections (Form 483) to the facility. This will be reviewed with facility management in a wrap-up meeting. After receiving the official EIR, a facility has 90 days to respond in writing to the agency. If the agency finds the answers satisfactory, the case file will be closed. The inspectors will issue requests to the facility on what may be required to address any deficiencies. These may be either "voluntary only" action or "mandatory action required" requests. Failure to comply with the GLPs or with mandatory action requests can result in a facility being disqualified. That is, its work will no longer be accepted by the agency. If the inspectors find reasonable grounds and suspect illicit activity (data falsification), the findings can be turned over to the Justice Department for possible criminal activity. Records can be sealed and impounded.

## Animal use and welfare

Toxicology has flowered in the twentieth century, largely as a result of rapidly expanding chemical technologies and the resulting social and legal pressures, but also as a result of increased scientific sophistication. The use of intact, live animals in toxicological research has been, and shall remain, a vital and necessary part of this progress. The practice of using animals has come under organized and vituperative attack by animal rights activists. This has culminated in the passage by Congress of the Animal Welfare Act in 1990 and the subsequent development by the Department of Agriculture of the Animal Welfare regulations. These are discussed in Chapter 10. Current influences across the entire realm of toxicology (safety assessment) include public concern with the use of animals in research and testing and the overstatement of cases of cruelty to animals and of the current status of alternative or *in vitro* models.

Contrary to the beliefs of some, total replacement of animals is not possible. While such replacement is readily possible for some uses, these are either when no regulatory submission of test results is required (in the US) or for some uses in meeting international regulatory requirements.

The key assumptions underlying modern toxicology are as follows:

1 Other organisms can serve as accurate predictive models of toxicity in humans.
2 Selection of an appropriate model is essential to accurate prediction in humans.
3 Understanding the strengths and weaknesses of any particular model is essential to understanding the relevance of specific findings to humans.

The nature of models and their selection in toxicologic research and testing have only recently become the subject of critical scientific review. Usually in toxicology when we refer to *models*, we actually mean *test organisms*, although in fact the ways in which parameters are measured (and in which parameters are measured to characterize an endpoint of interest) are also critical parts of the model (or, indeed, may actually constitute the "model").

Although there have been accepted principles for test organism selection, these have not generally been the final basis for such selection. It is a fundamental hypothesis of both historical and modern toxicology that adverse effects caused by chemical entities in higher animals are generally the same as those induced by those entities in humans. There are many who point to individual exceptions to this and conclude that the general principle is false. Yet, as our understanding of molecular biology advances and we learn more about the similarities of structure and function of higher organisms at the molecular level, it becomes clear that the mechanisms of chemical toxicity are largely identical in humans and animals. This increased understanding has caused some of the same people who question the general principle of predictive value to, in turn, suggest that our state of knowledge is such that mathematical models or simple cell culture systems could be used just as well as intact animals to predict toxicities in humans. This last suggestion also misses the point that the final expressions of toxicity in humans or animals are frequently the summation of extensive and complex interactions at cellular and biochemical levels. Zbinden has published extensively in this area, including a very advanced defense of the value of animal models. Lijinsky has reviewed the specific issues about the predictive value and importance of animals in carcinogenicity testing and research. Although it was once widely believed (and still is believed by many animal rights activists) that *in vitro* mutagenicity tests would entirely replace animal bioassays for carcinogenicity, this is clearly not the case on either scientific or regulatory grounds. Although there are differences in the responses of various species (including humans) to carcinogens, the overall predictive value of such results (when tempered by judgement) is clear. At the same time, well-reasoned use of *in vitro* or other alternative test model systems is essential to the development of a product safety assessment program which is both effective and efficient.

## Law versus regulation

The law is embodied in the documents written, debated, and passed by Congress and then approved by the President. It is then incorporated into the US Code of Federal Regulations. Part of the code will include the federal agency responsible for administering the law. The responsible agency will then devise and propose regulations that it believes will meet the intent of Congress. Regulations are the enforcing rules of the law and thus have an impact on the activities of toxicologists working in industrial settings. Both proposed (for comment) and final rules and regulations are published in the Federal Register. When final rules are published, they will include a response by the agency on the comments received concerning proposed regulations.

Regulations are organized in the Code of Federal Regulations using the following system:

- TITLE
  - CHAPTER (denoted by Roman numeral, upper case)
    - PART (denoted by Arabic numeral)
      - SUBPART (denoted by letter, upper case)
        - SECTION (decimal point, followed by Arabic numeral)

SUBSECTION (denoted by upper case Roman numeral)
PARAGRAPH (denoted by lower case letter in parentheses)

Further subdivisions are further denoted by parentheses, Arabic, and Roman numerals, respectively.

For example, Title 20, Chapter I of the Code of Federal Regulations covers all FDA Regulations. Subchapter F covers biologics. Subpart B covers establishment standards and Section 600.10 specifically covers personnel. Paragraph (c) covers restrictions on personnel, and part two of the paragraph covers the wearing of protective clothing.

Because parts and sections are enumerated without redundancy (there is no Part 600 in any other chapter of Title 20), it is customary not to specify subtitle, chapter, subchapter, or subpart in making references to the Code of Federal Regulations. Thus, the reference just described would be denoted 21 CFR 600.10 (c)(2).

Regulations are not forever. They are continually being added to, deleted, and modified. There is an entire industry based on keeping industry and the public aware of such changes, their proposal, and impacts by means of an array of newsletters. Some of the more prominent of these are summarized in Table 1.3.

A huge number of acronyms are now commonly used in both toxicology and regulatory actions. Appendix 1 presents an extensive listing of these. A great deal of information is now available from the Internet. Appendix 2 presents the addresses of key regulatory and government sites.

## Further reading


Holson JF, Kimmel CA, Hogue C, Carlo G. *Suitability of Experimental Studies for Predicting Hazards to Human Development.* Proceedings of the 1981 Annual Winter Meeting of the Toxicology Forum.

Huff J, Chemicals and cancer in humans; first evidence in experimental animals, *Environ Health Perspect* 1993;100:201–210.

Kimmel CA, Holson JF, Hogue CJ, Carlo GL, *Reliability of Experimental Studies for Predicting Hazards to Human Development*, NCTR Final report for experiment #6015, 1984.

Lijinsky W, Importance of animal experiments in carcinogenesis research, *Environ Mol Mutagen* 1988;11:307–314.

Schardein JL, *Chemically Induced Birth Defects.* New York: Marcel Dekker, 1993.

Zbinden G, *Predictive Value of Animal Studies in Toxicology.* Carshalton, UK: Center for Medicines Research, 1987.

# Chapter 2

# Human pharmaceutical products

*Shayne C. Gad and Christopher P. Chengelis*

## Introduction

The safety of pharmaceutical agents, medical devices, and food additives are the toxicology issues of the most obvious and longest-standing concern to the public. A common factor among the three is that any risk associated with a lack of safety of these agents is likely to affect a very broad part of the population, with those at risk having little or no option as to undertaking this risk. Modern drugs are essential for life in our modern society, yet there is a consistent high level of concern about their safety.

This chapter examines the regulations which establish how the safety of human pharmaceutical products are evaluated and established in the US and the other major international markets. As a starting place, the history of this regulation will be reviewed. The organizational structure of the Food and Drug Administration (FDA) will be briefly reviewed, along with the other quasi-governmental bodies that also influence the regulatory processes. The current structure and context of the regulations in the US and overseas will also be presented. From this point the general case of regulatory product development and approval will be presented. Toxicity assessment study designs will be presented. The broad special case of biotechnology-derived therapeutic products and environmental concerns associated with the production of pharmaceuticals will be briefly addressed. The significant changes in regulation brought about by harmonization are also reflected.

As an aid to the reader, appendices are provided at the end of this book: a codex of acronyms that are used in this field, followed by a glossary which defines some key terms.

## Brief history of US pharmaceutical law

A synopsis of the history of US drug legislation is presented in Table 2.1. Here we will review the history of the three major legislative acts covering pharmaceuticals.

### *1906: Pure Food and Drug Act*

The history of health product legislation in the US largely involves the passage of bills in Congress which were primarily in response to the public demand. In 1902, for example, Congress passed the Biologics Act in response to a tragedy in St. Louis where ten children had died after being given contaminated diphtheria toxins. Interestingly,

*Table 2.1* Important dates in US federal drug law[a]

| *Year* | *Event* |
|---|---|
| 1902 | Passage of the Virus Act, regulating therapeutic serums and antitoxins. Enforcement by the Hygienic Laboratory (later to become the National Institute of Health) and Treasury Department. |
| 1906 | Passage of Pure Food Act, including provisions for the regulations of drugs to prevent the sale of misbranded and adulterated products. Enforcement by the Chemistry Laboratory, Agriculture. |
| 1912 | Passage of the Sherley Amendment. Specifically outlawed any false label claims as to curative effect. |
| 1927 | Bureau of Chemistry renamed the Food, Drug and Insecticide Administration. |
| 1931 | Renamed again to Food and Drug Administration. |
| 1938 | Passage of the Food, Drug and Cosmetic Act. Superseded the law of 1906. Required evidence of safety, e.g., studies in animals. Included coverage of cosmetics and medical devices. Specifically excluded biologics. |
| 1944 | Administrative Procedures Act, codifying Public Health Laws: included provision that for a biological license to be granted, a product must meet standards for safety, purity, and potency. NIH also given the responsibility for developing biologics not developed by the private sector. |
| 1945 | Amendment to the 1936 Act requiring that the FDA examine and certify for release each batch of penicillin. Subsequently amended to include other antibiotics. |
| 1949 | Publication of the first set of criteria for animal safety studies. Following several revisions, guidelines published in 1959 as Appraisals Handbook. |
| 1951 | Passage of Durham-Humphrey Amendment. Provided the means for manufacturers to classify drugs as over-the-counter (not requiring prescription). |
| 1953 | Transfer of FDA to the Department of Health, Education and Welfare from Agriculture (now the Department of Health and Human Services). |
| 1962 | Passage of major amendments (the Kefauver Bill) to the 1938 FDCA, which required proof of safety and effectiveness (efficacy) before granting approval of New Drugs Applications. Required affirmative FDA approval. |
| 1968 | FDA placed under the Public Health Service of HEW. |
| 1970 | Controlled Substance Act and Controlled Substances Import and Export Act. Removed regulation of drug abuse from FDA (transferred to the Drug Enforcement Agency) and provided for stringent regulation of pharmaceuticals with abuse potential. |
| 1972 | Transfer of authority to regulate biologics transferred from NIH to FDA. The NIH retained the responsibility of developing biologics. |
| 1973 | Consumer Product Safety Act, leading to the formation of separate Consumer Product Safety Commission, which assumes responsibilities once handled by the FDA's Bureau of Product Safety. |
| 1976 | Medical Device Amendment to the FDCA requiring for devices that not only effectiveness be proven, but also safety. |
| 1979 | Passage of the Good Laboratory Practices Act. |
| 1983 | Passage of the first Orphan Drug Amendment to encourage development of drugs for small markets. |
| 1984 | Drug Price Competition and Patent Term Restoration Act intended to allow companies to recover some of the useful patent life of a novel drug lost due to the time it takes the FDA to review and approve. Also permits the marketing of generic copies of approved drugs. |
| 1985 | The "NDA rewrite" final rule. An administrative action streamlining and clarifying the New Drug Application process. Now embodied in 21 CFR 314. |
| 1986 | The United States Drug Export Amendment Act of 1986. Permitted the export of drugs outside the US prior to approval for the US market. |

made any therapeutic claim which was contrary to general medical practice and opinion. The definition of a drug was broadened to include devices used for medical purposes[1] *Adulteration* was defined as any drug product dangerous to health when used according to label directions. The patent manufacturers charged that no bill granted too much discretionary power to a federal agency – that no manufacturer could stay in business except by the grace of the Department of Agriculture, a charge that may have been correct. In response to the patent trade lobbying effort, the FDA launched its own educational drive of radio spots, displays (such as the sensationalized Chamber of Horrors exhibition, in which the toxicity of a variety of useless medicines was clearly displayed), mimeographed circulars, speaking engagements, posters, etc.

Ruth Lamb, FDA information officer at the time, was perhaps one of the hardest working and most quotable of the FDA staffers working the street at the time. For example, in reference to one of the counter-bills that had language similar to the original Copeland bill, but with extremely complicated enforcement provisions, Ruth Lamb called it "an opus for the relief of indigent and unemployed lawyers". She once described the Bailey amendment, which would have made proprietary drugs virtually immune to multiple seizures, as permitting the "sale of colored tap water as a cure for cancer unless arsenic was added to each dose making [it] immediately dangerous". After 1934, however, the educational efforts of the FDA were greatly attenuated by federal laws prohibiting lobbying by federal agencies.

The fall of 1937 witnessed the beginning of the often-told Elixir of Sulfanilamide incident, which remains one of the nation's worst drug tragedies. The Massengil Company was not one of the industry giants, but neither was it a "snake oil peddler". The company's chief chemist, Harold Watkins, was simply trying to develop a product and, in fact, did so in a manner consistent with the norms of the time. There was a perceived need for a liquid from of sulfanilamide, but it was difficult to dissolve. Watkins hit upon diethylene glycol. No toxicity tests were performed on the finished product, although the product did pass through the "control lab" where it was checked for appearance, fragrance, and consistency.

The first reports of human toxicity occurred in October 1937 when Dr. James Stevenson of Tulsa requested some information from the AMA because of the six deaths in his area that were attributable to the elixir. At the time, no product of Massengil stood accepted by the Council on Pharmacy and Chemistry, and the Council recognized no solution of sulfanilamide. The AMA telegraphed Massengil, requesting samples of the preparation for testing. Massengil complied. The test revealed the diethylene glycol to be the toxic agent and the AMA issued a general warning to the public on October 18, 1937. In the meantime, the FDA had become aware of the health risks and launched an investigation through its Kansas City station. By October 20, when at least fourteen people had died, Massengil wired the AMA to request an antidote for their own product. By the end of October, at least 73 people had died and another 20 suspicious deaths were linked to the drug. Had it not been for the response of the FDA, more deaths may have occurred. The Agency put its full force of field

1 The use of a broad definition of what constitutes a drug for regulatory purposes is a precedent that remains in place today. For example, the computer software used in diagnostic systems is considered to be a pharmaceutical for purposes of regulation.

investigators (239 members) on the problem and eventually recovered and accounted for 99.2 per cent of the elixir produced. Massengil fully cooperated with the investigation and in November published a public letter expressing regret over the matter, but further stating that no law had been broken. In fact, the company was eventually convicted on a long list of misbranding charges and fined a total of $26,000 (the largest fine ever levied under the 1906 law).

The Massengil incident made the limits of the 1906 law quite clear. Because there were no provisions against dangerous drugs, the FDA could move only on the technicality of misbranding. The term *elixir* was defined by the US Pharmacopoeia (USP) as "a preparation containing alcohol", which Elixir of Sulfanilamide was not. It was only this technicality that permitted the FDA to declare the "Elixir" misbranded, to seize the inventory, and to stop the sale of this preparation. If it had been called *Solution of Sulfanilamide*, no charges could have been brought.

The extensive press coverage of the disaster became part of the national dialogue. Letters poured in to congressmen demanding action to prevent another such tragedy. Medical and pharmacy groups and journals insisted that a new law was required. Congress was in special session in November 1937, and did not need to be told about the tragedy. Copeland and Representative Chapman (of Kentucky) pressed resolutions calling for a report from the FDA on the tragedy. When issued, the FDA report stunned Congress, not only because of the human disaster, but also because it made apparent that even had the bill then before Congress been law, the entire tragedy would still have occurred because there were no provisions for toxicity testing before new drugs entered the market. By December 1937 a new bill, S.3037, was introduced which stated that manufacturers seeking to place new drugs on the market would be required to supply records of testing, lists of components, descriptions of each manufacturing process, and sample labels. Drugs would require certification by the FDA before sale was permitted. A similar bill was introduced in the House by Chapman, although the issue of which agency was to control advertising of drugs was still festering in the House. In January 1938, debate started on the Wheeler–Lea Bill, which would ensure that all controls over drug advertising would remain in the Federal Trade Commission (FTC). Despite strong opposition by the FDA, the Wheeler–Lea Bill was signed into law March 1938. While the loss of advertising control was a blow to the FDA, the Wheeler–Lea Bill did facilitate the passage of the new Food and Drug Law.

With the issue of advertising controls settled, the Copeland–Chapman Bill faced one last hurdle. Section 701, which had been added in committee, provided for appeal suits that could be entered in any Federal District Court to enjoin the agency from enforcing new regulations promulgated as a result of the Act. Interestingly, this issue had more to do with foods than drugs, as its major focus was with acceptable tolerance limits for insecticides in food. The new bill defined an *adulterated food* as one containing any poison. However, because efforts to remove insecticides from fresh fruits and vegetables had never been completely successful, the Secretary of Agriculture needed this power to set tolerance levels. Allies of food producers tried to introduce provisions in the new bill that provided methods for stalling a tolerance regulation with rounds of appeals. The bill passed the House despite such provisions (Section 701) and despite the resistance of consumer groups and the FDA, and went into joint committee. Roosevelt, in one of his rare efforts to support the FDA, made it clear that he would not accept the bill with such a cumbersome appeals process. The resulting compromise was an appeals process which limited the new evidence that could be introduced into one of the ten circuit courts. Other provisions regarding labeling were also rectified in joint committee. In May 1938, S.3073 passed by unanimous vote. Both chambers ratified the joint committee report, and Franklin D. Roosevelt signed the new law in June of 1938.

A historical note to this story was that Royal Copeland did not live to see his measure passed. In May 1938, he collapsed on the Senate floor. His death occurred one month before President Roosevelt signed his bill into law.

### 1962: major amendment

The 1938 law very much changed the manner in which Americans purchased pharmaceutical agents. In effect, it changed the pharmaceutical industry from a traditional consumer product industry to one in which purchases were made as directed by a third party (the physician). In 1929, ethical pharmaceuticals (prescription drugs) comprised only 32 per cent of all medicines, while by 1969 this was up to 83 per cent (Temin, 1980). This led to a peculiar lack of competition in the ethical market. In 1959, Senator Estes Kefauver initiated his now-famous hearings on the drug industry. Interestingly, almost 30 years later, Senator Edward Kennedy had hearings on exactly the same matter. In 1961, Kefauver submitted a proposed legislation to amend the 1938 Act in such a way as to increase FDA oversight of the drug industry. The proposed amendment contained two novel propositions. The first was compulsory licensing, which would have required, for example, company "A" to license (with a royalty of no greater than 8 per cent of sales) company "B" to market a drug patented by company "A". Company "A" would have only 3 years' exclusivity with its patent. The second novel provision was that new drugs had to be not only "safe", but also "efficacious". There was not a ground swell of support for this legislation. When it was reported out of committee, it had been rewritten (including the removal of the licensing requirement) to the point that even Kefauver refused to support it. The Kennedy administration wanted new legislation but did not specifically support the Kefauver Bill; rather it introduced its own legislation, sponsored by Representative Orren Harris of Arkansas. It also had little support.

As in 1938, a tragic incident would intercede in the legislative process: 1961 would see the development of the thalidomide tragedy. An anti-anxiety agent marketed in Europe, thalidomide, was prescribed for pregnancy-related depression and taken by countless numbers of women. At about the same time, phocomelia, a birth defect marked by the imperfect development of arms and legs, appeared in Europe. Thalidomide was eventually determined to be the causative teratogen in 1961 and subsequently taken off the market in Europe. The William S. Merrill Company had applied for a New Drug Application (NDA) for thalidomide in the US in 1960. It was never approved because the FDA examiner, Dr. Frances Kelsey, had returned the application for lack of sufficient information. Eventually, the company withdrew the application. Senator Kefauver's staff had uncovered the thalidomide story as it was unfolding and had turned its findings over to the *Washington Post*. The *Post* reported the episode under the headline "Heroine of the FDA Keeps Bad Drug off the Market" in July 1962, 3 days after the Kefauver Bill was reported our of committee. Needless to say, the news created public support for the bill, which was sent back to committee and reported out again with new language in August 1962. The Kefauver–Harris bill was signed into law in October 1962. It was demonstrated after the fact that thalidomide was teratogenic in the rabbit; out of the episode grew the current practice that new human pharmaceuticals are tested for teratogenicity in two species, one generally being the rabbit.

The 1962 Drug Amendment made three major changes in the manner in which new drugs could be approved (Merrill, 1994). First, and perhaps the most important, was that it introduced the concept of effectiveness into the approval process. An NDA had to contain evidence that the drug was not only safe, but also effective. The 1938 law contained no such specification. The effectiveness requirement necessitated that a drug company had to do more extensive clinical trials. The new law required that companies apply to the FDA for approval of its clinical testing plan under an Investigational New Drug Application (INDA). No response from the FDA was deemed to be acceptance. As each level of clinical testing came to require FDA review and approval, the new law made the FDA an active partner in the development of all drugs.

The second major change enacted under the 1962 Drug Amendment was the change in the approval process from premarket notification to a premarket approval system. Under the terms of the 1938 law, an NDA would take effect automatically if the FDA did not respond. For example, the only reason thalidomide was not approved was because Dr. Kelsey returned the application to the sponsor with a request for more information. In contrast, the 1962 law required affirmative FDA action before a drug could be put on the market. Under the terms of the 1962 amendments, the FDA was also empowered to withdraw NDA approval and remove the drug from the market for a variety of reasons, including new evidence that the product was unsafe or that the sponsor had misrepresented or underreported data.

The third major change enlarged the FDA's authority over clinical testing of new drugs. Thus, not only was evidence of effectiveness required, but Section 505 (d) of the act specified the types of studies required. "Substantial evidence consisting of adequate and well-controlled investigations, including clinical investigations by qualified expert". In meeting the statutory requirement for setting standards of clinical evidence, the FDA has become highly influential in the design of drug testing regimens (Merrill, 1994). Interestingly, discussed in detail by Hutt (1987), the FDA was initially quite unprepared for this new level of responsibility. It was not until 1973 that audited regulations on the determination of safety and effectiveness were put into place (these were, in fact, approved by the Supreme Court). While there have been several procedural changes (e.g., the 1985 Investigational New Drug (IND) rewrite) and additions (e.g., the 1988 IND procedures for life-threatening disease treatment), there have actually been no major changes in the law through 1992 with the Prescription Drug Users Fee Act (PDUFA) and 1997 with the Food and Drug Administration Modernization Act (FDAMA).

We must interject an interesting historical side-light at this point. Despite its reputation, thalidomide has made a bit of a comeback in the 1990s (Blakeslee, 1998). Among other properties, thalidomide has been shown to have good anti-inflammatory properties, due to the fact that it apparently decreases the synthesis and/or release of tissue necrosis factor.

## 1992 and 1997: PDUFA and FDAMA

The history of pharmaceutical regulations has been dominated by two often-opposing schools of thought: the need to provide the citizenry with effective medicaments, and the need to protect the consumer from unsafe and misbranded products. The reader is referred to Peter B. Hutt's in-depth reviews (Hutt, 1983a,b) on the subject. For example, the very first federal drug legislation in the US was the Vaccine Act of 1813, which mandated the provision of the smallpox vaccine to the general public. In the modern era, legislative debate could be further defined as the constant swing back and forth on these two issues (see Hutt 1983a,b); i.e. safety versus development costs. In 1963, for example, Senator Hubert Humphrey presided over hearings on the FDA's implementation of the Drug Amendment of 1962. The FDA came under substantial criticism for failure to take strong action to protect the public from dangerous drugs. Eleven years later, Senator Edward Kennedy conducted hearings addressing exactly the same issue. Commissioner Schmidt pressed the point that the FDA is under constant scrutiny regarding the approval of "dangerous" drugs, but no hearing had ever been conducted (up to that time) on the failure of the FDA to approve an important new therapy.

The next decade and a half saw a proliferation of work that analyzed the impact of regulation on competitiveness and the introduction of new therapies (see Hutt, 1983b for a complete review). This included, for example, Grabowski and Vernon's work (Grabowski and Vernon, 1983), which concluded that regulation had significant adverse effect on pharmaceutical innovation. This examination of the cost of regulation continued into 1990. In a meticulous and well researched study DiMasi et al. (1994), reported that throughout the 1980s the number of INDAs were decreasing while the NDA success rate was also dropping, and the length of time between discovery and approval was increasing. Clearly this is a situation that could not go on forever. The cost of new drug development rose from $54 million (US) in 1976 to $359 million (US) in 1990. (Anon., 1998a). Members of the pharmaceutical industry and the biotechnology industry were becoming increasingly alarmed by the negative synergy caused by increased costs and increased time to market. In 1991, Dranove published an editorial examining the increased costs and decreased product flow that resulted from the 1962 amendment. He made the observation that European requirements are less stringent than those of the US, yet the Europeans do not seem to be afflicted by a greater number of dangerous drugs (see Table 2.2).

In an age of decreasing regulatory recourses, the FDA (as well as Congress) was under increasing pressure to review and release drugs more quickly. In response, Congress passed the 1992 Prescription Drug Users Fee Act (PDUFA). Under the terms of this act, companies would pay a fee to the agency to defray costs associated with application review. They would supposedly provide the FDA with the resources available to decrease application review time. In return, companies were guaranteed a more rapid review time. By all accounts, PDUFA has been successful. In 1992 (the year PDUFA was passed), 26 NDAs were approved, requiring on average 29.9 months for data review; while in 1996, 53 new drug (or biological) products were approved, each requiring an average of 17.8 months of review time. PDUFA was successful in decreasing review times, but has not really streamlined the procedures.

The AIDS activist community was particularly vocal and effective in demanding

*Table 2.2* Post-approval adverse side effects and related drug withdrawals in the 1990s (51% of approved drugs have serious post-approval identified adverse side effects, FDAMA passed in 1997)

| | *Withdrawals* | | |
|---|---|---|---|
| *Year* | *Drug* | *Indication/class* | *Causative side effect* |
| 1991 | Enkaid (4 years on market) | Antiarrhythmic | Cardiovascular (sudden cardiac death) |
| 1992 | Temafloxacin | Antibiotic | Blood and kidneyencainade, alpidem |
| 1997 | Fenfluramine*/dexafluramine (combo used since 1984) (*24 years on market) | Diet | Heart valve abnormalities |
| | Seldane (terfenadine) (12 years on market) | Antihistamine | Ventricular arrhythmias |
| 1998 | Posicormibefradil (1 year on market) | $Ca^{2+}$ channel blocker | Lethal drug interactions (inhibited liver enzymes) |
| | Duract (bronfemic sodium) (early preapproval warning of ↑ liver enzymes) | Pain relief | Liver damage |
| | Verdia (application withdrawn after submission) | Angiotension II receptor blocker | Liver damage |
| 1999 | Tronan (severely restricted) | Antibiotic | Liver/kidney |
| | Rotashield | Rotavirus vaccine | Bowel obstruction |
| 2000 | Renzulin (approved December 1996) | Type II diabetes | Liver damage |
| | Propulsid | Heartburn | Cardiovascular irregularities/death |

more rapid approvals and increased access to therapies. There was also demand for FDA reform on a number of other fronts (e.g., medical devices, pediatric claims, women and minority considerations, manufacturing changes, etc.). In 1993 the House Commerce Committee on Oversight and Investigations, chaired by John Dingel (D-MI) released a comprehensive investigation and evaluation of the FDA entitled, *Less than the Sum of its Parts*. The report was highly critical of the FDA, and made a number of recommendations (Pilot and Waldemann, 1998). The mid 1990s also saw the reinventing government initiatives (RIGO) chaired by Vice-President Al Gore. Under RIGO, the FDA sought to identify and implement administrative reform. The RIGO report issued was entitled *Reinventing Regulation of Drugs and Medical Devices.*The 104th Congress started hearings on FDA reform again in the winter of 1995. Two bills were introduced that provided the essential outline of what would become FDAMA. Senator Nancy Kassebaum (R-KS), chair of the Senate Committee on Labor and Human Resources, introduced S-1477. The second was H.R.3201, introduced by Representative Joe Barton (R-TX). Other bills were introduced by Senator Paul Wellstone (D-MN) and Representative Ron Weyden (D-OR), which focused more on medical devices but sill paved the way for bipartisan support of FDA reform (Pilot and Waldemann, 1998). Eventually, the 105th Congress passed the Food and Drug Administration Modernization Act (FDAMA), which was signed into law by President Clinton in November 1997. The various sections of FDAMA are listed in Table 2.3. By any measure, it was a very broad and complex, if not over-deep, piece of legislation. In 1998, Marwick (1998) observed, "a measure of the extent of the task is that implementation of the act will require 42 new regulations, …23 new guidance notices, and 45 reports and other tasks". The FDA has identified these various tasks, regulations and guidances necessary for the implementation of FDAMA. (FDA's FDAMA Implementation Chart is available at http://www.fda.gov/po/modact97.html), and the reader is urged to explore this site. There is a FDAMA icon on the FDA home page, and both Center for Biologic Evaluation and Research (CBER) and the Center for Drug Evaluation and Research (CDER) have issued various guidance documents. Some of the more interesting sections of the act that may be of interest to toxicologists include:

- Renewal of PDUFA for another 5 years.
- Fast track for breakthrough products.
- Changes in the fashion Biologicals are regulated (elimination of the Establishment and Product licenses, both replaced with a Biological License Application (BLA)).
- Changes in the fashion Antibiotics are developed and regulated.
- Incentives for the development of pediatric claims.
- Companies will be permitted to disseminate information about approved uses for their products.

*Table 2.3* Summary of the contents of the 1997 Food and Drug Administration Modernization Act

| *Title/subtitle* | *Section* |
|---|---|
| I. Improving regulatory drugs | |
| A. Fees relating to drugs | 101. Findings |
| | 102. Definitions |
| | 103. Authority to assess and use drug fees |
| | 104. Annual reports |
| | 105. Savings |
| | 106. Effective date |
| | 107. Termination of effectiveness |
| B. Other improvements | 111. Pediatric studies of drugs |
| | 112. Expanding study and approval of fast track drugs |
| | 113. Information program on trials for serious disease |
| | 114. Healthcare economic information |
| | 115. Manufacturing changes for drugs |
| | 116. Streamlining clinical research for drugs |
| | 118. Data requirements for drugs and biologics |
| | 119. Content and review of applications |
| | 120. Scientific advisory panels |
| | 121. Positron emission tomography |
| | 122. Requirements for radiopharmaceuticals |
| | 123. Modernization of regulation |
| | 124. Pilot and small scale manufacture |
| | 125. Insulin and antibiotics |
| | 126. Elimination of certain labeling requirements |
| | 127. Application of federal law to pharmacy compounding |
| | 128. Reauthorization of clinical pharmacology program |
| | 129. Regulation of sunscreen products |

Table 2.3 (*continued*)

| *Title/subtitle* | *Section* |
|---|---|
| | 130. Report of post-marketing approval studies |
| | 131. Notification of discontinuance of a life-saving product |
| II. Improving regulation of devices | 201. Investigational device exemptions |
| | 202. Special review for certain devices |
| | 203. Expanding humanitarian use of devices |
| | 204. Device standards |
| | 205. Collaborative determinations of device data requirements |
| | 206. Premarket notification |
| | 207. Evaluation of automatic class iii designation |
| | 208. Classification panels |
| | 209. Certainty of review timeframes |
| | 210. Accreditation of person for review of premarket notification reports |
| | 211. Device tracking |
| | 212. Postmarket notification |
| | 213. Reports |
| | 214. Practice of medicine |
| | 215. noninvasive blood glucose meter |
| | 216. Data relating to premarket approval: product development protocol |
| | 217. Number of required clinical investigations for approval |
| III. Improving regulation of food | 301. Flexibility for regarding claims |
| | 302. Petitions for claims |
| | 303. Health claims for food products |
| | 304. Nutrient content claims |
| | 305. Referral statements |
| | 306. Disclosure of radiation |
| | 307. Irradiation petition |
| | 308. Glass and ceramic ware |
| | 309. Food contact substance |
| IV. General provisions | 401. Dissemination of information new uses |
| | 402. Expanded access of investigational therapies and diagnostics |
| | 403. Approval of supplemental applications for approved products |
| | 404. Dispute resolution |
| | 405. Informal agency statements |
| | 406. FDA mission and annual report |
| | 407. Information system |
| | 408. Education and training |
| | 409. Centers for education and research on therapeutics |
| | 410. Mutual recognition of agreements and global harmonization |
| | 411. Environmental impact review |
| | 412. National uniformity for nonprescription drugs and cosmetics |
| | 413. FDA study of mercury in drugs and foods |
| | 414. Interagency collaboration |
| | 415. Contracts for expert review |
| | 416. Product classification |
| | 417. Registration of foreign establishments |
| | 418. Clarification of seizure authority |
| | 419. Interstate commerce |
| | 420. Safety report disclaimers |
| | 421. Labeling and advertising compliance with statutory requirements |
| | 422. Rule of construction |
| V. Effective date | 501. Effective date |

- FDAMA requires that the FDA establishes a clinical trials data base for drugs used to treat serious and life-threatening diseases, other than AIDS and cancers (data bases for these diseases have already been established).

The full impact of FDAMA in the pharmaceutical industry in general, and on toxicology within this industry in particular remains to be established.

This is a debate that has continued to the present day and has been highlighted by the demands for anti-HIV chromotherapeutic agents.

While it is not possible to review the history of regulations worldwide, it is possible to point out some differences. We will highlight specific differences where appropriate throughout the remainder of the text.

The strength of the US regulatory system was highlighted at the BioEurope 1993 Conference. David Holtzman stated:

> the main subject of the conference was regulation, and the US was perceived to have the superior regulatory agency. It may be more difficult to satisfy but it is more predictable and scientifically based. (Holtzman, 1993)

this predictability has not stultified the growth and biotechnology industry in the US and has, in fact, made the US a more inciting target for investment than Europe. It is also a system that, while not perfect, has permitted very few unsafe products on the market.

## FDAMA

### *Summary: consequences and other regulations*

In summary, federal regulation of the safety of drugs has had three major objectives:

1. Requiring testing to establish safety and efficacy.
2. Establishing guidelines as to which tests are required and how they are designed.
3. Promulgating requirements of data recording and reporting.

The first of these objectives was served by the 1906 Act, which required that agents be labeled appropriately. This was amended in 1938, in response to the tragedies associated with Elixir of Sulfanilamide and Lash Lure®[2], to require that drugs and marketed formulations of drugs be shown to be safe when used as intended. In the aftermath of the thalidomide tragedy, the 1962 Kefauver–Harris Amendment significantly tightened requirements for preclinical testing (the INDA) and pre-market approval (the NDA) of new drugs. The regulations pertaining to INDAs and NDAs have been modified (most recently in 1988), but essentially remain as the backbone of regulations of the toxicity evaluation of new human pharmaceutical agents.

The Good Laboratories Practice (GLP) Act, which specifies standards for study planning, personnel training, data recording, and reporting, came out in 1978 in response to perceived shoddy practices of the operations of laboratories involved in the conduct of preclinical safety studies. It was revised in 1985 and is discussed elsewhere in this book.

[2] An ethylenediamine containing cosmetic which caused facial disfigurement and blindness.

The final major regulatory initiative on how drugs will be preclinically evaluated for safety arose out of the Acquired Immune Deficiency Syndrome (AIDS) crisis. To that point, the process of drug review and approval had very generally been perceived as slowing down, the FDA pursuing a conservative approach to requiring proof of safety and efficacy before allowing new drugs to become generally available. In response to AIDS, in 1988 the Expedited Delivery of Drugs for Life-Threatening Diseases Act established a basis for less rigorous standards (and more rapid drug development) in some limited cases.

In the UK, the Committee on Safety of Medicines (reporting to the Minister of Health) regulates drug safety and development under the Medicines Act of 1968 (which has replaced the Therapeutic Substances Act of 1925). Details on differences in drug safety regulations in the international marketplace can be found in Alder and Zbinden (1988), but key points are presented in this chapter.

## Overview of US regulations

### *Regulations: general considerations*

The US federal regulations that govern the testing, manufacture, and sale of pharmaceutical agents and medical devices are covered in Chapter 1, Title 21 of the Code of Federal Regulations (21 CFR). These comprise nine 6 × 8 inch (printing on both sides of the pages) volumes which stack 8 inches high. This title also covers foods, veterinary products, and cosmetics. As these topics will be discussed elsewhere in this book, here we will briefly review those parts of 21 CFR that are applicable to human health products and medicinal devices.

Of most interest to a toxicologist working in this arena would be Chapter 1, Subchapter A (parts 1–78), which cover general provisions, organization, etc. The GLPs are codified in 21 CFR 58.

General regulations that apply to drugs are in Subchapter C (parts 200–299). This covers topics such as labeling, advertising, commercial registration, manufacture, and distribution. Of most interest to a toxicologist is a section on labeling (Part 201, Subparts A–G, which covers Sections 201.1–201.317 of the regulations) as much of the toxicological research on a human prescription drug goes toward supporting a label claim. For example, specific requirements on content and format of labeling for human prescription drugs are covered in Section 201.57. Directions for what should be included under the "Precautions" section of a label are listed in 201.57(f). This includes 201.57(f)(6), which covers categorization of pregnancy risk, and the reliance upon animal reproduction studies in making these categorizations is made quite clear. For example, a drug is given a pregnancy category B if "animal reproduction studies have failed to demonstrate a risk to the fetus". The point here is not to give the impression that the law is most concerned with pregnancy risk. Rather, we wish to emphasize that much basic toxicological information must be summarized on the drug label (or package insert). This section of the law is quite detailed as to what information is to be presented and the format of the presentation. Toxicologists working in the pharmaceutical arena should be familiar with this section of the CFR.

### *Regulations: human pharmaceuticals*

The regulations specifically applicable to human drugs are covered in Subchapter D, Parts 300–399. The definition of a new drug is covered in Part 310(g):

> A new drug substance means any substance that when used in the manufacture, processing or packaging of a drug causes that drug to be a new drug but does not include intermediates used in the synthesis of such substances.

The regulation then goes on to discuss "newness with regard to new formulations, indications, or in combinations". For toxicologists, the meat of the regulations can be found in Section 312 (INDA) and Section 314 (applications for approval to market a new drug or antibiotic drug or NDA). The major focus for a toxicologist working in the pharmaceutical industry is on preparing the correct toxicology "packages" to be included to "support" these two types of applications. (The exact nature of these packages will be covered below).

In a nutshell, the law requires solid scientific evidence of safety and efficacy before a new drug will be permitted in clinical trials or (later) on the market. The INDA (covered in 21CFR310) is for permission to proceed with clinical trials on human subjects. Once clinical trials have been completed, the manufacturer or "sponsor" can then proceed to file an NDA (covered in 21 CFR 314) for permission to market the new drug.

As stated in 321.21, "A sponsor shall submit an IND if the sponsor intends to conduct a clinical investigation with a new drug... [and] shall not begin a clinical investigation until... an IND... is in effect". Similar procedures are in place in other major countries. In the UK, for example, a Clinical Trials Certificate (CTC) must be filed or a Clinical Trial Exemption (CTX) obtained before clinical trials may proceed. Clinical trials are divided into three phases, as described in 312.21. Phase I trials are initial introductions into healthy volunteers primarily for the purposes of establishing tolerance (side effects), bioavailability, and metabolism. Phase II clinical trials are "controlled studies... to evaluate effectiveness of the drug for a particular indication or disease". The secondary objective is to determine common short-term side effects; hence the subjects are closely monitored. Phase III studies are expanded clinical trials. It is during this phase that the definitive, large-scale, double-blind studies are performed.

The toxicologist's main responsibilities in the IND process are to design, conduct, and interpret appropriate toxicology studies (or "packages") to support the initial IND and then design the appropriate studies necessary to support each additional phase of investigation. Exactly what may constitute appropriate studies is covered elsewhere in this chapter. The toxicologist's second responsibility is to prepare the toxicology summaries for the (clinical) investigator's brochure (described in 312.23(a)(8)(ii)). This is an integrated summary of the toxicological effects of the drug in animals and *in vitro*. The FDA has prepared numerous guidance documents covering the content and format of INDs. It is of interest that in the Guidance for Industry (Lumpkin, 1995) an in-depth description of the expected contents of the Pharmacology and Toxicology sections was presented. The document contains the following self-explanatory passage:

> Therefore, if final, fully quality-assured individual study reports are not available at the time of IND submission, an integrated summary report of toxicological findings based on the unaudited draft toxicologic reports of the completed animal studies may be submitted.

If unfinalized reports are used in an initial IND, the finalized report must be submitted within 120 days of the start of the clinical trial. The sponsor must also prepare a document identifying any differences between the preliminary and final reports, and the impact (if any) on interpretation.

Thus, while the submission of fully audited reports is preferable, the agency does allow for the use of incomplete reports.

Once an IND or CTC/X is opened, the toxicologists may have several additional responsibilities. First, to design, conduct, and report the additional tests necessary to support a new clinical protocol or an amendment to the current clinical protocol (Section 312.20). Second, to bring to the sponsor's attention any finding in an ongoing toxicology study in animals "suggesting a significant risk to human subjects, including any finding of mutagenicity, teratogenicity or carcinogenicity", as described in 21 CFR 312.32. The sponsor has a legal obligation to report such findings within 10 working days. Third, to prepare a "list of the preclinical studies, completed or in progress during the past year" and a summary of the major preclinical findings. The sponsor is required (under Section 312.23) to file an annual report (within 60 days of the IND anniversary date) describing the progress of the investigation. INDs are never "approved" in the strict sense of the word. Once filed, an IND can be opened 30 days after submission, unless the FDA informs the sponsor otherwise. The structure of an IND is outlined in Table 2.4.

If the clinical trials conducted under an IND are successful in demonstrating safety and effectiveness (often established at a pre-NDA meeting, described in 21 CFR 312.47(b)(2)), the sponsor can then submit an NDA. Unlike an IND, the NDA must be specifically approved by the Agency. The toxicologist's responsibility in the NDA/Marketing Authorization Application (MAA) process is to prepare an integrated summary of all the toxicology and/or safety studies performed and be in a position to present and review the toxicology findings to the FDA or its advisory bodies. The approval process can be exhausting, including many meetings, hearings, appeals, etc. The ground rules for all of these are described in Part A of the law. For example, all NDAs are reviewed by an "independent" (persons not connected with either the sponsor or the Agency) Scientific Advisory Panel which will review the findings and make recommendations as to approval. MAAs must be reviewed by and reported on by

*Table 2.4* Composition of standard investigational new drug application[a]

1. IND cover sheets (form FDA-1571)
2. Table of contents
3. Introductory statement
4. General (clinical) investigation plan
5. (Clinical) investigators brochure
6. (Proposed) clinical protocol(s)
7. Chemistry, manufacturing, and control information
8. Pharmacology and toxicology information (includes metabolism and pharmacokinetic assessments done in animals)
9. Previous human experience with the investigational drug
10. Additional information
11. Other relevant information

[a] Complete and thorough reports on all pivotal toxicological studies must be provided with the application.

an expert recognized by the cognizant regulatory authority. Final statutory approval in the US lies with the Commissioner of the FDA. It is hoped that few additional studies will be requested during the NDA review and approval process. When an NDA is approved, the Agency will send the sponsor an approval letter and will issue a Summary Basis of Approval (SBA) (312.30), which is designed and intended to provide a public record on the Agency's reasoning for approving the NDA while not revealing any proprietary information. The SBA can be obtained through Freedom of Information Act, and can provide insights into the precedents for which types of toxicology studies are used to support specific types of claims.

### *Regulations: environmental impact*

Environmental impact statements, while once important only for animal drugs, must now accompany all MDAs. This assessment must also be included in the Drug Master File (DMF). The procedures, formats, and requirements are described in 21 CFR 2531. This requirement has grown in response to the National Environmental Policy Act, the heart of which required that federal agencies evaluate every major action that could effect the quality of the environment. In the INDs, this statement can be a relatively short section claiming that relatively small amounts will post little risk to the environment. The EEC has similar requirements for drug entities in Europe, though data requirements are more strenuous. With NDAs, this statement must be more substantial, detailing any manufacturing and/or distribution process that may result in release into the environment. Environmental fate (e.g., photohydrolysis) and toxicity (e.g., fish, daphnia, and algae) studies will be required. While not mammalian toxicology in the tradition of pharmaceutical testing, preparing an environmental impact statement will clearly require toxicological input. The FDA has published a technical bulletin covering the tests it may require. (FDA, 1987).

### *Regulations: antibiotics*

The NDA law (safety and effectiveness) applies to all drugs, but antibiotic drugs were treated differently until the passage of FDAMA in 1997. Antibiotic drugs had been treated differently by the FDA since the development of penicillin revolutionized medicine during World War II. The laws applicable to antibiotic drugs were covered in 21 CFR 430 and 431. Antibiotics such as penicillin or doxorubicin are drugs derived (in whole or in part) from natural sources (such as moulds or plants) which have cytotoxic or cytostatic properties. They were treated differently from other drugs as the applicable laws required batch-to-batch certification process. Originally passed into law in 1945 specifically for penicillin, this certification process was expanded by the 1962 amendment (under Section 507 of the Food, Drug, and Cosmetic Act) to require certification of all antibiotic drugs, meaning that the FDA would assay each lot of antibiotic for purity, potency, and safety. The actual regulations were covered in 21 CFR Subchapter D, Parts 430–460 (over 600 pages), which describes the standards and methods used for certification for all approved antibiotics. Section 507 was repealed by FDAMA (Section 125). As a result of the repeal of Section 507, the FDA is no longer required to publish antibiotic monographs. In addition, the testing, filing and reviewing of antibiotic applications are now handled under Section 505 of the act like any other new therapeutic agent. The FDA has published a guidance document to which the reader is referred for more details (Anon, 1998a,b).

### Regulations: biologics

Biological products are covered in Subchapter F, Parts 600–680. As described in 21 CFR 600.3(h), "biological product means any virus, therapeutic serum, toxin, antitoxin or analogous product applicable to the prevention, treatment or cure of diseases or injuries of man". In other words, these are vaccines and other protein products derived from animal sources. Clearly the toxicological concerns with such products are vastly different than those involved with low-molecular-weight synthetic molecules. There is little rational basis, for example, for conducting a 1-year, repeated dose toxicity study with a vaccine or a human blood product. The FDA definition for safety with regard to these products is found in 21 CFR 603.1(p): "Relative freedom from harmful effect to persons affected, directly or indirectly, by a product when prudently administered". Such safety consideration has more to do with purity, sterility, and adherence to good manufacturing standards than with the toxicity of the therapeutic molecule itself. The testing required to show safety is stated in the licensing procedures 21 CFR 601.25(d)(1): "Proof of safety shall consist of adequate test methods reasonably applicable to show the biological product is safe under the prescribed conditions". Once a license is granted, each batch or lot of biological product must be tested for safety, and the methods of doing so are written into the law. A general test for safety (i.e. required in addition to other safety tests) is prescribed using guinea pigs as described in 610.11. Additional tests are often applied to specific products. For example, 21 CFR 630.35 describes the safety tests required for measles vaccines, which includes tests in mice and *in vitro* assays with tissue culture. Many new therapeutic entities produced by biotechnology are seeking approval as biologics with the results being FDA approval of a Product License Application (PLA). Table 2.5 presents general guidance for the basis of deciding if an individual entity falls under the CDER or the CBER authority for review.

The International Conferences on Harmonization has published its document S6 Preclincial Safety Evaluation of Biotechnology-Derived Pharmaceuticals. The FDA (both the CDER, and the CBER jointly) has published the document as a Guidance for Industry (Anon, 1997a,b).

A current list of regulatory documents available by e-mail (including the most recent PTCs, or points to consider) can be found at doc_list@a1.fda.gov.

### Regulations versus law

A note of caution must be inserted here. The law (the document passed by Congress) and the regulations (the documents written by the regulatory authorities to enforce the laws) are separate documents. The sections in the law do not necessarily have numerical correspondence. For example, the regulations on the NDA process is described in 21 CFR 312, but the law describing the requirement for an NDA process is in Section 505 of the FDCA. Because the regulations rather than the laws themselves have a greater impact on toxicological practice, greater emphasis is placed on regulation in this chapter. For a complete review of FDA law, the reader is referred to the monograph by the Food and Drug Law Institute in 1984.

*Table 2.5* Product class review responsibilities

| |
|---|
| *Center for drug evaluation and review* |
| Natural products purified from plant or mineral sources |
| Products produced from solid tissue sources (excluding procoagulants, venoms, blood products, etc.) |
| Antibiotics, regardless of method of manufacture |
| Certain substances produced by fermentation |
| Disaccharidase inhibitors |
| Hmg-CoA inhibitors |
| Synthetic chemicals |
| Traditional chemical synthesis |
| Synthesized mononuclear or polynuclear products including antisense chemicals |
| Hormone products |
| *Center for biologics evaluation and review* |
| Vaccines, regardless of manufacturing method |
| In vivo diagnostic allergenic products |
| Human blood products |
| Protein, peptide, and/or carbohydrate products produced by cell culture(other than antibiotics and hormones) |
| Immunoglobulin products |
| Products containing intact cells or microorganisms |
| Proteins secreted into fluids by transgenic animals |
| Animal venoms |
| Synthetic allergens |
| Blood banking and infusion adjuncts |

*Table 2.6* Congressional committees responsible for FDA oversight

| | |
|---|---|
| *Authorization* | |
| Senate | All public health service agencies are under the jurisdiction of the Labor and Human Resources Committee |
| House | Most public health agencies are under the jurisdiction of the Health and the Environmental Subcommittee of the House Energy and Commerce Committee |
| *Appropriation* | |
| Senate | Unlike most other public health agencies, the FDA is under the jurisdiction of the Agriculture, Rural Development, and Related Agencies Subcommittee of the Senate Appropriations Committee |
| House | Under the jurisdiction of the Agriculture, Rural Development, and Related Agencies Subcommittee of the House Appropriations Committee |

Laws authorize the activities and responsibilities of the various federal agencies. All proposed laws before the US Congress are referred to committees for review and approval. The committees responsible for FDA oversight are summarized in Table 2.6. This table also highlights the fact that authorizations and appropriations (the funding necessary to execute authorizations) are handled by different committees.

## Organizations regulating drug and device safety in the US

The agency formally charged with overseeing the safety of drugs in the US is the FDA. It is headed by a commissioner who reports to the Secretary of the Department of Health and Human Services (DHHS) and has a tremendous range of responsibilities. Drugs are overseen primarily by the CDER (though some therapeutic or healthcare entities are considered biologics and are overseen by the corresponding CBER). Figure 2.1 presents the organization of CDER. The organization of CBER is shown in Figure 2.2.

Most of the regulatory interactions of toxicologists are with the two offices of Drug Evaluation, which have under them a set of groups focused on areas of therapeutic claim (cardiorenal, neuropharmacological, gastrointestinal and coagulation, oncology and pulmonary, metabolism and endocrine, antiinfective and antiviral). Within each of these are chemists, pharmacologists/toxicologists, statisticians, and clinicians. When an INDA is submitted to the offices of Drug Evaluation, it is assigned to one of the therapeutic groups based on its area of therapeutic claim. Generally, it will remain with that group throughout its regulatory approval "life". INDs, when allowed, grant investigators the ability to go forward into clinical (human) trials with their drug candidate in a predefined manner, advancing through various steps of evaluation in human (and in additional preclinical or animal studies) until an NDA can be supported, developed, and submitted. Likewise for biological products, the PLA or other applications (INDA, IND) are handled by the offices of Biological Products Review of the CBER.

For drugs, there is at least one nongovernmental body which must review and approve various aspects – the USP – which maintains (and revises) the compendia of the same name. This volume sets forth standards for purity of products in which residues may be present and tests for determining various characteristics of drugs, devices, and biologics. The USP also contains significant "guidance" for the evaluation of safety for devices.

## Process of pharmaceutical product development and approval

Except for a very few special cases (treatments for life-threatening diseases such as cancer or AIDS), the safety assessment of new drugs as mandated by regulations proceeds in a rather fixed manner. The IND is filed to support this clinical testing. An initial set of studies (typically, studies of appropriate length by the route intended for humans are performed in both a rodent (typically rat) and a nonrodent (usually a dog or a primate)) are required to support Phase I clinical testing. Such Phase I testing is intended to evaluate the safety ("tolerance" in clinical subjects), pharmacokinetics, and general biological effects of a new drug, and is conducted in normal volunteers (almost always males).

Successful completion of Phase I testing allows, with the approval of the FDA, progression into Phase II clinical testing. Here, selected patients are enrolled to evaluate therapeutic efficacy, dose ranging, and more details about the pharmacokinetics and metabolism. Longer-term systemic toxicity studies must be in conformity with the guidelines that are presented in the next section. Once a sufficient understanding of the actions, therapeutic dose response, and potential risk-to-benefit ratio of the drug is in hand (once again, with FDA approval), trials move into Phase III testing.

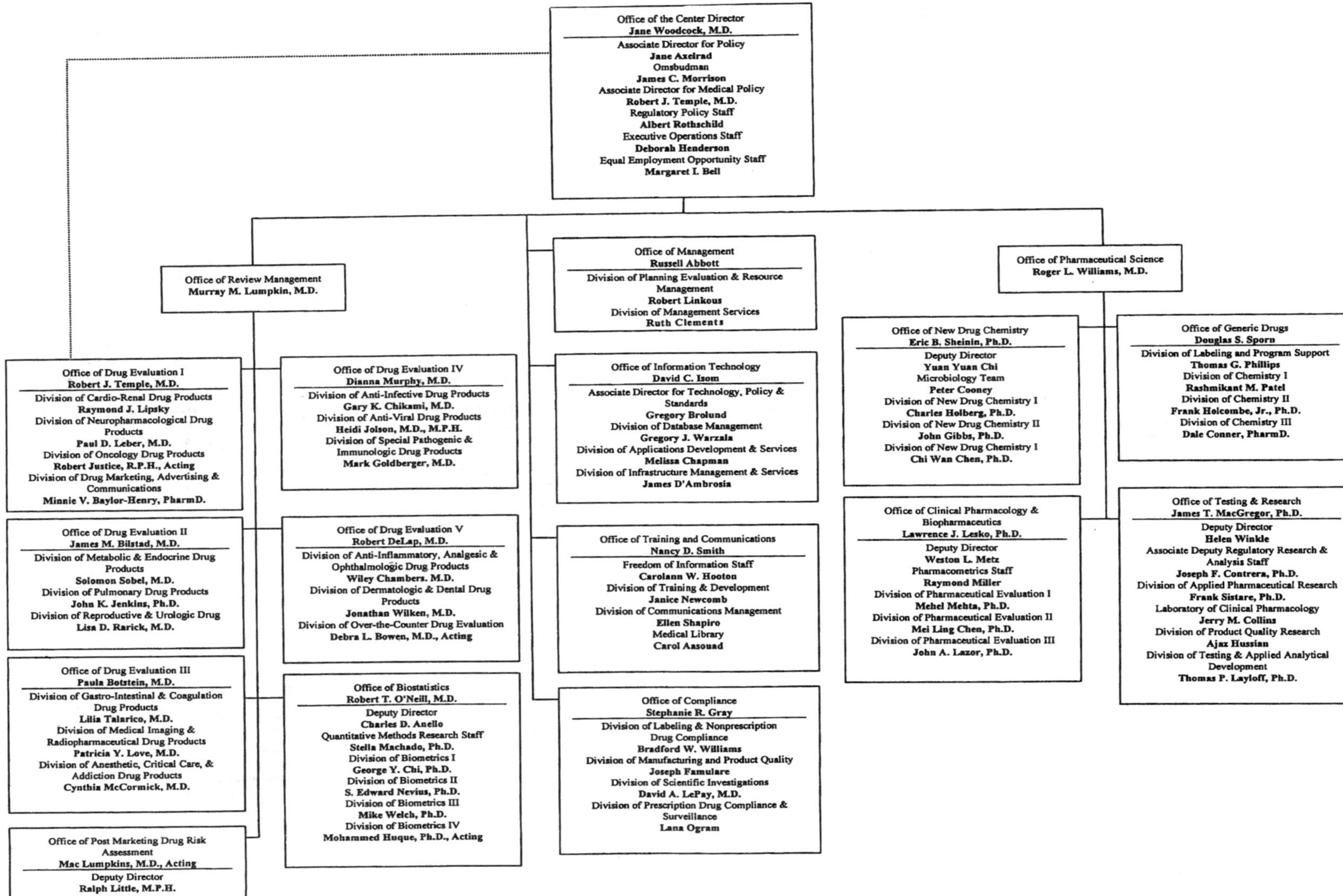


*Figure 2.1* The organization of CDER.

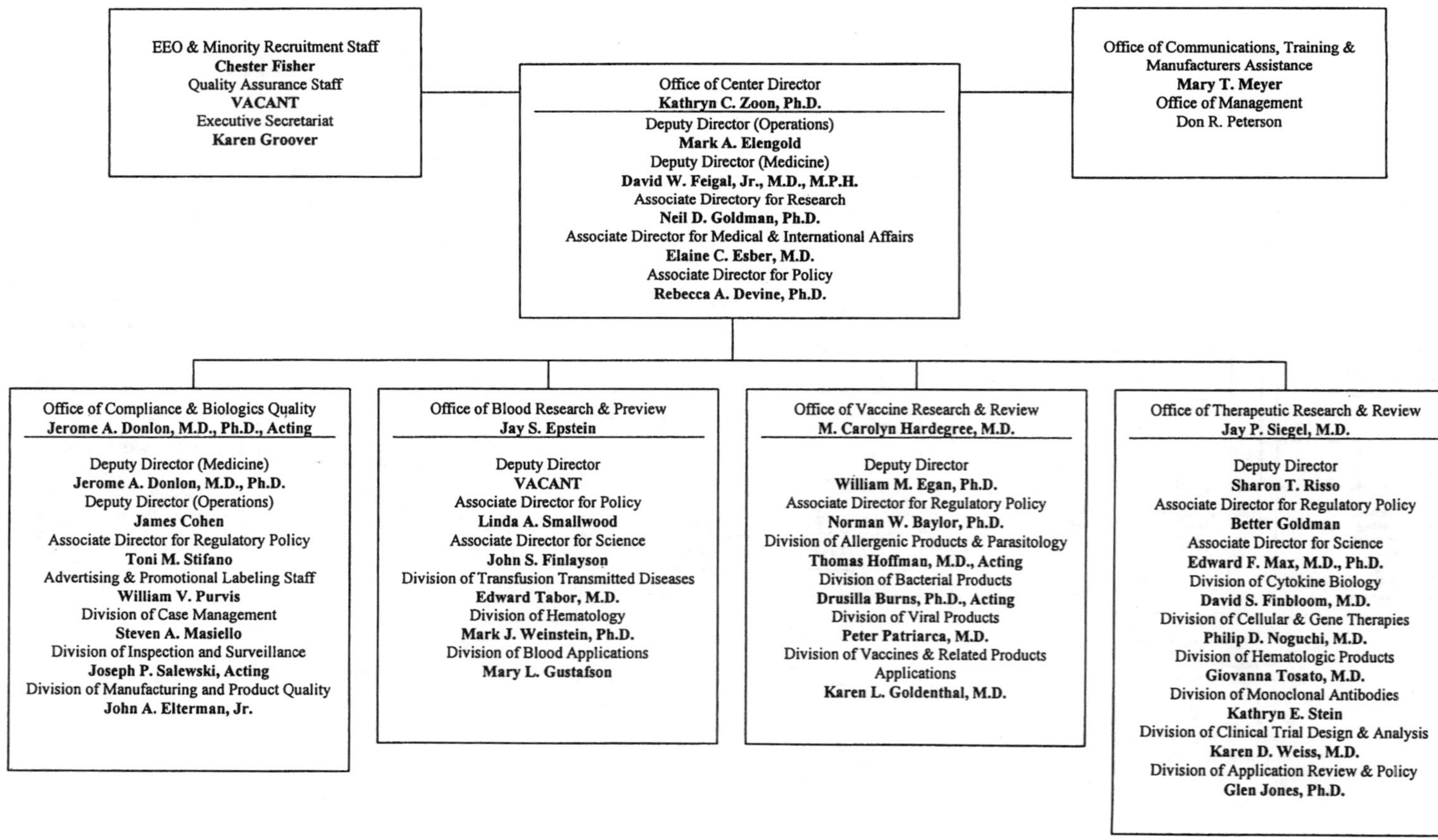


*Figure 2.2* The organization of CBER.

Phase III tests are large, long, and expensive. They are conducted using large samples of selected patients and are intended to produce proof of safety and efficacy of the drug. Two studies providing statistically significant proof of the claimed therapeutic benefit must be provided. All the resulting data from preclinical and clinical animal studies are organized in a specified format in the form of a NDA, which is submitted to the FDA.

By the time that Phase III testing is completed, some additional preclinical safety tests must also generally be in hand. These include the three separate reproductive and developmental toxicity studies (Segments I and III in the rat, and Segment II in the rat and rabbit) and carcinogenicity studies in both rats and mice (unless the period of therapeutic usage is intended to be very short). Some assessment of genetic toxicity will also be expected.

The ultimate product of the pharmaceutical toxicologist will thus generally be the toxicology summaries of the IND and NDA (or PLA). For medical devices, the equivalents are the IDE and Product Development Notification (PDN). Data required to support each of these documents is specified in a series of guidelines, as will be discussed below.

Acceptance of these applications is contingent not only upon adherence to guidelines and good science, but also adherence to GLPs (see Chapter 14 in this volume).

## Testing guidelines

### *Toxicity testing: traditional pharmaceuticals*

Although the 1938 Act required safety assessment studies, no consistent guidelines were available. Guidelines were first proposed in 1949 and published in the *Food, Drug and Cosmetic Law Journal* that year (Burns, 1983). Following several revisions, these guidelines were issued as *The Appraisal Handbook* in 1959. While never formally called a *guideline,* it set the standard for preclinical toxicity test design for several years. The current basic guidelines for testing required for safety assessment in support of the phases of clinical development of drugs were first outlined by Goldenthal (1968) and later incorporated into a 1971 FDA publication entitled *FDA Introduction to Total Drug Quality.*

### *General or systematic toxicity assessment*

Table 2.7 presents an overview of the current FDA toxicity testing guidelines for human drugs. They are misleading in their apparent simplicity, however. First, each of the systemic toxicity studies in these guidelines must be designed and executed in a satisfactory manner. Sufficient animals must be used to have confidence in finding and characterizing any adverse drug actions that may be present. In practice, as the duration of the study increases, small doses are administered and larger numbers of animals must be employed per group. These two features – dosage level and group size – are critical to study designs. Table 2.8 presents general guidance on the number of animals to be used in systemic studies. These and other technical considerations for the safety assessment of pharmaceuticals are present in detail in Gad (1994).

The protocols discussed thus far have focused on general or systemic toxicity assessment. The agency and, indeed, the lay public have a special set of concerns

*Table 2.7* Synopsis of general guidelines for animal toxicity studies for drugs

| *Category* | *Duration of human administration* | *Clinical phase* | *Subacute or chronictoxicity* | *Special studies* |
|---|---|---|---|---|
| Oral or parenteral | Several days | I, II, III, NDA | Two species; 2 weeks | For parentally administered drugs; compatibility with blood where applicable |
| | Up to 2 weeks | I | Two species; 4 weeks | |
| | | II | Two species; up to 4 weeks | |
| | | III, NDA | Two species; up to 3 months | |
| | Up to 3 months | I, II | Two species; 4 weeks | |
| | | III | Two species; 3 months | |
| | | NDA | Two species; up to 6 months | |
| | 6 Months to unlimited | I, II | Two species; 3 months | |
| | | III | Two species; 6 months or longer | |
| | | NDA | Two species; 9 months (non-rodent) and 12 months (rodent) + two rodent species for CA; 18 months (mouse) 24 months (rat)v | |
| Inhalation (general anesthetics) | | I, II, III, NDA | Four species; 5 days (3 h/day) | |
| Dermal | Single application | I | One species; single 24 h exposure followed by 2-week observation | Sensitization |
| | Single or short-term application | II | One species; 20-day repeated exposure (intact and abraded skin) | |

Table 2.7 (*continued*)

| *Category* | *Duration of human administration* | *Clinical phase* | *Subacute or chronictoxicity* | *Special studies* |
|---|---|---|---|---|
| | Short-term application | III | As above | |
| | Unlimited application | NDA | As above, but intact skin study extended up to 6 months | |
| Ophthalmic | Single application | I | | Eye irritation tests with graded doses |
| | Multiple application | I, II, III | One species; 3 weeks daily applications, as in clinical use | |
| | | NDA | One species; duration commensurate with period of drug administration | |
| Vaginal or rectal | Single application | I | | Local and systematic toxicity after vaginal or rectal application in two species |
| | Multiple application | I, II, III, NDA | Two species; duration and number of applications determined by proposed use | |
| Drug combinations | (4) | I | | Lethality by appropriate route, compared to components run concurrently in one species |
| | | II, III, NDA | Two species; up to 3 months | |

with reproductive toxicity, fetal/embryo toxicity, and developmental toxicity (also called *teratogenicity*). Collectively, these concerns often go by the acronyms *DART* (Developmental and Reproductive Toxicity) or *RTF* (Reproduction, Teratogenicity, Fertility). Segment II studies are more designed to detect developmental toxicity. Only pregnant females are dosed during the critical period of organogenesis. Generally, the first protocol DART test (exclusive of range-finding studies) is a Segment I study of rats in fertility and general reproductive performance. This is generally done while the drug is in Phase II clinical trials. Alternatively, many companies are now performing the Segment II teratology study in rats before the Segment I study because the former is less time- and resource-intensive. One or both should be completed before including women of child-bearing potential in clinical trials. The FDA requires teratogenicity testing in two species – a rodent (rat or mouse) and the rabbit. The use of the rabbit was instituted as a result of the finding that thalidomide was a positive teratogen in the rabbit but not in the rat. On occasion, when a test article is not compatible with the rabbit, teratogenicity data in the mouse may be substituted. There are also some specific classes of therapeutics (the quinalone antibiotics, for example) where Segment II studies in primates are effectively required prior to product approval. Both should be completed before entering Phase III clinical trials. The most complicated of the DART protocols – Segment III – is generally commenced during Phase III trials and should be part of the NDA. There are differences in the different national guidelines (as discussed later with international considerations) regarding the conduct of these studies. The large multinational drug companies try to design their protocols to be in compliance with as many of the guidelines as possible to avoid duplication of testing while allowing the broadest possible approval and marketing of therapeutics.

### Genetic toxicity assessment

Genetic toxicity testing generally focuses on the potential of a new drug to cause mutations (in single-cell systems) or other forms of genetic damage. The tests, generally short in duration, often rely on *in vitro* systems and generally have a single endpoint of effect (point mutations, chromosomal damage, etc.). For a complete review of proto-

*Table 2.8* Numbers of animals per dosage group in systemic toxicity studies

| *Study duration (per sex)* | *Rodents (per sex)* | *Non-rodents* |
|---|---|---|
| 2–4 weeks | 5 | 3 |
| 13 weeks | 20[a] | 6 |
| 26 weeks | 30 | 8 |
| Chronic | 50 | 10 |
| Carcinogenicity | 60[b] | Applies only to contraceptives |
| Bioassays | | Applies only to contraceptives |

[a] Starting with 13-week studies, one should consider adding animals (particularly to the high dose) to allow evaluation of reversal of effects.

[b] In recent years there have been decreasing levels of survival in rats on 2-year studies. What is required is that at least 20–25 animals/sex per group survive at the end of the study. Accordingly, practice is beginning to use 70 or 75 animals per sex, per group.

cols, technology, etc., the reader is referred to Brusick (1987). It is of interest that the FDA has no standard or statutory requirement for genetic toxicity testing but generally expects to see at least some such tests performed and will ask for them if the issue is not addressed. If one performs such a study, any data collected, of course, must be sent to the Agency as part of any INDA, PLA, or NDA. These studies have yet to gain favor with the FDA (or other national regulatory agencies) as substitutes for *in vivo* carcinogenicity testing. However, even with completed negative carcinogenicity tests, at least some genetic toxicity assays are generally required. Generally, pharmaceuticals in the US are evaluated for mutagenic potential (e.g., the Ame's assay) or for chromosomal damage (e.g., the In Vivo Mouse Micronucleus Test). In general, in the US, pharmaceutical companies apply genetic toxicity testing in the following fashion:

- *As a screen*: an agent that is positive in one or more genetic toxicity tests may be more likely than one that is negative to be carcinogenic and, therefore, may not warrant further development.
- *As an adjunct*: an agent that is negative in carcinogenicity testing in two species and also negative in a genetic toxicity battery is more likely than not to be noncarcinogenic in human beings.
- *To provide mechanistic insight*: for example, if an agent is negative in a wide range of genetic toxicity screens, but still produces tumors in animals, then one could hypothesize that an epigenetic mechanism was involved.

While not officially required, the FDA does have the authority to request, on a case-by-case basis, specific tests it feels may be necessary to address a point of concern. A genetic toxicity test could be part of such a request. In general, therefore, companies deal with genetic toxicity (after "screening") on a case-by-case basis, dictated by good science. If more than a single administration is intended, common practice is to perform the tests prior to submitting an IND.

### Toxicity testing: biotechnology products

As mentioned, the regulation of traditional pharmaceuticals (small molecules such as aspirin or digitalis) and biologicals (proteins such as vaccines and antitoxins derived from animal sources) have very different histories. See the discussion on biologics earlier in this chapter. Until 1972, the NIH (or its forerunning agency, the Hygienic Laboratory of the Department of the Treasury) was charged with the responsibilities of administering the Virus Act of 1902. With the passage of the Food and Drug Laws of 1906, 1938, and 1962, there was reoccurring debate about whether these laws applied or should apply to biologicals. This debate was resolved when the authority for the regulation of biologics was transferred to the FDA's new Bureau of Biologics (now the CBER) in 1972. Since then, there appears to have been little difference in the matter of regulation for biologics and pharmaceuticals. The FDA essentially regulates biologics as described under the 1902 Act, but then uses the rule-making authority granted under the Food and Drug Act to "fill in the gaps".

The Bureau of Biologics was once a relatively "sleepy" agency, primarily concerned with the regulation of human blood products and vaccines used for mass immunization programs. The authors of the 1902 law could hardly have foreseen the explosion in biotechnology that occurred in the 1980s. New technology created a welter of new biological products, such as recombinant-DNA-produced proteins (e.g., tissue plasminogen activator), biological response modifiers (cytokinins and colony-stimulating factors), monoclonal antibodies, antisense oligonucleotides, and self-directed vaccines (raising an immune response to self-proteins such as gastrin for therapeutic reasons). The new products raised a variety of new questions on the appropriateness of traditional methods of evaluating drug toxicity that generated several points-to-consider documents. For the sake of brevity, this discussion will focus on the recombinant DNA proteins. Some of the safety issues that have been raised over the years are as follows:

- The appropriateness of testing a human-specific peptide hormone in nonhuman species.
- The potential that the peptide could breakdown due to nonspecific metabolism, resulting in products that had no therapeutic value or even a toxic fragment.
- The potential sequelae to an immune response (formation of neutralizing antibodies, provoking an autoimmune or a hypersensitivity response), and pathology due to immune precipitation, etc.
- The presence of contamination with oncogenic virus DNA (depending on whether a bacterial or mammalian system was used on the synthesizing agent) or endotoxins.
- The difficulty interpreting the scientific relevance of response to supraphysiological systemic doses of potent biological response modifiers.

The intervening last few years have shown that some of these concerns were more relevant than others. The "toxic peptide fragment" concern, for example, has been shown to be without merit. The presence of potentially oncogenic virus DNA and endotoxins is a quality assurance concern and is not truly a toxicological problem. Regardless of the type of synthetic pathway, all proteins must be synthesized in compliance with Good Manufacturing Practices. Products must be as pure as possible, not only free of rDNA but also free of other types of cell debris (endotoxin). Batch-to-batch consistency with regard to molecular structure must also be demonstrated using appropriate methods (e.g., amino acid). The regulatory thinking and experience over the last 15 years has come together in the document, "S6 Preclincial Safety Evaluation of Biotechnology-Derived Pharmaceuticals" prepared by the International Conferences on Harmonization. The FDA (both the CDER, and the Center for Biologics Evaluation and Research jointly) has published the document as a Guidance for Industry (Anon, 1997a,b). The document intended to provide basic guidance for the preclinical evaluation of biotechnology derived products, including proteins and peptides, either produced by cell culture using rDNA technology, but did not cover antibiotics, allergenic extracts, heparin, vitamins, cellular drug products vaccines, or other products regulated as biologics. Items covered are summarized as follows:

- *Test article specifications*: in general, the product that is used in the definitive pharmacology and toxicology studies should be comparable to the product proposed for the initial clinical studies.
- *Animal species/model selection*: safety evaluation should include the use of relevant species, in which the test article is pharmacologically active due, for example to the expression of the appropriate receptor molecule. These can be screened with *in vitro* rector binding assays. Safety evaluation should normally include two appro-

priate species, if possible and/or feasible. The potential utility of gene knockout and/or transgenic animals in safety assessment is discussed.

- *Group size*: no specific numbers are given, but it does state that a small sample size may lead to failure to observe toxic events.
- *Administration*: the route and frequency should be as close as possible to that proposed for clinical use. Other routes can be used when scientifically warranted.
- *Immunogenicity*: it has also been clearly demonstrated in the testing of rDNA protein products that animals will develop antibodies to foreign proteins. This response has been shown to neutralize (rapidly remove from circulation) the protein, but no pathological conditions have been shown to occur as a sequelae to the immune response. Bear in mind, however, that interleukins have powerful effects on immune response, but these are due to their physiological activity and not due to an antigen–antibody response. The first has to do with "neutralizing antibodies"; i.e. is the immune response so great that the test article is being removed from circulation as fast as it is being added? If this is the case, does long-term testing of such a chemical make sense? In many cases, it does not. The safety testing of any large molecule should include the appropriate assays for determining whether the test system has developed a neutralizing antibody response. Depending on the species, route of administration, intended therapeutic use, and development of neutralizing antibodies (which generally takes about 2 weeks), it is rare for a toxicity test on an rDNA protein to be longer than 4 weeks duration. However, if the course of therapy in humans is to be longer than 2 weeks, formation of neutralizing antibodies must be demonstrated or longer-term testing performed. The second antigen–antibody formation concern is that a hypersensitivity response will be elicited. Traditional preclinical safety assays are generally adequate to guard against this if they are 2 weeks or longer in duration and the relevant endpoints are evaluated.
- *Safety pharmacology*: it is important to investigate the potential for unwanted pharmacological activity in appropriate animal models and to incorporate monitoring for these activities in the toxicity studies.
- *Exposure assessment*: single and multiple dose pharmacokinetics, toxicokinetics and tissue distribution studies in relevant species are useful. Proteins are not given orally, demonstrating absorption and mass balance is not typically a primary consideration. Rather, this segment of the test should be designed to determine half-life (and other appropriate pharmacokinetic descriptor parameters), the plasma concentration associated with biological effects, and potential changes due to the development of neutralizing antibodies
- *Reproductive performance and developmental toxicity studies* will be dictated by the product, clinical indication and intended patient population.
- *Genotoxicity studies*: the S6 document states that the battery of genotoxicity studies routinely conducted for traditional pharmaceuticals is not appropriate for biotechnology derived pharmaceuticals. In contrast to small molecules, genotoxicity testing with a battery of *in vitro* and *in vivo* techniques of protein molecules has not become common US industry practice. Such tests are not formally required by the FDA but, if performed, have to be reported. They are required by European and Japanese regulatory authorities. This has sparked a debate as to whether or not genotoxicity testing is necessary or appropriate for rDNA protein molecules. It is

the authors' opinion that such testing is, scientifically, of little value. First, large protein molecules will not easily penetrate the cell wall of bacteria or yeast, and (depending on size, charge, lipophilicity, etc.) penetration across the plasma lemma of mammalian cells will be highly variable. Second, if one considers the well-established mechanism(s) of genotoxicity of small molecules, it is difficult to conceive how a protein can act in the same fashion. For example, proteins will not be metabolized to be electrophilic active intermediates that will crosslink guanine residues. In general, therefore, genotoxicity testing with rDNA proteins is a waste of resources. It is conceivable, however, that some proteins, because of their biological mechanism of action, may stimulate the proliferation of transformed cells. For example, it is a feasible hypothesis that a colony-stimulating factor could stimulate the proliferation of leukemic cells (it should be emphasized that this is a hypothetical situation, presented here for illustrative purposes). Again, this is a question of a specific pharmacological property, and such considerations should be tested on a case-by-case basis.
- *Carcinogenicity studies* are generally inappropriate for biotechnology derived pharmaceuticals; however, some products may have the potential to support or induce proliferation of transformed cells... possibly leading to neoplasia. When this concern is present, further studies in relevant animal models may be needed.

These items are covered in greater detail in the S6 guidance document and in a review by Yaes and Ryffel (1997).

So, given the above discussion, what should the toxicology testing package of a typical rDNA protein resemble? Based on the products that have successfully wound their way through the regulatory process, the following generalizations can be drawn:

- The safety tests look remarkably similar to those for traditional tests. Most have been done on three species: the rat, the dog, or the monkey. The big difference has to do with the length of the test. It is rare for a safety test on a protein to be more than 13 weeks long.
- The dosing regimens can be quite variable and at times very technique-intensive. These chemicals are almost always administered by a parenteral route of administration; normally intravenously or subcutaneously. Dosing regimens have run the range from once every 2 weeks for an antihormone "vaccine" to continuous infusion for a short-lived protein.
- As reviewed by Ryffel (1996) most side effects in man of a therapy with rDNA therapy may be predicted by data from experimental toxicology studies, but there are exceptions. IL-6 for example, induced a sustained increase in blood platelets and acute phase proteins, with no increase in body temperature. In human trials, however, there were increases in temperature.
- The S6 document also mentions monoclonal antibody products. Indeed, many of the considerations for rDNA products are also applicable to monoclonal antibodies (including hybridized antibodies). With monoclonal antibodies, there is the additional concern of crossreactivity with nontarget molecules.

As mentioned, the rapid development in the biotechnology industry has created some confusion as to what arm of the FDA is responsible for such products. In October 1992, the two major reviewing groups, CBER and CDER, reached a series of agreements to explain and organize the FDA's position on products that did not easily fall into its traditional classification schemes. CDER will continue to have responsibility for traditional chemically synthesized molecules as well as those purified from mineral or plant sources (except allergenics), antibiotics, hormones (including insulin, growth hormone, etc.), most fungal or bacterial products (disaccharidase inhibitors), and most products from animal or solid human tissue sources. CBER will have responsibility for products subject to licensure (BLA), including all vaccines, human blood or blood-derived products (as well as drugs used for blood banking and transfusion), immunoglobulin products, products containing intact cells, fungi, viruses, proteins produced by cell culture or transgenic animals, and synthetic allergenic products. This situation was further simplified by the introduction of the concept of "well-characterized biologics". When introduced during the debate on FDA reform in 1996, the proposed section of S-1447 stated that "Biological products that the secretary determines to be well-characterized shall be regulated solely under the Federal Food, Drug and Cosmetic Act". Under this concept, highly purified, well-characterized therapeutic rDNA proteins would be regulated by CDER, regardless of therapeutic target (Anon., 1996).

### *Toxicity/safety testing: cellular and gene therapy products*

Human clinical trials of cellular and gene therapies involve administration to patients of materials considered investigational biological, drug or device products. Somatic cell therapy refers to the administration to humans of autologous, allogenic or xenogenic cells which have been manipulated or processed *ex vivo*. Gene therapy refers to the introduction into the human body of genes or cells containing genes foreign to the body for the purposes of prevention, treatment, diagnosing or curing disease.

Sponsors of cellular or gene therapy clinical trials must file an IND or in certain cases an Investigational Device Exemption (IDE) with the FDA before initiation of studies in humans. It is the responsibility of the CBER to review the application and determine if the submitted data and the investigational product meet applicable standards. The critical parameters of identity, purity, potency, stability, consistency, safety and efficacy relevant to biological products are also relevant to cellular and gene therapy products.

In 1991, FDA first published a Points to Consider document on human somatic cell and gene therapy. At this time virtually all gene therapies were retroviral and were prepared as *ex vivo* somatic cell therapies. This was subsequently reviewed by Kessler et al. (1993). While the data for certain categories of information such as the data regarding the molecular biology were defined in previous guidance documents relating to recombinant DNA products, the standards for preclinical and clinical development were less well-defined. Over the past 5 years, the field has advanced to include not only new vectors but also novel routes of administration. The PTC on Human Somatic Cell and Gene Therapy (1996) has thus been recently amended to reflect both the advancements in product development and more importantly the accumulation of safety information over the past 5 years.

FDA regulations state that the sponsor must submit, in the IND, adequate information about pharmacological and toxicological studies of the drug including laboratory animals or *in vitro* studies on the basis of which the sponsor has considered that it is reasonably safe to conduct the proposed clinical investigation. For cellular and gene therapies, designing and conducting relevant preclinical safety testing has been a challenge to both the FDA and the sponsor. For genes delivered using viral vectors, the safety of the vector system per se must be considered and evaluated.

The preclinical knowledge base is initially developed by designing studies to answer fundamental questions. The development of this knowledge base is generally applicable to most pharmaceuticals as well as biopharmaceuticals, and include data to support: (1) the relationship of the dose to the biological activity, (2) the relationship of the dose to the toxicity, (3) the affect of route and/or schedule on activity or toxicity and (4) identification of the potential risks for subsequent clinical studies. These questions are considered in the context of indication and/or disease state. In addition there are often unique concerns relating to the specific category or product class.

For cellular therapies safety concerns may include development of a data base from studies specifically designed to answer questions relating to growth factor dependence, tumorigenicity, local and systemic toxicity, and affects on host immune responses including immune activation and altered susceptibility to disease. For viral-mediated gene therapies, specific questions may relate to the potential for over expression of the transduced gene, transduction of normal cells/tissues, genetic transfer to germ cells and subsequent alterations to the genome, recombination/rescue with endogenous virus, reconstitutions of replication competence, potential for insertional mutagenesis/malignant transformation, altered susceptibility to disease, and/or potential risk(s) to the environment.

To date cellular and gene therapy products submitted to FDA have included clinical studies indicated for bone marrow marking, cancer, cystic fibrosis, AIDS, and inborn errors of metabolism and infectious diseases. Of the current active INDs approximately 78 per cent have been sponsored by individual investigators or academic institutions and 22 per cent have also been industry sponsored. In addition to the variety of clinical indications the cell types have also been varied. Examples include Tumor Infiltrating Lymphocytes (TIL) and Lymphocyte Activated Killer (LAK) cells, selected cells from bone marrow and peripheral blood lymphocytes, e.g., stem cells, myoblasts, tumor cells and encapsulated cells (e.g., islet cells and adrenal chromaffin cells).

### *Cellular therapies*

Since 1984 CBER has reviewed close to 300 somatic cell therapy protocols. Examples of the specific categories include manipulation, selection, mobilization, tumor vaccines and other.

- *Manipulation*: autologous, allogenic, or xenogenic cells which having been expanded, propagated, manipulated or had their biological characteristics altered *ex vivo* (e.g., TIL or LAK cells; islet cells housed in a membrane).
- *Selection*: products designed for positive or negative selection if autologous or allogenic cells intended for therapy (e.g., purging of tumor from bone marrow, selection of CD34 + cells).
- *Mobilization*: *in vivo* mobilization of autologous stem cells intended for transplantation.
- *Tumor vaccines*: autologous or allogenic tumor cells which are administered as vaccine (e.g., tumor cell lines, tumor cell lysates, primary explant). This group

also includes autologous antigen presenting cells pulsed with tumor specific peptides or tumor cell lysates.
- *Other*: autologous, allogenic, and xenogenic cells which do not specifically fit above. This group includes cellular therapies such as extracorporeal liver assist devices.

### *Gene therapies*

The types of vectors that have been used, or proposed, for gene transduction include retrovirus, adenovirus, adeno-associated viruses, other viruses (e.g., herpes, vaccinia, etc.), and plasmid DNA. Methods for gene introduction include *ex vivo* replacement, drug delivery, marker studies, and others and *in vivo*, viral vectors, plasmid vectors and vector producer cells.

*Ex vivo*

- *Replacement*: cells transduced with a vector expressing a normal gene in order to correct or replace the function of a defective gene.
- *Drug delivery*: cells transduced with a vector expressing a gene encoding a therapeutic molecule which can be novel or native to the host.
- *Marker studies*: cells (e.g., bone marrow, stem cells) transduced with a vector expressing a marker or reporter gene used to distinguish it from other similar host tissues.
- *Other*: products which do not specifically fit under above (e.g., tumor vaccines in which cells are cultured or transduced *ex vivo* with a vector).

*In vivo*

- *Viral vectors*: the direct administration of a viral vector (e.g., retrovirus, adenovirus, adeno-associated virus, herpes, vaccinia) to patients.
- *Plasmid vectors*: the direct administration of plasmid vectors with or without other vehicles (e.g., lipids) to patients
- *Vector producer cells*: the direct administration of retroviral vector producer cells (e.g., murine cells producing HTK vector) to patients.

### *Preclinical safety evaluation*

The goal of preclinical safety evaluation includes: recommendation of an initial safe starting dose and safe dose-escalation scheme in humans, identification of potential target organ(s) of toxicity, identification of appropriate parameters for clinical monitoring and identification of "at risk" patient population(s). Therefore, when feasible, toxicity studies should be performed in relevant species to assess a dose-limiting toxicity. General considerations in study design include selection of the model (e.g., species, alternative model, animal model or disease), dose (e.g., route, frequency and duration) and study endpoint (e.g., activity and/or toxicity).

The approach to preclinical safety evaluation of biotechnology-derived products, including novel cellular and gene therapies, has been referred to as the 'case-by case' approach. This approach is science-based, data-driven and flexible. The major distinction from past practices of traditional pharmaceuticals is that the focus is directed at asking specific questions across various product categories. Additionally, there is a consistent reevaluation of the knowledge base to reassess real or theoretical safety concerns and hence reevaluation of the need to answer the same questions across all product categories. In some cases there may even be conditions which may not need specific toxicity studies. For example, when there is a strong efficacy model which is rationally designed to answer specific questions and/or there is previous human experience with a similar product with respect to dose and regimen.

### *Basic principles for preclinical safety evaluation of cellular and gene therapies*

Biotechnology-derived products in general

- Use of product in animal studies that is comparable to that of the product proposed for clinical trial(s).
- Adherence to basic principles of GLP to ensure quality of the study including a detailed protocol prepared prospectively.
- Use of the same or similar route and method of administration as proposed for clinical trials (whenever possible).
- Determination of appropriate doses delivered based upon preliminary activity obtained from both *in vitro* and *in vivo* studies (i.e. finding a dose likely to be effective and not dangerous, no observed adverse effect level, and a dose causing dose limiting toxicity).
- Selection of one or more species sensitive to the endpoint being measured, e.g., infections or pathologic sequelae and/or biological activity or receptor binding.
- Consideration of animal model(s) of disease may be better to assess the contribution of changes in physiologic or underlying physiology to safety and efficacy.
- Determination of affect on host immune response.
- Localization/distribution studies – evaluation of target tissue, normal surrounding tissue and distal tissue sites and any alteration in normal or expected distribution.
- Local reactogenicity.

### *Additional considerations for cellular therapies*

- Evaluation of cytopathogenicity
- Evaluation of signs of cell transformation/growth factor dependence-effect on animal cells, normal human cells and cells prone to transform easily
- Determination of alteration in cell phenotype, altered cell products and/or function
- Tumorigenicity

### *Additional considerations for gene therapies*

- Determination of phenotype/activation state of effector cells
- Determination of vector/transgene toxicity

- Determination of potential transfer to germ line
- *In vitro* challenge studies – evaluation of recombination or complementation, potential for "rescue" for subsequent infection with wild-type virus
- Determination of persistence of cells/vector
- Determination of potential for insertional mutagenesis (malignant transformation)
- Determination of environmental spread (e.g., viral shedding)

## Toxicity testing: special cases

On paper, the general case guidelines for the evaluation of the safety of drugs are relatively straightforward and well understood. However, there are also a number of special case situations under which either special rules apply or some additional requirements are relevant. The more common of these are summarized below.

### *Oral contraceptives*

Oral contraceptives are subject to special testing requirements. These have recently been modified so that in addition to those preclinical safety tests generally required, the following are also required (Berliner, 1974):

- A 3-year carcinogenicity study in beagles (this is a 1987 modification in practice from earlier FDA requirements and the 1974 publication).
- A rat reproductive (Segment III) study including a demonstration of return to fertility.

### *Life-threatening diseases (compassionate use)*

Drugs to treat life-threatening diseases are not strictly held to the sequence of testing requirements as put forth in Table 2.3 because the potential benefit on any effective therapy in these situations is so high. In the early 1990s, this situation applied to AIDS-associated diseases and cancer. The development of more effective HIV therapies (protease inhibitors) has now made cancer therapy more the focus of these considerations. Though the requirements for safety testing prior to initial human trials is unchanged, subsequent requirements are flexible and subject to negotiation and close consultation with FDA's Division of Oncology (within CDER) (FDA, 1988a–c). The more recent thinking on anticancer agents has been reviewed by DeGeorge et al. (1998). The preclinical studies that will be required to support clinical trials and marketing of new anticancer agents will depend on the mechanism of action and the target clinical population. Toxicity studies in animals will be required to support initial clinical trials. These studies have multiple goals:

- Determine a starting dose for clinical trials
- Identify target organ toxicity and assess recovery
- Assist in the design of clinical dosing regimens

The studies should generally confirm to the protocols recommended by the National Cancer Institute. as discussed by Grieshaber (1991) Greishaber (1991). In general, it can be assumed that most antineoplastic cytotoxic agents will be highly toxic. Two studies are essential to support initial clinical trials (IND phase) in patients with advanced disease. These are studies of 5–14 days in length, but with longer recovery periods. A study in rodents is required that identifies those doses that produce either life-threatening or nonlife threatening toxicity. Using the information from this first study, a second study in nonrodents (generally the dog) is conducted to determine if the tolerable dose in rodents produces life-threatening. Doses are compared on a $mg/m^2$ basis. The staring dose in initial clinical trails is generally one-tenth of that required to produce severe toxicity in rodents (STD10) or one-tenth the highest dosed in nonrodents that does not cause severe irreversible toxicity. While not required, pharmacokinetic information, especially information comparing the plasma concentration associated with toxicity in both species is virtually essential. Special attention is paid to organs with high cell-division rates, bone marrow, tests, lymphoid tissue testing, and GI tract. As these agents are almost always given intravenously, special attention needs to be given relatively early in development to intravenous irritation and blood compatibility studies. Subsequent studies to support the NDA will be highly tailored, depending on the following:

- Therapeutic indication and mechanism of action
- The results of the initial clinical trials
- The nature of the toxicity
- Proposed clinical regimen

Even at the NDA stage, toxicity studies with more than 28 days of dosing are rarely required. While not required for the IND, assessment of genotoxicity and developmental toxicity will need to be addressed. For genotoxicity, it will be important to establish the ratio between cytotoxicity and mutagenicity. In vivo models, e.g., the mouse micronucleus test, can be particularly important in demonstrating the lack of genotoxicity at otherwise sub-toxic doses. For developmental toxicity, International Conference on Harmonization (ICH) stage C–D studies (traditionally known as Segment II studies for teratogenicity in rat and rabbits) will also be necessary.

The emphasis of this discussion has been on purely cytotoxic neoplastic agents. Additional considerations must be given to cytotoxic agents that are administered under special circumstances: those that are photo-activated, delivered as liposomal emulsions, or delivered as antibody conjugates. These types of agents will require additional studies. For example, a liposomal agent will need to be compared to the free agent and a blank liposomal preparation. There are also studies that may be required for a particular class of agent. For example, anthracylcines are known to be cardiotoxic, so comparison of a new anthracylcine agent to previously marketed anthracylines will be expected.

In addition to antineoplastic, cytotoxic agents, there are cancer therapeutic or preventative drugs that are intended to be given on a chronic basis. This includes chemopreventatives, hormonal agents, immunomodulators, etc. The toxicity assessment studies on these will more closely resemble those of more traditional pharmaceutical agents. Chronic toxicity, carcinogenicity and full developmental toxicity (ICH A–B, C–D, E–F) assessments will be required. For a more complete review, the reader is referred to DeGeorge et al. (1998).

### *Optical isomers*

The FDA (and like regulatory agencies, as reviewed by Daniels et al., 1997) has become increasingly concerned with the safety of stereoisomeric or chiral drugs. Stereoisomers are molecules that are identical to one another in terms of atomic formula and covalent bonding, but differ in the three-dimensional projections of the atoms. Within this class are those molecules that are nonsuperimposable mirror images of one another. These are called enantiomers (normally designated as R- or S-). Enantiometric pairs of a molecule have identical physical and chemical characteristics except for the rotation of polarized light. Drugs have generally been mixtures of optical isomers (enantiomers), because of the difficulties in separating the isomers. It has become apparent in recent years, however, that these different isomers may have different degrees of both desirable therapeutic and undesirable toxicologic effects. Technology has also improved to the extent that it is now possible to perform chiral specific syntheses, separations, and/or analyses. It is now highly desirable from a regulatory (FDA, 1988a–c, Anon 1992a) basis to develop a single isomer unless all isomers have equivalent pharmacological and toxicologic activity. The FDA has divided enantiometric mixtures into the following categories.

- Both isomers have similar pharmacologic activity, which could be identical, or they could differ in the degrees of efficacy.
- One isomer is pharmacologically active, while the other is inactive.
- Each isomer has completely different activity.

During preclinical assessment of an enantiometric mixture, it may be important to determine to which of these three classes it belongs. The pharmacological and toxicological properties of the individual isomers should be characterized. The pharmacokinetic profile of each isomer should be characterized in animal models with regard to disposition and interconversion. It is not at all unusual for each enantiomer to have a completely different pharmacokinetic behavior.

If the test article is an enantiomer isolated from a mixture that is already well characterized (e.g., already on the market), then appropriate bridging guides need to be performed which compare the toxicity of the isomer to that of the racemic mixture. The most common approach would be to conduct a subchronic (3 months) and a Sement II type teratology study with an appropriate "positive" control group which received the racemate. In most instances no additional studies would be required if the enantiomer and the racemate did not differ in toxicity profile. If, on the other hand, differences are identified, the reasons for this difference need to be investigated and the potential implications for human subjects need to be considered.

### *Special populations: pediatric and geriatric claims*

Relatively few drugs marketed in the US (approximately 20 per cent) have pediatric dosing information available. Clinical trials have rarely been done specifically on pediatric patients. Traditionally, dosing regimens for children have been derived empirically by extrapolating on the basis of body weight or surface area. This approach assumes that the pediatric patient is a young adult, which simply may not be the case. There are many examples of how adults and children differ qualitatively in metabolic and/or pharmacodynamic responses to pharmaceutical agents. In their review, Schacter and DeSantis (1998) state

> The benefit of having appropriate usage information in the product label is that healthcare practitioners are given the information necessary to administer drugs and biologics in a manner that maximizes safety, minimizes unexpected adverse events, and optimizes treatment efficacy. Without specific knowledge of potential drug effects, children may be placed at risk. In addition, the absence of appropriate proscribing information, drugs and biologics that represent new therapeutic advances may not be administered to the pediatric population in a timely manner.

In response to the need for pediatric information, the FDA developed a pediatric plan. This two-phase plan called first for the development of pediatric information on marketed drugs. The second phase focused on new drugs. The implementation of the plan was to be coordinated by the Pediatric Subcommittee of the Medical Policy Coordinating Committee of CDER. The Pediatric Use Labeling Rule was a direct result of phase 1 in 1994. (Anon, 1998b). Phase 2 resulted in 1997 from a proposed rule entitled "Pediatric Patients, Regulations Requiring Manufacturers to Assess the Safety and Effectiveness of New Drugs and Biologics". Soon after this rule was proposed, the FDA Modernization Act of 1997 was passed. FDAMA contained provisions that specifically addressed the needs and requirements for the development of drugs for the pediatric population.

The FDAMA bill essentially codified and expanded several regulatory actions initiated by the FDA during the 1990s. Among the incentives offered by the bill, companies will be offered an additional 6 months of patent protection for performing pediatric studies (clinical trials) on already approved products. In fact, the FDA was mandated by FDAMA to develop a list of over 500 drugs for which additional information would produce benefits for pediatric patients. The FDA is supposed to provide a written request for pediatric studies to the manufacturers (Hart, 1999).

In response to the pediatric initiatives, the FDA has published policies and guidelines and conducted a variety of meetings. CDER has established a web site (http//www.fda.-gov/cder/pediatric) which lists three pages of such information. Interestingly, the focus has been on clinical trials, and almost no attention has been given to the preclinical toxicology studies that may be necessary to support such trials. There are three pages of documents on the pediatric web site. None appear to address the issue of appropriate testing. This is a situation that is just now being addressed and is in a great deal of flux.

In the absence of any guidelines from the Agency for testing drugs in young or "pediatric" animals, one must fall back on the maxim of designing a program that makes the most scientific sense. As a guide, the FDA designated levels of post-natal human development and the approximate equivalent ages (in the author's considered opinion) in various animal models are given in Table 2.9. The table is somewhat inaccurate, however, because of difference in the stages of development at birth. A rat is born quite underdeveloped when compared to a human being. A 1-day-old rat is not equivalent to a 1-day-old full-term human infant. A 4-day-old rat would be more appropriate. In terms of development, the pig may be the best model of those listed; however, one should bear in mind that different organs have different developmental schedules in different species.

*Table 2.9* Comparison of post-natal development stages

| *Stage* | *Human* | *Rat* | *Dog* | *Pig* |
|---|---|---|---|---|
| Neonate | Birth to 1 month | Birth to 1 week | Birth to 3 weeks | Birth to 2 weeks |
| Infant | 1 month to 2 years | 1–3 weeks | 3–6 weeks | 2–4 weeks |
| Child | 2–12 years | 3–9 weeks | 6 weeks to 5 months | 4 weeks to 4 months |
| Adolescent | 12–16 years | 9–13 weeks | 5–9 months | 4–7 months |
| Adult | Over 16 years | Over 13 weeks | Over 9 months | Over 7 months |

Table 2.9 can be used as a rough guide in designing toxicity assessment experiments in developing animals. In designing the treatment period, one needs to consider, not only the dose and the proposed course of clinical treatment, but also the proposed age of the patient, and whether or not an equivalent dosing period in the selected animal model covers more than one developmental stage. For example, if the proposed patient population is human infants, initiating a toxicity study of the new pharmaceutical agent in 3-day-old rats is not appropriate. Furthermore, if the proposed course of treatment in adult children is 2 weeks, it is unlikely that this would crossover into a different developmental stage. A 2-week treatment initiated in puppies, however, might easily span two developmental stages. Thus, in designing an experiment in young animals one must carefully consider the length of the treatment period balancing the developmental age of the animal model and the proposed length of clinical treatment. Where appropriate (infant animals), one needs to also assess changes in standard developmental landmarks, (e.g., eye opening, pinae eruption, external genitalia development, etc.) as well as the more standard indicators of target organ toxicity. The need for maintaining the experimental animals past the dosing period, perhaps into sexual maturity, to assess recovery or delayed effects needs also to be carefully considered

To summarize, the current status of assessment of toxicity in post-natal mammals, in response to the pediatric initiatives covered in FDAMA, is an extremely fluid situation. One needs to carefully consider a variety of factors in designing the study, and should discuss proposed testing programs with the appropriate office at CDER.

Drugs intended for use in the elderly, like those intended for the very young, may also have special requirements for safety evaluation, but geriatric issues were not addressed in the FDAMA of 1997. The FDA has published a separate guidance document for geriatric labeling. As was the case with pediatric guidance, this document does not address preclinical testing. With the elderly, the toxicological concerns are quite different than the developmental concerns associated with pediatric patients. With the elderly, one must be concerned with the possible interactions between the test article and compromised organ function. The FDA had previously issued a guidance for clinically examining clinical safety of new pharmaceutical agents in patients with compromised renal and/or hepatic function (CDER, 1989). The equivalent ICH guideline (S5A) was issued in 1994. Whether this type of emphasis will require toxicity testing in animal models with specifically induced organ insufficiency remains to be seen. In the interim, we must realize that there is tacit evaluation of test article-related toxicity in geriatric rodents for those agents that undergo 2-year carcinogenicity testing. As the graying of America continues, labeling for geriatric use may become more of an issue in the future.

## Orphan drugs

The development of sophisticated technologies, coupled with the rigors and time required for clinical and preclinical testing has made pharmaceutical development very expensive. In order to recoup such expenses, pharmaceutical companies have tended to focus on therapeutic agents with large potential markets. Treatment for rare but life-threatening diseases have been "orphaned" as a result. An *orphan product* is defined as one targeted at a disease which affects 200,000 or fewer individuals in the US. Alternatively, the therapy may be targeted for more than 200,000 but the developer would have no hope of recovering the initial investment without exclusivity. The Orphan Drug Act of 1983 was passed in an attempt to address this state of affairs. Currently applicable regulations were put in place in 1992 (Anon, 1992b). In 1994, there was an attempt in Congress to amend the act, but it failed to be passed into law. The current regulations are administered by the office of Orphan Product Development (OPD). The act offers the following incentives to encourage the development of products to treat rare diseases:

- 7 years exclusive market following the approval of a product for an orphan disease
- Written protocol assistance from the FDA
- Tax credits for up 50 per cent of qualified clinical research expenses
- Available grant to support pivotal clinical trials

As reviewed by Haffner (1998), other developed countries have similar regulations.

The ODA did not change the requirements of testing drug products. The nonclinical testing programs are similar to those used for more conventional products. They will undergo the same FDA review process. A major difference, however, is the involvement of the OPD. A sponsor must request OPD review. Once OPD determines that a drug meets the criteria for orphan drug status it will work with the sponsor to provide the assistance required under the Act. The ODA does not review a product for approval. The IND/NDA process is still handled by the appropriate reviewing division (e.g., cardiovascular) for formal review. The Act does not waive the necessity for submission of an IND, not for the responsibility of toxicological assessment. As always, in cases where there is ambiguity, a sponsor may be well-served to request a pre-IND meeting at the appropriate division to discuss the acceptability of a toxicology assessment plan.

## Botanical drug products

There is an old saying, "What goes around, comes around": and so it is with botanicals. At the beginning of the twentieth century, most marketed pharmaceutical agents were botanical in origin. For example, aspirin was first isolated from willow bark. These led the way in the middle part of the century, for reasons having to do with patentability, manufacturing costs, standardization, selectivity, and potency. The dawning of the twenty-first century has seen a grass-roots return to botanical preparations (also sold as herbals or dietary supplements). These preparations are being marketed to the lay public as "natural" supplements to the nasty synthetic chemicals now proscribed as pharmaceutical products. In 1994, the Dietary Supplement Health and Education Act was passed which permitted the marketing of dietary supplements (including botanicals) with limited submissions to the FDA. (Wu et al., 2000). If a producer makes a claim that an herbal preparation is beneficial to a specific part of the body (e.g., enhanced memory), then it may be marketed after a 75-day period of FDA review but without formal approval. On the other hand, if any curative properties are claimed, then the botanical will be regulated as a drug and producers will be required to follow the IND/NDA process. In 1997 and 1998 combined, some 26 INDs were filed for botanical products (Wu et al., 2000).

The weakness in the current regulation has to do with its ambiguity. The line between a beneficial claim and a curative claim is sometimes difficult to draw. What is the difference, for example, between an agent that enhances memory and one that prevents memory loss? Given the number of products and claims hitting the shelves every day, this situation will probably demand increased regulatory scrutiny in the future.

## International pharmaceutical regulation and registration

### *International Conference on Harmonization*

The International Conference on Harmonization (ICH) of Technical Requirements for Registration of Pharmaceuticals for Human Use was established to make the drug-regulatory process more efficient in the US, Europe, and Japan. The US involvement grew out of the fact that the US is party to the General Agreement on Tariffs and Trade, which included the Agreement on Technical Barriers to Trade, negotiated in the 1970s, to encourage reduction of nontariff barriers to trade (Barton, 1998). The main purpose of ICH is, through harmonization, to make new medicines available to patients with a minimum of delay. More recently, the need to harmonize regulation has been driven, according to ICH, by the escalation of the cost of R&D. The regulatory systems in all countries have the same fundamental concerns about safety, efficacy, and quality, yet sponsors had to repeat many time-consuming and expensive technical tests to meet country-specific requirements. Second, there was a legitimate concern over the unnecessary use of animals. Conference participants include representatives from the drug-regulatory bodies and research-based pharmaceutical industrial organizations of three regions; the European Union, the US and Japan comprise over 90 per cent of the world's pharmaceutical industry. Representation is summarized in Table 2.10. The biennial conference has met four times, beginning in 1991, rotating between sites in the US, Europe, and Japan. The next meeting is scheduled for the year 2001 and will be held on the West Coast of the US. The precise venue has yet to be named.

The ICH meets its objectives by issuing guidelines for the manufacturing, development, and testing of new pharmaceutical agents that are acceptable to all three major parties. For each new guideline, the ICH Steering Committee establishes an expert working group with representation from each of the six major participatory ICH bodies. Each new draft guideline goes through the five various steps of review and revision summarized in Table 2.11. So far, the ICH has proposed or adopted over 40 safety, efficacy, and quality guidelines (listed in Table 2.12) for use by the drug-regulatory agencies in the US, Europe, and Japan. The guidelines are organized under broad categories: the "E" series dealing with clinical trials, the "Q" series dealing with quality (including chemical manufacturing and control as wells as traditional GLP issues), and the "S" series dealing with safety. Guidelines may be obtained from the

*Table 2.10* ICH representation[a]

| *Country/region* | *Regulatory* | *Industry* |
|---|---|---|
| European Union | European Commission (2) | European Federation of Pharmaceutical Industries Associations (2) |
| Japan | Ministry of Health and Welfare (2) | Japanese Pharmaceutical Manufactures Association (2) |
| United States | Food and Drug Administration (2) | Pharmaceutical Research and Manufacturers of America (2) |
| Observing organizations | World Health Organization, European Free Trade Area, Canadian Health Protection Branch | International Federation of Pharmaceutical Manufactures Associations (2): also provides the secretariat |

[a] Numbers in parens = number of representatives on the ICH steering Committee.

ICH secretariat, c/o IFPMA, 30 rue de St.-Jean, PO Box 9, 1211 Geneva 18, Switzerland, or may be down-loaded from a website set up by Ms. Nancy McClure (http://www.mcclurenet.com/index.html). They are also published in the Federal Register. It is the guidelines of the "S" series that will have the most impact on toxicologists. The biggest changes dealing with toxicological assessment are summarized as follows:

*Carcinogenicity studies*: carcinogenicity studies are covered in Guidelines S1A, S1B and S1C. The guidelines are almost more philosophical than they are technical. In comparison to the EPA guidelines for example, the ICH guidelines contain little in the way of concrete study criteria (e.g., the number of animals, the necessity for clinical chemistry, etc.). There is discussion on when carcinogenicity studies should be done, whether two species are more appropriate than one, and how to set dosages on the basis of human clinical PK data. The major changes being wrought by these guidelines are:

- Only one 2-year carcinogenicity study should be generally required. Ideally, the species chosen should be the one most like man in terms of metabolic transformations of the test article.
- The traditional second long-term carcinogenicity study can be replaced by a shorter-term alternative model. In practical terms, this guideline is beginning to result in sponsors conducting a 2-year study in the rat and a 6-month study in an alternative mouse model, such as the P53 or the TG.AC genetically manipulated mouse strains.
- In the absence of target organ toxicity with which to set the high dose at the

*Table 2.11* Steps in ICH guideline development and implementation

| *Step* | |
|---|---|
| 1 | Building scientific consensus in joint regulatory/industry expert working groups |
| 2 | Agreement by the steering committee to release the draft consensus text for wider consultation |
| 3 | Regulatory consultation in the three regions. Consolidation of the comments |
| 4 | Agreement on a harmonized ICH guideline; adopted by the regulators |
| 5 | Implementation in the three ICH regions |

*Table 2.12* International conference on harmonization guidelines

| *Ref.* | *Guideline* | *Date* |
|---|---|---|
| E1 | The extent of population exposure to assess clinical safety | Oct. 1994 |
| E2A | Clinical safety data management: definitions and standards for expedited reporting | Oct. 1994 |
| E2B | Guideline on clinical safety data management; notice | Jan. 1998 |
| E2C | Clinical safety data management: periodic safety update reports for marketed drugs | May 1997 |
| E3 | Structure and content of clinical study reports | Nov. 1995 |
| E4 | Dose response information to support drug registration | March 1994 |
| E5 | Ethnic factors in the acceptability of foreign clinical data | Feb. 1998 |
| E6 | Good clinical practice: consolidated guideline; notice of availability | May 1997 |
| E6A | GCP addendum on investigator's brochure | March 1995 |
| E6B | GCP: Addendum on essential documents for the conduct of a clinical trial | Oct. 1994 |
| E7 | Studies in support of special populations: geriatrics | June 1993 |
| E8 | Guidance on general considerations for clinical trials; notice | Dec. 1997 |
| E9 | Draft guideline on statistical principles for clinical trials; notice of availability | May 1997 |
| M3 | Guidance on nonclinical safety studies for the conduct of human clinical trials for pharmaceuticals; notice | Nov. 1997 |
| Q1A | Stability testing of new drug substances and products | Oct. 1993 |
| Q1B | Stability testing | |
| Q1C | Stability testing | |
| Q2A | Validation of analytical procedures: definitions and terminology | Oct. 1994 |
| Q2B | Validation of analytical procedures: methodology | |
| Q3A | Guideline on impurities in new drug substances | Mar. 1995 |
| Q3B | Guideline on impurities in new drug products; availability; notice | May. 1997 |
| Q3C | Guideline on impurities: residual solvents; availability; notice | Dec. 1997 |
| Q5A | Quality of biotechnological products viral safety evaluation of biotechnology products derived from cell lines of human or animal origin | |
| Q5B | Quality of biotechnology products analysis of the expression construct in cells used for production of r-DNA derived protein product | Nov. 1995 |
| Q5C | Quality of biotechnological products: stability testing of biotechnological /biology products | 3 Nov. 1995 |
| Q5D | Availability of draft guideline on quality of biotechnological/biological products: derivation and characterization of cell substrates used for production of biotechnological/biological products; notice | May 1997 |
| Q6A | Draft guidance on specifications: test procedures and acceptance criteria for new drug substances and new drug products: chemical substances; notice | Nov. 1997 |
| Q6B | Specifications: test procedures and acceptance criteria for biotechnology products | Feb. 1998 |
| S1A | Guidance on the need for carcinogenicity studies of pharmaceuticals | Nov. 1995 |
| S1B | Draft guideline on testing for carcinogenicity of pharmaceuticals; notice | Nov. 1998 |
| S1C | Dose selection for carcinogenicity studies of pharmaceuticals | Oct. 1994 |
| S1Ca | Guidance on dose selection for carcinogenicity studies of pharmaceuticals: addendum on a limit dose and related notes; availability; notice | Dec. 1997 |
| S2A | Genotoxicity: guidance on specific aspects of regulatory genotoxicity tests for pharmaceuticals | July 1995 |

Table 2.12 (*continued*)

| *Ref.* | *Guideline* | *Date* |
|---|---|---|
| S2B | Guidance on genotoxicity: a standard battery for genotoxicity testing of pharmaceuticals; availability; notice | Nov 1997 |
| S3A | Toxicokinetics: guidance on the assessment of systemic exposure in toxicity studies | Oct. 1994 |
| S3B | Pharmacokinetics: guidance for repeated dose tissue distribution studies | Oct. 1994 |
| S4 | Single dose acute toxicity testing for pharmaceuticals; revised guidance; availability; notice | Aug. 1996 |
| S4A | Draft guidance on the duration of chronic toxicity testing in animals (rodent and nonrodent toxicity testing); availability; notice | Nov. 1997 |
| S5A | Detection of toxicity to reproduction for medicinal products | June 1993 |
| S5B | Reproductive toxicity to male fertility | |
| S6A | Guidance on preclinical safety evaluation of biotechnology-derived pharmaceuticals; availability | Nov. 1997 |

maximally tolerated dose, the high dose can be set at the dose that produces an area under the curve (AUC). This is 25-fold higher than that obtained in human subjects.

*Chronict toxicity*: traditionally, chronic toxicity of new pharmaceuticals in the US was assessed in studies of 1-year duration in both the rodent and the nonrodent species of choice. The European view was that studies of 6 months are generally sufficient. The resulting guideline (S4A) was a compromise. Studies of 6 months duration were recommended for the rodent, as rodents would also be examined in 2-year studies. For the nonrodent (dog, nonhuman primate, and pig) studies of 9 months duration were recommended.

*Developmental and reproductive toxicity*: this was an area in which there was considerable international disagreement and the area in which ICH has promulgated the most technically detailed guidelines (S5A and S5B). Some of the major changes include:

- The traditional Segment I, II, and III nomenclature has been replaced with different nomenclature, as summarized in Table 2.13.
- The dosing period of the pregnant animals during studies on embryonic development (traditional Segment (II studies) has been standardized.
- New guidelines for fertility assessment (traditional Segment I) studies that have shortened the pre-mating dosing schedule (for example, in male rats from 10 weeks to 4 weeks). There has been an increased interest in assessment of spermatogenesis and sperm function.
- The new guidelines allow for a combination of studies in which the end-point typically assessed in the traditional Segment II and III studies are now examined under a single protocol.

For a more complete review of the various study designs, the reader is refereed to the review by Manson (1994).

While they were not quite as sweeping in approach as the aforementioned guidelines, a toxicologist working in pharmaceutical safety assessment should become familiar with the all the other ICH guidelines in the S series.

*Table 2.13* Comparison of traditional and ICH guidelines for reproductive and developmental toxicology

| *Traditional protocol* | *Stages covered* | *ICH protocol* | *Dosing regimen* |
|---|---|---|---|
| Segment I (rats) | A. Premating to conception<br>B. Conception to implantation | Fertility and early embryonic development, including implantation | Males: 4 weeks pre-mating, mating (1–3 weeks) plus 3 weeks post-mating<br>Females: 2 weeks pre-mating, mating through day 7 of gestation |
| Segment II (rabbits) | C. Implantation to closure of hard palate<br>D. Closure of hard palate to the end of pregnancy | Embryo-fetal development | Female rabbits: day 6–20 of pregnancy |
| Study title | Termination | Endpoints: in-life | Endpoints: postmortem |
| Fertility and early embryonic development, including implantation | Females: day 13–15 of pregnancy | Clinical signs and mortality | Macroscopic exam + histo on gross lesions |
| | Males: day after completion of dosing | Body weights and feed intake<br>Vaginal cytology | Collection of reproductive organs for possible histology<br>Quantitation of corpa lutea and implantation sites |

Table 2.13 (*continued*)

| *Traditional protocol* | *Stages covered* | *ICH protocol* | *Dosing regimen* |
|---|---|---|---|
| | | | Seminology (ocunt, motility and morphology) |
| Embryo-fetal development | | Clinical signs and mortality | Macroscopic exam + histo on gross lesions |
| | | Body weights and changes | Quantitation of corpa lutea and implantation sites |
| | | Feed in-take | Fetal body weights<br>Fetal abnormalities |
| Pre- and post-natal development, including maternal function | | Clinical signs and mortality body weights and changes | Macroscopic exam + histo on gross lesions |
| | | Feed in-take | Implantation |
| | | Duration of pregnancy | Abnormalities (including terata) |
| | | Parturition | Live/dead offspring at birth<br>Pre-and post-weaning survival and growth ($F_1$)<br>Physical development($F_1$)<br>Sensory functions and reflexes ($F_1$)<br>Behavior ($F_1$) |

In an interesting recent article, Ohno (1998) discussed not only the harmonization of nonclinical guidelines, but also the need to harmonize the timing of nonclinical tests in relation to the conduct of clinical trials. For example, there are regional differences in the inclusion of women of childbearing potential in clinical trials. In the US, including women in such trials is becoming more important, and therefore evaluation of embryo-fetal development will occur earlier in the drug development process than in Japan. Whether or not such timing or staging of nonclinical tests becomes part of an ICH guideline in the near future remains to be established.

### Other international considerations

The US is the single largest pharmaceutical market in the world. But the rest of the world (particularly, but not limited to the second and third largest markets, Japan and the European Union) represents in aggregate a much larger market, so no one develops a new pharmaceutical for marketing in just the US. The effort at harmonization (exemplified by the ICH) has significantly reduced differences in requirements for these other countries, but certainly not obliterated them. Though a detailed understanding of their regulatory schemes is beyond this volume, the bare bones and differences in toxicology requirements are not.

#### *European Union*

The standard European Union toxicology and pharmacologic data requirements for a pharmaceutical include

1 Single-dose toxicity
2 Repeat-dose toxicity (subacute and chronic trials)
3 Reproduction studies (fertility and general reproductive performance, embryotoxicity and peri/postnatal toxicity)
4 Mutagenic potential (*in vitro* and *in vivo*)
5 Carcinogenicity
6 Pharmacodynamics

- Effects related to proposed drug indication
- General pharmacodynamics
- Drug interactions

7 Pharmacokinetics

- Single dose
- Repeat dose
- Distribution in normal and pregnant animals
- Biotransformation

8 Local tissue tolerance
9 Environmental toxicity

In general, the registration process in the EU allows one to either apply to an overall medicines authority or to an individual national authority. Either of these steps is supposed to lead to mutual recognition by all the individual members.

### *Japan*

In Japan, the Koseisho is the national regulatory body for new drug.

The standard $LD_{50}$ test is no longer a regulatory requirement for new medicines in the US, the EU, or Japan. The Japanese guidelines were the first to be amended in accordance with this agreement, with the revised guidelines becoming effective in August 1993. The Japanese may still anticipate that single dose (acute) toxicity studies should be conducted in at least two species, one rodent and one nonrodent (the rabbit is not accepted as a nonrodent). Both males and females should be included from at least one of the species selected: if the rodent, then a minimum of five per sex; if the nonrodent, at least two per sex. In nonrodents, both the oral and parenteral routes should be used, and normally the clinical route of administration should be employed. In nonrodents, only the intended route of administration need be employed; if the intended route of administration in humans is intravenous, then use of this route in both species is acceptable. An appropriate number of doses should be employed to obtain a complete toxicity profile and to establish any dose–response relationship. The severity, onset, progression, and reversibility of toxicity should be studied during a 14-day follow-up period, with all animals being necropsied. When macroscopic changes are noted, the tissue must be subjected to histological examination.

Chronic and subchronic toxicity studies are conducted to define the dose level, when given repeatedly, that causes toxicity, and the dose level that does not lead to toxic findings. In Japan, such studies are referred to as repeated-dose toxicity studies. As with single-dose studies, at least two animal species should be used, one rodent and one nonrodent (rabbit not acceptable). In rodent studies, each group should consist of at least ten males and ten females; in nonrodent species, three of each sex are deemed adequate. Where interim examinations are planned, however, the numbers of animals employed should be increased accordingly. The planned route of administration in human subjects is normally explored. The duration of the study will be dictated by the planned duration of clinical use (Table 2.14).

At least three different dose groups should be included, with the goals of demonstrating an overtly toxic dose and a no-effect dose, and establishing any dose–response relationship. The establishment of a nontoxic dose within the framework of these studies is more rigorously adhered to in Japan than elsewhere in the world. All surviving animals should also be necropsied, either at the completion of the study or during its extension recovery period, to assess reversal of toxicity and the possible appearance of delayed toxicity. Full histological examination is mandated on all nonrodent animals

*Table 2.14* MHW preclinical study length requirement before human studies

| *Duration of dosing in toxicity study* | *Duration of human exposure* |
|---|---|
| 1 month | Single dose or repeated dosage not exceeding 1 week |
| 3 months | Repeated dosing exceeding 1 week and to a maximum of 4 weeks |
| 6 months | Repeated dosing exceeding 4 weeks and to a maximum of 6 months |
| 12 months[a] | Repeated dosing exceeding 6 months or where this is deemed to be appropriate |

[a] Where carcinogenicity studies are to be conducted, the Koseisho had agreed to forego chronic dosage beyond 6 months. Source: New Drugs Division Notification No. 43, June 1992.

used in a chronic toxicity study; at a minimum, the highest-dose and control groups of rodents must be submitted to a full histological examination.

While the value of repeated-dose testing beyond 6 months has been questioned (Lumley, 1992), such testing is a regulatory requirement for a number of agencies, including the US FDA and the Koseisho. In Japan, repeated-dose testing for 12 months is required only for new medicines expected to be administered to humans for periods in excess of 6 months (Anon, 1994). At the First ICH held in Brussels, the consensus was that 12-month toxicity studies in rodents could be reduced to 6 months where carcinogenicity studies are required. While not yet adopted in the Japanese guidelines, 6-month repeated-dose toxicity studies have been accepted by the agencies of all three regions. Japan – like the EU – accepts a 6-month duration if accompanied by a carcinogenicity study. The US still requires a 9-month nonrodent study.

With regard to reproductive toxicology, as a consequence of the first ICH, the US, EU, and Japan agreed to recommend mutual recognition of their respective current guidelines. A tripartite, harmonized guideline on reproductive toxicology has achieved ICH Step 4 status and should be incorporated into the local regulations of all three regions soon. This agreement represents a very significant achievement that should eliminate many obstacles to drug registration.

**Preclinical male fertility studies**. Before conducting a single-dose male volunteer study in Japan, it is usually necessary to have completed a preclinical male fertility study (Segment 1) that has an in-life phase of 10 or more weeks (i.e. 10 weeks of dosing, plus follow-up). Although government guidelines do not require this study to be completed before Phase 1 trials begin, the responsible Institutional Review Board or the investigator usually imposes this condition. Japanese regulatory authorities are aware that the Segment 1 male fertility study is of poor predictive value. The rat, which is used in this study, produces a marked excess of sperm. Many scientists therefore believe that the test is less sensitive than the evaluation of testicular weight and histology that constitute part of the routine toxicology assessment

**Female reproductive studies**. Before entering a female into a clinical study, it is necessary to have completed the entire reproductive toxicology program, which consists of the following studies:

- Segment 1: fertility studies in the rat or mouse species used in the Segment 2 program;
- Segment 2: teratology studies in the rat or mouse, and the rabbit; and
- Segment 3: late gestation and lactation studies in a species used in the Segment 2 studies.

Such studies usually take approximately 2 years. Although the US regulations state the need for completion of Segments 1 and 2, and the demonstration of efficacy in male patients, where appropriate, before entering females into a clinical program, the current trend in the US is toward relaxation of the requirements to encourage investigation of the drug both earlier and in a larger number of females during product development. Growing pressure for the earlier inclusion of women in drug testing may encourage selection of this issue as a future ICH topic. The trend in the US and the EU toward including women earlier in the critical program has not yet been embraced in Japan, however.

The three tests required in Japan for genotoxicity evaluation are a bacterial gene mutation test, *in vitro* cytogenetics, and *in vivo* tests for genetic damage. The Japanese regulations state these tests to be the minimum requirement and encourage additional tests. Currently, Japanese guidelines do not require a mammalian cell gene mutation assay. Harmonization will likely be achieved by the Koseisho recommending all four tests, which will match requirements in the US and EU; at present, this topic is at Step 1 in the ICH harmonization process. The mutagenicity studies should be completed before the commencement of Phase 2 clinical studies.

Guidelines presented at the second ICH are likely to alter the preclinical requirements for registration in Japan; they cover toxicokinetics and when to conduct repeated-dose tissue distribution studies. The former document may improve the ability of animal toxicology studies to predict possible adverse events in humans; currently, there are no toxicokinetic requirements in Japan, and their relevance is questioned by many there. Although there is general agreement on the registration requirement for single-dose tissue distribution studies, implementation of the repeated-dose study requirement has been inconsistent across the three ICH parties.

### *Safety pharmacology*

Japan was the first major country to require extensive pharmacological profiling on all new pharmaceutical agents as part of the safety assessment profile. Prior to commencement of initial clinical studies, the drug's pharmacology must be characterized in animal models. In the US and Europe, these studies have been collectively called safety pharmacology studies. For a good general review of the issues surrounding safety pharmacology the reader is referred to Hite (1997). The Japanese guidelines for such characterizations were published in 1991. They include

- Effects on general activity and behavior
- Effects on the central nervous system
- Effects on the autonomic nervous system and smooth muscle
- Effects on the respiratory and cardiovascular systems
- Effects on the digestive system
- Effects on water and electrolyte metabolism
- Other important pharmacological effects

Source: New Drugs Division Notification No. 4, January 1991.

In the US, pharmacological studies in demonstration of efficacy have always been required, but specific safety pharmacological studies have never been required. Special situational or mechanistic data would be requested on a case-by-case basis. This is a situation that is changing. In the US the activities of the Safety Pharmacology Discussion Group, for example, have helped bring attention to the utility and issues surrounding safety pharmacology data. In 1999 and 2000, the major toxicological and pharmacological societal meetings had symposia on safety pharmacological testing. Many major US Pharmaceutical companies are in the process of implementing programs in safety pharmacology. The issue has been taken up by ICH and the draft guideline is currently at the initial stages of review. This initial draft (Guideline S7) includes core tests in the assessment of CNS, cardiovascular and respiratory function. Studies will be expected to be performed under GLP guidelines.

## Combination products

Recent years have seen a vast increase in the number of new therapeutic products which are not purely drug, device or biologic, but rather a combination of two or more of these. This leads to a problem of deciding which of the three centers shall have ultimate jurisdiction.

The Center for Devices and Radiological Health is designated the center for major policy development and for the promulgation and interpretation of procedural regulations for medical devices under the Act. The Center for Devices and Radiological Health regulates all medical devices inclusive of radiation-related devices, that are not assigned categorically or specifically to CDER. In addition, Center for Devices and Radiological Health (CDRH) will independently administer the following activities (references to "Sections" are the provisions of the Act):

1 Small business assistance programs under Section 10 of the amendments (see PL 94-295). Both CDER and CDRH will identify any unique problems relating to medical device regulation for small business;
2 Registration and listing under Section 510 including some CDER administered device applications. The CDER will receive printouts and other assistance, as requested;
3 Color additives under Section 706, with review by CDER, as appropriate;
4 Good Manufacturing Practices (GMPs) Advisory Committee. Under Section 520(f) (3), CDER will regularly receive notices of all meetings, with participation by CDER, as appropriate; and
5 Medical device reporting. The manufacturers, distributors, importers, and users of all devices, including those regulated by CDER, shall report to CDRH under Section 519 of the Act as required. The CDRH will provide monthly reports and special reports as needed to CDER for investigation and follow-up of those medical devices regulated by CDER.

### *Device programs that CDER and CBRH each will administer*

Both CDER and CDRH will administer and, as appropriate, enforce the following activities for medical devices assigned to their respective centers (references to "Sections" are the provisions of the Act):

1 Surveillance and compliance actions involving general controls violations, such as misbranded or adulterated devices under Sections 301, 501, and 502;
2 Warning letters, seizures, injunctions, and prosecutions under Sections 302, 303, and 304;
3 Civil penalties under Section 303(f) and administrative restraint under Section 304(g);
4 Nonregulatory activities, such as educational programs directed at users, participation in voluntary standards organizations, etc.;
5 Promulgation of performance standards and applications of special controls under Section 514;
6 Premarket notification, investigational device exemptions including humanitarian exemptions, premarket approval, product development protocols, classification,

device tracking, petitions for reclassification, post market surveillance under Sections 510(k), 513, 515, 519, 520(g)&(m), and 522, and the advisory committees necessary to support these activities;

7 Banned devices under Section 516;
8 FDA-requested and firm-initiated recalls whether under Section 518 or another authority and other Section 518 remedies such as recall orders;
9 Exemptions, variances and applications of CGMP regulations under Section 520(f);
10 Government-Wide Quality Assurance Program; and
11 Requests for export approval under Sections 801(e) and 802.

### *Coordination*

The centers will coordinate their activities in order to assure that manufacturers do not have to independently secure authorization to market their product from both centers unless this requirement is specified in Section VII.

### *Submissions*

Submissions should be made to the appropriate center, as specified herein, at the addresses provided below:

Address update: Food and Drug Administration, Center for Drug Evaluation and Research, Central Document Room (Room #2-14), 12420 Parklawn Drive, Rockville, MD 20852, USA

or

Food and Drug Administration, Center for Devices and Radiological Health, Document Mail Center (HFZ-401), 1390 Piccard Drive, Rockville, MD 20850, USA

For submissions involving medical devices and/or drugs that are not clearly addressed in this agreement, sponsors are referred to the product jurisdiction regulations (21 CFR Part 3). These regulations have been promulgated to facilitate the determination of regulatory jurisdiction but do not exclude the possibility for a collaborative review between the centers.

### *Center jurisdiction*

The following subsections provide details concerning status, market approval authority, special label/regulatory considerations, investigational options, and inter-center consultations for the categories of products specified. Section VII provides the general criteria that CDRH and CDER will apply in reaching decisions as to which center will regulate a product.

#### DEVICE WITH PRIMARY PURPOSE OF DELIVERING OR AIDING IN THE DELIVERY OF A DRUG THAT IS DISTRIBUTED WITHOUT A DRUG (i.e., UNFILLED)

Examples:

- Devices that calculate drug dosages
- Drug delivery pump and/or catheter infusion pump for implantation iontophoreses device

- Medical or surgical kit (e.g., tray) with reference in instructions for use with specific drug (e.g., local anesthetic)
- Nebulizer
- Small Particle Aerosol Generator (SPAG) for administering drug to ventilated patient
- Splitter block for mixing nitrous oxide and oxygen
- Syringe, jet injector, storage and dispensing equipment

*Status*: device and drug, as separate entities.

*Market approval authority*: CDRH and CDER, respectively, unless the intended use of the two products, through labeling, creates a combination product.

*Special label/regulatory considerations*: the following specific procedures will apply depending on the status of the drug delivery device and drugs that will be delivered with the device:

1 It may be determined during the design or conduct of clinical trials for a new drug that it is not possible to develop adequate performance specification data on those characteristics of the device that are required for the safe and effective use of the drug. If this is the case, then drug labeling cannot be written to contain information that makes it possible for the user to substitute a generic, marketed device for the device used during developments to use with the marketed drug. In these situations, CDER will be the lead center for regulation of the device under the device authorities.
2 For a device intended for use with a category of drugs that are on the market, CDRH will be the lead center for regulation for the device under the device authorities. The effects of the device use on drug stability must be addressed in the device submission, when relevant. An additional showing of clinical effectiveness of the drug when delivered by the specific device will generally not be required. The device and drug labeling must be mutually conforming with respect to indication, general mode of delivery (e.g., topical, I.V.), and drug dosage/schedule equivalents.
3 For a drug delivery device and drug that are developed for marketing to be used together as a system, a lead center will be designated to be the contact point with the manufacturer(s). If a drug has been developed and marketed and the development and studying of device technology predominates, the principle mode of action will be deemed to be that of the device, and CDRH would have the lead. If a device has been developed and marketed and the development and studying of drug predominates then, correspondingly, CDER would have the lead. If neither the drug nor the device is on the market, the lead center will be determined on a case-by-case basis.

*Investigation options*: IDE or IND, as appropriate.

*Inter-center consultation*: CDER, when lead center, will consult with CDRH if CDER determines that a specific device is required as part of the NDA process. CDRH as lead center will consult with CDER if the device is intended for use with a marketed drug and the device creates a significant change in the intended use, mode of delivery (e.g., topical, I.V.), or dose/schedule of the drug.

#### DEVICE WITH PRIMARY PURPOSE OF DELIVERING OR AIDING IN THE DELIVERY OF A DRUG AND DISTRIBUTED CONTAINING A DRUG (i.e., "PRE-FILLED DELIVERY SYSTEM")

Examples:

- Nebulizer
- Oxygen tank for therapy and OTC emergency use
- Pre-filled syringe
- Transdermal patch

*Status*: combination product
*Market approval authority*: CDER using drug authorities and device authorities, as necessary.
*Special label/regulatory considerations*: none.
*Investigation options*: IND.
*Inter-center consultations*: Optional.

#### DEVICE INCORPORATING A DRUG COMPONENT WITH THE COMBINATION PRODUCT HAVING THE PRIMARY INTENDED PURPOSE OF FULFILLING A DEVICE FUNCTION

Examples:

- Bone cement containing antimicrobial agent
- Cardiac pacemaker lead with steroid-coated tip
- Condom, diaphragm, or cervical cap with contraceptive or antimicrobial agent (including virucidal) agent
- Dental device with fluoride
- Dental wood wedge with hemostatic agent
- Percutaneous cuff (e.g., for a catheter or orthopedic pin) coated/impregnated with antimicrobial agent
- Skin closure or bandage with antimicrobial agent
- Surgical or barrier drape with antimicrobial agent
- Tissue graft with antimicrobial or other drug agent
- Urinary and vascular catheter coated/impregnated with antimicrobial agent
- Wound dressing with antimicrobial agent

*Status*: combination product.
*Market approval authority*: CDRH using device authorities.
*Special label/regulatory considerations*: these products have a drug component that is present to augment the safety and/or efficacy of the device.
*Investigation options*: IDE.
*Inter-center consultation*: required if a drug or the chemical form of the drug has not been legally marketed in the US as a human drug for the intended effect.

#### DRUG INCORPORATING A DEVICE COMPONENT WITH THE COMBINATION PRODUCT HAVING THE PRIMARY INTENDED PURPOSE OF FULFILLING A DRUG FUNCTION

Examples:

- Skin-prep pads with antimicrobial agent
- Surgical scrub brush with antimicrobial agent

*Status*: combination product.

*Market approval authority*: CDER using drug authorities and, as necessary, device authorities.

*Special label/regulatory considerations*: marketing of such a device requires a submission of an NDA with safety and efficacy data on the drug component or it meets monograph specifications generally recognized as safe (GRAS) and generally recognized as effective (GRAE). Drug requirements, e.g., CGMPs, registration and listing, experience reporting, apply to products.

*Investigation options*: IND.

*Inter-center consultation*: Optional.

#### DEVICE USED IN THE PRODUCTION OF A DRUG EITHER TO DELIVER DIRECTLY TO A PATIENT OR FOR USE IN THE PRODUCING MEDICAL FACILITY (EXCLUDING USE IN A REGISTERED DRUG MANUFACTURING FACILITY)

Examples:

- Oxygen concentrators (home or hospital)
- Oxygen generator (chemical)
- Ozone generator

*Status*: device.

*Market approval authority*: CDER, applying both drug and device authorities.

*Special label/regulatory consideration*: may also require an NDA if the drug produced is a new drug. Device requirements, e.g., CGMPs, registration and listing, experience reporting will apply to products.

*Investigation options*: IDA; or NDA, as appropriate.

*Inter-center consultation*: optional.

#### DRUG/DEVICE COMBINATION PRODUCT INTENDED TO PROCESS A DRUG INTO A FINISHED PACKAGE FORM

Examples:

- Device that uses drug concentrates to prepare large volume parenterals.
- Oxygen concentrator (hospital) output used to fill oxygen tanks for use within that medical facility.

*Status*: combination product.

*Market approval authority*: CDER, applying both drug and device authorities.

*Special label/regulatory considerations*: respective drug and device requirements, e.g., CGMPs, registration and listing, experience reporting will apply.

*Investigation options*: IDE or NDA, as appropriate.

*Inter-center consultation*: optional, but will be routinely obtained.

#### DEVICE USED CONCOMITANTLY WITH A DRUG TO DIRECTLY ACTIVATE OR TO AUGMENT DRUG EFFECTIVENESS

Examples:

- Biliary lithotriptor used in conjunction with dissolution agent

- Cancer hyperthermia used in conjunction with chemotherapy
- Current generator used in conjunction with an implanted silver electrode (drug) that produces silver ions for an antimicrobial purpose
- Materials for blocking blood flow temporarily to restrict chemotherapy drug to the intended site of action
- UV and/or laser activation of oxsoralen for psoriasis or cutaneous T-cell lymphoma

*Status*: device and drug, as separate entities.

*Market approval authority*: CDRH and CDER, respectively.

*Special label/regulatory considerations*: the device and drug labeling must be mutually conforming with respect to indications, general mode of delivery (e.g., topical, I.V.), and drug dosage/schedule equivalence. A lead center will be designated to be the contact point with the manufacturer. If a drug has been developed and approved for another use and the development and studying of device technology predominates, then CDRH would have the lead. If a device has been developed and marketed for another use and the development and studying of drug action predominates, then CDER would have the lead. If neither the drug nor the device is on the market, the lead center will be determined on a case-by-case basis. If the labeling of the drug and device creates a combination product, as defined in the combination product regulations, then the designation of the lead center for both applications will be based upon a determination of the product's primary mode of action.

*Investigation options*: IDE or IND, as appropriate.

*Inter-center consultations*: required.

#### DEVICE KITS LABELED FOR USE WITH DRUGS THAT INCLUDE BOTH DEVICE(S) AND DRUG(S) AS SEPARATE ENTITIES IN ONE PACKAGE WITH THE OVERALL PRIMARY INTENDED PURPOSE OF THE KIT FULFILLING A DEVICE FUNCTION

Example:

- Medical or surgical kit (e.g., tray) with drug component

*Status*: combination product.

*Market approval authority*: CDRH, using device authorities is responsible for the kit if the manufacturer is repackaging a market drug. Responsibility for overall packaging resides with CDRH. CDER will be consulted as necessary on the use of drug authorities for the repackaged drug component.

*Special label/regulatory consideration*: device requirements, e.g., CGMPs, registration and listing, experience reporting apply to kits. Device manufacturers must ensure that manufacturing steps do not adversely affect drug components of the kit. If the manufacturing steps do affect the marketed drug (e.g., the kit is sterilized by irradiation), and ANDA or NDA would also be required with CDRH as lead center.

*Investigation options*: IDA or IND, as appropriate.

*Inter-center consultation*: optional if ANDA or NDA not required.

### LIQUIDS, GASES OR SOLIDS INTENDED FOR USE AS DEVICES (E.G., IMPLANTED, OR COMPONENTS, PARTS, OR ACCESSORIES TO DEVICES)

Examples:

- Dye for tissues used in conjunction with laser surgery, to enhance absorption of laser light in target tissue
- Gas mixtures for pulmonary function testing devices
- Gases used to provide "physical effects"
- Hemo-dialysis fluids
- Hemostatic devices and dressings
- Injectable silicon, collagen, and Teflon
- Liquids functioning through physical action applied to the body to cool or freeze tissues for therapeutic purposes
- Liquids intended to inflate, flush, or moisten (lubricate) in dwelling device (in or on the body)
- Lubricants and lubricating jellies
- Ophthalmic solutions for contact lenses
- Organ/tissue transport and/or perfusion fluid with antimicrobial or other drug agent, i.e., preservation solutions
- Powders for lubricating surgical gloves
- Sodium hyaluronate or hyaluronic acid for use as a surgical aid
- Solution for use with dental "chemical drill"
- Spray on dressings not containing a drug component

*Status*: device.
*Market approval authority*: CDRH.
*Special label/regulatory considerations*: none.
*Investigation options*: IDE.
*Inter-center consultation*: required if the device has direct contact with the body and the drug or the chemical form of the drug has not been legally marketed as a human drug.

### PRODUCTS REGULATED AS DRUGS

Examples:

- Irrigation solutions
- Purified water or saline in pre-filled nebulizers for use in inhalation therapy
- Skin protectants (intended for use on intact skin)
- Sun screens
- Topical/internal analgesic-antipyretic

*Status*: drug.
*Market approval authority*: CDER.
*Special label/regulatory considerations*: none.
*Investigation options*: IND.
*Inter-center consultations*: optional.

AD HOC JURISDICTIONAL DECISIONS

| *Examples* | *Status* | *Center* |
|---|---|---|
| Motility marker constructed of radiopaque plastic brachytherapy capsules, needles, etc., that are radioactive and may be removed from the body after radiation | Device | CDRH |
| Therapy has been administered | Device | CDRH |
| Skin markers | Device | CDRH |

*Status*: device or drug.
*Market approval authority*: CDRH or CDER as indicated.
*Special label/regulatory considerations*: none.
*Investigation options*: IDE or IND, as appropriate.
*Inter-center consultation*: required to ensure agreement on drug/device status.

## General criteria affecting drug/device determination

The following represent the general criteria that will apply in making device/drug determinations.

## Device criteria

- A liquid, powder, or other similar formulation intended only to serve as a component, part, or accessory to a device with a primary mode of action that is physical in nature will be regulated as a device by CDRH.
- A product that has the physical attributes described in Section 201(h) (e.g., instrument, apparatus) of the Act and does not achieve its primary intended purpose through chemical action within or on the body, or by being metabolized, will be regulated as a device by CDRH.
- The phrase "within or on the body" as used in Section 201(h) of the Act does not include extra corporeal systems or the solutions used in conjunction with such equipment. Such equipment and solutions will be regulated as devices by CDRH.
- An implant, including an injectable material, placed in the body for primarily a structural purpose even though such an implant may be absorbed or metabolized by the body after it has achieved its primary purpose will be regulated as a device by CDRH.
- A device containing a drug substance as a component with the primary purpose of the combination being to fulfill a device function is a combination product and will be regulated as a device by CDRH.
- A device (e.g., machine or equipment) marketed to the user, pharmacy, or licensed practitioner that produces a drug will be regulated as a device or combination product by CDER. This does not include equipment marketed to a registered drug manufacturer.
- A device whose labeling or promotional materials make reference to a specific drug or generic class of drugs unless it is prefilled with a drug ordinarily remains a device regulated by CDRH. It may, however, also be subject to the combination products regulation.

## Drug criteria

- A liquid, powder, tablet, or other similar formulation that achieves its primary intended purpose through chemical action within or on the body, or by being metabolized, unless it meets one of the specified device criteria, will as regulated as a drug by CDER.
- A device that serves as a container for a drug or a device that is a drug delivery system attached to the drug container where the drug is present in the container is a combination product that will be regulated as a drug by CDER.
- A device containing a drug substance as a component with the primary purpose of the combination product being to fulfill a drug purpose is a combination product and will be regulated as a drug by CDER.
- A drug whose labeling or promotional materials makes reference to a specific device or generic class of devices ordinarily remains a drug regulated by CDER. It may, however, also be subject to the combination products regulation.

## Conclusions

In summary, we have touched upon the regulations that currently control the types of preclinical toxicity testing done on potential human pharmaceuticals and medical device products. We have reviewed the history, the law, the regulations themselves, the guidelines, and common practices employed to meet regulatory standards. Types of toxicity testing were discussed, as were the special cases pertaining to, for example, biotechnology products.

## References


Alder S, Zbinden G. *National and International Drug Safety Guidelines.* Switzerland: MTC Verlag Zoblikon, 1988.

Anderson O. *The Health of a Nation; Harvey Wiley and the Fight for Pure Food.* Chicago: University of Chicago, 1958.

Anon. FDA's Policy Statement for the Development of New Stereoisomeric Drugs, 1992a. http://www.fda.gov/cder/guidance/stereo.htm

Anon. Orphan Drug Regulations 21 CFR PART 316, 1992b. http://www.fda.gov/orphan/about/odreg.htm

Anon. Guidance for Industry and Reviewers: Repeal of Section 507 of the Federal Food, Drug and Cosmetic Act, 1997a. http://www.fda.gov/cder/guidance/index.htm

Anon. Points to Consider in the Manufacture and Testing of Monoclonal Antibody Products for Human Use, 1997b. http://www.fda.gov/cber/cberftp.html

Anon. Annual Report of the Pharmaceutical Research and Manufacturer Association (Priority #2: Improved FDA Regulation of Drug Development), 1998a.

Anon. Guidance for Industry: Content and Format for Geriatric Labeling (Draft Guidance), 1998b. http://www.fda.gov/cder/guidance/2527dft.pdf

Anon. *Guidelines for Toxicity Study of Drugs Manual.* Tokyo: Ykuji Nippo, Ltd., 1994.

Anon. Well-characterized biologics would be regulated under FD&C Act by Kassebaum FDA Reform Bill: *FDC Reports (pink sheets)* 1996;58:11–12.

Barton B. International Conference on Harmonization – good clinical practices update. *Drug Info J* 1998;32:1143–1147.

Berliner VR. US Food and Drug Administration requirements for toxicity testing of contraceptive products. In: Briggs MH, Diczbalusy, E, editors. Pharmacological models in contraceptive development. *Acta Endocrinol (Copenhagen)* 1974:Supp. 185:240–253.

Blakeslee D. Thalidomide: a Background Briefing. JAMA HIV/AIDS Information Center Newsline, 1998. http://www.amaassn.org/special/hiv/newsline/briefing/briefing.htm

Brusick, D. *Principles of Genetic Toxicology* Plenum Press, New York, 1987.

Burns J. Overview of safety regulations governing food, drug and cosmetics. In: Homberger F, editor. *The United States in Safety and Evaluation and Regulation of Chemicals 3: Interface Between Law and Science.* New York: Karger, 1983.

Daniels J, Nestmann E, Kerr A. Development of stereoisomeric (chiral) drugs: a brief review of the scientific and regulatory considerations. *Drug Info J* 1997;50:639–646.

DeGeorge J, Ahn C, Andrews P, Bower M, Giorgio D, Goheer M, Lee-Yam D, McGuinn W, Schmidt W, Sun C, Tripathi S. Regulatory considerations for the preclinical development of anticancer drugs. *Cancer Chemother Pharmacol* 1998;41:173–185.

DiMasi J, Seibring M, Lasagna L. New drug development in the United States from 1963 to 1992. *Clin Pharmacol Ther* 1994;55:609–622.

FDA. *Environmental Assessment Technical Assistance Handbook.* NTIS No. PB87-175345, 1987.

FDA. Good Laboratory Practices, *CFR* 1988a;21(58).

FDA. Investigational new drug, antibiotic, and biological drug product regulations, procedures for drug intended to treat life-threatening and severely debilitating illnesses. *Fed Register* 1988b;53(204):41516–41524.

FDA. FDA perspective on development of stereoisomers. *AAPS* 1988c;May 16:12–16, .

Gad, SC. *Safety assessment for pharmaceuticals.* New York: Van Nostrand Reinhold, 1994.

Goldenthal E. Current view on safety evaluation of drugs. *FDA Papers,* 1968;May:13–18.

Grabowski HG, Vernon JM. *The Regulation of Pharmaceuticals.* Washington, DC: American Enterprise Institute, 1983.

Grieshaber C. Prediction of human toxicity of new antineoplastic drugs from studies in animals. In: Powis G, Hacker M, editors. *The Toxicity of Anticancer Drugs.* New York: Pergamon Press, 1991. pp. 10–24.

Haffner M. Orphan drug development – international program and study design issues. *Drug Info J* 1998;32:93–99.

Hart C. Getting past the politics and paperwork of pediatric drug studies. *Mod Drug Dis* 1999;2:15–16.

Hutt PB. Investigations and reports respecting FDA regulations of new drugs (Part I). *Clin Pharmacol Ther* 1983a;33:537–548.

Hutt PB. Investigations and reports respecting FDA regulations of new drugs (Part II). *Clin Pharmacol Ther* 1983b;33:174–187.

Hutt PB. Progress in new drug regulation. *Clin Res Prac Reg Affairs* 1987;5:307–318.

Kessler D, Siegel J, Noguchi P, Zoon K, Feidon K, Woodcock J. Regulation of somatic-cell therapy and gene therapy by the Food and Drug Administration. *New Engl J Med* 1993;329:1169–1173.

Lumley CE, Walker, SE. An international appraisal of the minimum duration of chronic animal toxicity studies. *Hum Exp Toxicol* 1992:11:155–162.

Lumpkin M. Guidance for industry: content and format of investigational new drug applications (INDs) for Phase 1 studies of drugs, including well-characterized, therapeutic, and biotechnology-derived products, 1995. http://www.fda.gov/cder/guidance/phase1.pdf

Manson J. Testing of pharmaceutical agents for reproductive toxicity in developmental toxicology. In: Kimmel C, Buelke-Sam J, editors. *Developmental Toxicology,* 2nd Ed. New York: Raven Press, 1994. pp. 379–402.

Marwick C. Implementing the FDA Modernization Act. (Medical News and Perspectives). *J Am Med Assoc* 1998;279: 815–816.

Merrill R. Regulation of drugs and devices: an evolution. *Health Aff* 1994;13:47–69.

Pilot L, Waldemann D. Food and Drug Modernization Act of 1997: medical device provisions. *Food Drug Law J* 1998;53:267–295.
Ryffel B. Unanticipated human toxicology of recombinant proteins. *Arch Toxicol* 1996; Suppl. 18:333–341.
Ryffel B. Safety of human recombinant proteins. *Biomed Environ Sci* 1997;10:65–71.
Schacter E, DeSantis P. Labeling of drug and biologic products for pediatric use. *Drug Info J* 1998;32: 299–303.
Wu K-M, DeGeorge J, Atrachi A, Barry E, Bigger A, Chen C, Du T, Freed L, Geyer H, Goheer A, Jacobs A, Jean D, Rhee H, Osterburg R, Schmidt W, Farrelly J. Regulatory science: a special update from the United States Food and Drug Administration. Preclinical issues and status of investigations of botanical drug products in the United States. *Toxicol Lett* 2000;111:199–202.

## Further reading

Buesing-McSweegan M. Submitting biologics applications to the Center for Biologics Evaluation and Research electronically. *Drug Info J* 1999;33:1–15.
Cadden S. *New Drug Approval in Canada*, Parexel, Waltham, MA, 1999.
CDER. Guideline for the study of drugs likely to be used in the elderly. *FDA*, November 1989.
CDER. Premarket notifications: 510 (K) regulatory requirements for medical devices. HHS *FDA* 1990:4158.
Currie WJC. *New Drug Approval in Japan*, Parexel, Waltham, MA, 1995.
Dranove D. The cost of compliance with the 1962 FDA Amendments. *J Health Econ* 1991;19:235–238.
Evers PT. *New Drug Approval in the European Union.* Waltham, MA: Parexel, 1985.
FDA. *FDA Introduction to Total Drug Quality.* US Government Printing Office, Washington, DC, 1971.
FDA, Points to consider in human somatic cell therapy and gene therapy, 1991.
FDA, Points to consider in the characterization of cell lines used to produce biologicals, 1993.
FDA, Points to consider in the production and testing of new drugs and biologicals produced by recombinant DNA technology, 1985.
FDA, Draft Addendum to the PTC in the Human Somatic Cell and Gene Therapy, 1999.
Gussin R. Recent development in biotechnology: the influence on drug development. *Drug Info J* 1989;20:115–119.
Hayes T, Ryffel B. Symposium in writing: safety considerations of recombinant protein therapy; introductory comments. *Clin Immunol Immunother* 1997;83:1–4.
Herrick A. *New Drugs.* New York: Revere Publishing, 1946.
Hess G. FDA accelerates its approval of new drugs. *Chem Market Newslett* 1999;255:21.
Hite M. Safety pharmacology approaches. *Int J Toxicol* 1997;16: 23–31.
Holtzman DC. European biotech firms invest in US facilities. *BioWorld Today*, 1994;June:2.
ICH. International Conference on Harmonization Safety Steps 4/5 Documents. Buffalo Grove, IL: Interpharm Press, Inc., 1997.
Ohno Y. Harmonization of the timing of the nonclinical tests in relation to the conduct of clinical trials. *J Control Release* 1999;62:57–63.
Wessinger J. Pharmacologic and toxicologic considerations for evaluating biologic products. *Reg Toxicol Pharmacol* 1989;10:255–263.
Yakuji Nippo, Ltd. *Drug Registration Requirements in Japan.* Tokyo: Yakuji Nippo, Ltd., 1991.
Young F, Nightingale S, Mitchell W, Beaver L. The United States Drug Export Amendment Act of 1986: perspectives from the Food and Drug Administration. *Int Digest Health Legislation* 1989;40:246–252.

Chapter 3

# Animal health products

*Patricia Frank and James H. Schafer*

From a regulatory perspective, veterinary animal health products are divided into two groups: those given to food-producing animals, and those administered to non-food (companion) animals. The safety assessment and regulatory issues for each group of products are unique. This chapter examines the testing required to bring these veterinary products to market in the United States. Regulations governing veterinary products, as well as the organization of the Center for Veterinary Medicine (CVM), are presented. Available CVM guidelines for direction in planning, executing, and submitting studies are listed. The contents of the initial filing to the CVM (an Investigational New Animal Drug Application), and the organization and contents of the New Animal Drug Application are discussed. Finally, the risk assessment process is presented for products that leave residues in tissues subsequently consumed by humans.

## Center for Veterinary Medicine (CVM)

### *Regulatory history*

Under the federal Food, Drug, and Cosmetic Act, the Center for Veterinary Medicine of the US Food and Drug Administration (USFDA or FDA) is responsible for reviewing, evaluating, and approving all drug products for use in animals prior to introducing the drugs into the marketplace. Chapter 1 presents the overall history of the FDA ("Agency") and the laws that enable it. Only the aspects relative to animal products are given here.

Massachusetts enacted the first food law in the US in 1784, but the first attempt to control animal health occurred in 1891 when Congress passed an act requiring the inspection of animals for disease prior to slaughter.

Dr. Harvey Wiley was appointed chief chemist, US Department of Agriculture (USDA) in 1883 and is credited with establishing the regulation of foods and drugs; his efforts resulted in the enactment of the first Food and Drug Act ("the Act") in 1906. The Act was extended in 1938 to cover cosmetics and medical devices and required all drugs to be shown safe for use. Additional amendments such as for food color additives were added in 1958–1962, continuing with the Kefauver–Harris Amendment requiring that all drugs be shown both safe and effective prior to distribution and sale. This amendment contained a clause (the Delaney Clause, i.e., "DES exemption") permitting the use of potentially cancer-causing drugs in production animals provided that no residues were found in edible tissues.

In 1968 the Act was amended to include sections specifically addressing animal drugs. The modifications were designed to ensure that animal drugs are safe and effective for their intended uses and that no unsafe residues are present in foods at the time of consumption. The 1968 amendment consolidated the veterinary drug review process of the FDA by imposing a unified review method applicable to animal drugs, particularly those used in production animals. Three significant changes were promulgated. First, the amendment designated five named antibiotics as *new animal drugs* by statutory definition. This had the effect of superimposing batch certification requirements on new drugs rather than allowing these certifications to be used as an alternative regulatory system, as had been done previously. Second, the amendment required that each animal drug approval be published in the Federal Register as a specific regulation, identifying the name of the drug, its sponsor, and the prescribed conditions of use. Third, the amendment extended the Delaney Clause to all animal drugs.

In 1987 a rule establishing the criteria and procedures for evaluating the safety of carcinogenic residues was enacted. In 1988 the Generic Animal Drug and Patent Term Restoration Act established the procedure for an Abbreviated New Animal Drug Application (ANADA) by a sponsor. In 1996, the Animal Drug Availability Act was enacted to facilitate the approval and marketing of new animal drugs and medicated feeds.

During the first 50 years of its existence, the CVM protected the consumer almost entirely through law enforcement actions. During the last 25 years or so, it has relied more on pre-marketing approvals, balanced by post-marketing surveillance and compliance. Because safety studies supporting the use of products in animals are regulated by Good Laboratory Practices (Code of Federal Regulations, Title 21, Part 58), animal drug sponsors must be certain that effective quality assurance and data audit systems are in place and that all raw data are carefully checked before submission to CVM. Top enforcement priority is now given to illegal drug residues in meat, milk, eggs, and more recently, in fish.

## Organization and administration

Along with the office of the Center Director, the CVM is comprised of four additional groups of staff members: management, pre-marketing, post-marketing, and research. The section of primary interest to toxicologists involved in veterinary product development is that of New Animal Drug Evaluation (pre-marketing staff), including five divisions: Therapeutic Drugs for Non-food Animals, Biometrics and Production Drugs, Therapeutic Drugs for Food Animals, Manufacturing Technologies, and Human Food Safety.

When a submission is filed with the CVM, the manufacturing, composition, and control sections will be reviewed by the Division of Manufacturing Technologies. Environmental data will be reviewed by the Environmental Assessment Team of the Division of Manufacturing Technologies. Efficacy and safety data will be reviewed by one of the other divisions, depending on the class of drug. Tissue residue and toxicology data (related to drug residues) will be reviewed by the Division of Human Food Safety. A current listing of the organization, including the personnel in each division and their telephone numbers, can be obtained from the CVM by written request or at the FDA/CVM web site, http://www.fda.gov/cvm/fda/.

The Code of Federal Regulations (CFR) is published by the Office of the Federal Register, National Archives and Records Administration, as a special edition of the Federal Register, and is available for purchase from the Superintendent of Documents, US Government Printing Office, Washington, DC 20402. Included in the CFR, Title 21, are the volumes pertaining to the FDA, which are updated each year effective April 1st. Title 21, parts 500–599, includes regulations governing animal drugs, animal feed, and related products. Requirements such as registration, listing, labeling, and current Good Manufacturing Practices (cGMPs) are contained in Title 21, Parts 200–299. Good Laboratory Practice (GLP) for Nonclinical Studies are listed in 21 CFR 58.

The CVM makes available a series of educational materials to assist toxicologists, veterinarians, and individuals in the animal drug and feed industries in understanding these laws and regulations. These materials may be acquired from the CVM website at http://www.fda.gov/cvm/fda/, or through the CVM Industry Information Staff, The Center for Veterinary Medicine, Communications and Education Branch, 7500 Standish Place, HFV-12, Rockville, MD 20855, +1-301-594-1755. Additional information may be obtained from CVM's Office of New Animal Drug Evaluation, HFV-100, 7500 Standish Place, Rockville, MD 20855, +1-301-594-1620, fax +1-301-594-2297.

The Center for Veterinary Medicine Memos (CVMMs), published by the Agency, are intended to help regulated industry, scientists, veterinary professionals, and the general public better understand laws and regulations enforced by the FDA and to improve the safety and effectiveness of animal drugs. CVMMs are available from the FDA Industry Information Staff.

The CVM Guidelines provide procedures for collecting the research data that are necessary to support new animal drug approval requirements. These guidelines describe procedures acceptable to the FDA; however, they do not preclude alternative methods, provided that the drug sponsor believes they may be applicable. Discussion with the CVM *prior* to undertaking such studies is advisable. A listing of pertinent guidelines that may be helpful can be found at the end of this chapter.

A current organization chart may be found at http://www.fda/gov/cvm/mappgs/center.pdf.

### Freedom of information

In addition to CVMMs and CVM Guidelines which are usually available at no cost and often accessible on the internet, the Freedom of Information Act permits an individual or corporation to procure a "Freedom of Information (FOI) Summary" of an approved drug. This document is requested from the FDA Freedom of Information Staff, HFI-35, Room 12-A-20, 5600 Fishers Lane, Rockville, MD 20857, or through private companies such as FOI Services, Inc. (11 Firstfield Road, Gaithersburg, MD 20878-1703) who typically charge a nominal fee for this service. Information ordered directly from the FDA may take longer to arrive, and such requests are published in the Freedom of Information Log, which is available to the public. The request for a drug FOI Summary should contain the trade name, the generic name, and the manufacturer, and must contain the following statement: "Under the Freedom of Information Act and implementing regulations, please forward to us the following...".

An FOI Summary for each drug is prepared by the sponsor and approved by CVM as a required part of the New Animal Drug Application (NADA) or ANADA. The summary is released for public view and made available for distribution when the drug is approved. The FOI document must provide a summary of each study in sufficient detail to demonstrate the safety and effectiveness upon which the CVM based its approval. Corroborative or supportive (non-pivotal) studies are not included in the FOI document.

The contents of an FOI Summary are presented in moderate detail and illustrate the studies needed to gain approval for a veterinary drug. For example, an FOI Summary may contain the following elements:

1 *General information*: NADA number, sponsor, product name.
2 *Indications for use.*
3 *Dosage forms, routes of administration, and recommended dosages.*
4 *Effectiveness*: The dose rationale, including studies conducted to establish a dose or a dose range are presented; field studies are also described.
5 *Animal safety*: Pivotal studies would normally include tolerance test (usually conducted at 10× the recommended dose administered once) and a target animal safety study (usually conducted at 1×, 3×, and 5× the recommended dose for three times the recommended duration). If the product is to be used in breeding animals, reproductive safety studies will be needed. An option is to label the product " not for use in breeding animals." Additional safety studies could be required depending on known or suspected properties of the test substance.
6 *Human food safety*: In food animals, tissue tolerance levels must be established, as well as withdrawal times following treatment, before the tissues can be used for food.
7 *Agency conclusions*: This is the last page of an NADA and is left blank by the sponsor. The Agency places its conclusion here; that is, based on the evidence, the drug is approved for the conditions listed in the Indications section. In addition, labels for all packaging including samples, the finding of no significant environmental impact, and the Environmental Assessment document are included.

## Regulatory process and procedures

### *Investigational new animal drugs*

To initiate clinical field trials with a new animal drug or to obtain a withdrawal period to allow treated food-producing animals to enter the human food chain, the drug must be listed with the CVM as an Investigational New Animal Drug Application (INAD). If no withdrawal period has been established, no production animals may be consumed by humans; the carcasses must be destroyed. Depending on the compound and the available information, CVM may grant an extended (conservative) withdrawal period. The drug sponsor should submit a "Notice of Claimed Exemption for an Investigational New Animal Drug" (21 CFR 511.1) to the CVM, along with a summary of information known about the drug. Also included should be the intended use for the drug, the species in which it will be tested, a summary of available data and literature (including foreign studies), and a summary of the toxicity testing performed to date with particular emphasis on the proposed target species.

It is recommended that the sponsor meet with CVM (product development meeting) to clearly identify the regulatory approval requirements for the specific compound being developed. Protocols for efficacy and safety studies to be conducted should be submitted for CVM review prior to initiation of studies. An INAD number must be assigned by the CVM before interstate shipment of an investigational drug for clinical trials is allowed.

Data must be presented for a dose rationale. These data may be from traditional dose titration testing or from other host animal studies in which a dose response is demonstrated. A target animal safety study at multiples of the efficacious dose is performed (see Target Animal Safety Testing section). In some instances, often at the discretion of the sponsor, distribution and metabolism studies in the target animal are completed.

Clinical testing for efficacy is performed in the field by qualified investigators (usually practicing veterinarians with client-owned animals) selected by the drug sponsor. At least one well-controlled pivotal clinical study must be submitted. The number of animals per test group should be determined statistically using sample size calculations. Clinical studies should generally be done in several geographically separate areas of the US to show that the drug or product works in a variety of different situations.

## New drugs

### New animal drug application (NADA)

#### GENERAL

An NADA is compiled according to 21 CFR 514; three copies must be submitted. In *Phasing the Review of Data Submissions* (Policy and Procedures Manual, FDA Guide 1240.3040, March 1989) the procedure is outlined for the phased submission of specific data and information to be used to support an NADA by sections in order to facilitate CVM review of the data.

#### SECTIONS OF THE NADA

The NADA contains the following sections:

1 *Identification*: This section defines the NADA. Form FDA 356V and a cover letter are included.
2 *Table of contents and summary*: A detailed table of contents is presented to allow the various FDA reviewers easy access to information in the NADA. The summary is concise, yet contains all the salient points which need to be highlighted for the reviewers.
3 *Labeling*: A copy of the sponsor's draft label for the product is required. A request to the CVM for a label from a product which has been approved recently may be helpful in formatting the label according to current FDA style.
4 *Components and composition*: A complete list of articles used for the production of the new animal drug and a full list of each article used in the composition of the drug product should be provided.
5 *Manufacturing information*: This section contains a complete description of the facilities, equipment, and manufacturing procedures used to prepare the drug substance and finished dosage form.

6 *Drug samples*: Reference samples of the drug and a sample of the finished dosage form are to be submitted on request.
7 *Residue information*: Toxicology data required to establish human safety are presented, if required, and residue information, including analytical methods for the residues, is detailed for drugs indicated for production animals.
8 *Safety and efficacy*: Target animal safety data, toxicity data, and the results of clinical field trials are presented. Compatibility studies with other drugs may be required. If the drug is an antibiotic for use in food-producing animals, microbial-resistance and *Salmonella* shedding studies may be necessary.
9 *Good laboratory practices*: A statement of compliance or non-compliance with good laboratory practices is presented for each safety study.
10 *Environmental impact*: This section contains a summary of all the environmental studies conducted with the product, as well as the complete studies themselves. The effect of the drug on the environment is based on estimated sales and use patterns of the drug and the residence time of the drug in the environment. Categorical exclusion from the need to provide an Environmental Assessment may be requested in some cases (21 CFR 25.33 (a), (c), (d), or (e)).
11 *Freedom of information summary*: The FOI summary prepared by the sponsor is included and will be supplied to the public upon request.

### SUPPLEMENTS AND AMENDMENTS

Supplements or amendments to the NADA should contain only the sections that apply. The CVM no longer requires supplemental applications for minor changes such as extension or expiration dates, updates in specifications or methods to bring them into compliance with official methods, or minor label changes. The notification necessary for various changes is provided in 21 CFR 514.8.

### TARGET ANIMAL SAFETY GUIDELINES

Target animal safety guidelines have been established for dogs, cats, horses, ruminants, swine, and poultry (see CVM Guideline 33). The guidelines detail how to design an acceptable protocol. Studies must be performed at one, three, and five times (1×, 3×, 5×) the proposed clinical dose for three times (3×) the proposed duration of use. Thus, definitive safety studies may not be possible until after establishment of a dose. The sponsor should evaluate the likelihood of toxicity with a short-term study to decide whether a drug tolerance study will be needed. A control group (untreated, placebo, or vehicle) must be included in all target animal safety studies.

The age and sex of the test animals should mimic as closely as possible those of the animals receiving the drug in the clinical setting. For example, antimastitis drugs must be tested in cows of the appropriate age and worming agents should be tested in puppies as well as mature dogs.

A drug tolerance test (usually performed as a single 10× dose) is required to demonstrate the drug's margin of safety. If signs and symptoms of toxicity are observed in the 1×, 3×, 5× study, a separate drug tolerance study may be needed. For drugs with a wide margin of safety, evaluating the drug at 10× would be the maximum dose required.

*Group size* An appropriate number of animals in each group must be selected based on statistical considerations; the CVM is concerned with the needless use of animals. The number of animals per group used for target animal safety studies will depend on the drug and the intended target species. Refer to the current CVM guideline for the specific target animals. A protocol should be submitted to the CVM before the study starts to achieve agreement on study design and parameters to be measured to prevent unnecessary duplication of testing. It should contain a complete statistics section which includes both the significance level and the power of the test.

*Study duration* Dosing duration is determined by the clinical use of the drug. For drugs to be administered long-term, i.e., over 14 days, the safety studies should be at least as long as the clinical treatment course. For short-term drugs such as antibiotics or anthelmintics, the safety studies should continue for at least three times the clinical treatment plan. That is, for a drug which will be given for 3–5 days, 15 days would be an appropriate length for the safety studies. The sponsor should bear in mind that if the drug might be used repeatedly during a relatively short time interval, the safety studies should extend longer than the one-time use would indicate. For chronic medications to be used in dogs and cats, such as diuretics or cardiac drugs, studies of 6 months or longer might be needed. It is recommended that protocols be submitted to CVM for review, and the protocol be revised according to their comments prior to study start. CVM's Document and Submission Information (formerly called Submission Tracking and Reporting System – STARS document) lists an anticipated protocol review time of 50 days (100 days if the protocol is submitted with data).

*Reproduction studies* If the drug is intended for use in breeding animals, reproductive safety testing in breeding animals is required. Requirements for reproductive studies are outlined in CVM Guideline 33. The parental, in utero, and postnatal exposure are adjusted to suit the specific species. As an alternative, the drug label may exclude breeding animals.

*Special studies* Additional tests such as irritation studies for injectable and topical drugs should be performed. Drug interaction studies with commonly used medications may be required in the target animal. These may include testing for interactions with commonly used food supplements and vitamins.

## Generic drugs

### Abbreviated New Animal Drug Application (ANADA)

The 1988 Generic Animal Drug Act provided legislation that permits approval for generic versions of off-patent animal drugs. It also requires that NADAs contain patent numbers and expiration dates of patents covering each component of the application. Bioequivalence of generic versions of drugs against the FDA-designated reference product must be established. For true solutions, and for some topical products, the need to conduct animal bioequivalence studies might be waived. Animal drug pharmacokinetic studies, pharmacologic end-point studies, or clinical end-point studies are required to establish bioequivalence for the filing of an ANADA.

Bioequivalence is preferably demonstrated through "blood level" studies which determine the concentration of the parent compound(s) and/or metabolite(s) in serum or plasma following the administration of the drug product. If a blood level study is not feasible (e.g., when a satisfactory analytical method is not available), a physiological endpoint study may be substituted. Also acceptable for some drugs, such as anthelmintic products, are clinical endpoint studies in which improvement in a disease state (e.g., parasite burden) is measured. For drugs used in food-producing animals the FDA requires tissue residue depletion studies in addition to blood level or end-point equivalence studies.

## Risk assessment

Risk assessment for animal health products concerns risk to the target animal, the environment, the drug handler, and the human consumer relative to drug residues in meat, milk and eggs.

Risk to the target animal species is addressed by the target animal safety studies. For a drug used in food-producing animals, the risk to the environment and to the human handler of the product must be assessed. When an additive is administered to food-producing animals in their feed, or when a dosage form is directly administered to such animals either in a free-range situation or in a feedlot, the producer of such an additive or drug must calculate the potential effects on the environment. Drug and/or metabolites can reach the soil, plants, and the water table through feedlot runoff, use of manure to fertilize crop fields, and the raising of crops on pasture land. The impact of compounds entering the soil and water needs to be assessed.

The types of studies that should be conducted to prepare an environmental assessment report are described in detail in the *Environmental Assessment Technical Handbook* (National Technical Information Services, 5285 Port Royal Road, Springfield, VA 22161). These generally assess toxicity to wildlife, aquatic life, soil life, and plant life. The binding of drug to soil and its half-life in the environment must be calculated. This requires an assessment of photodegradation, hydrolysis, soil adsorption and desorption, and microbial breakdown of the drug and its major metabolites.

In addition to effects on the environment, when feed additives are prepared for inclusion in the diet of food-producing animals, the exposure of humans can be substantial. Most diet supplements are added to the food at the use site, i.e., the farm or ranch where the animals are housed. The farmer who mixes the additive into the feed is not constrained by worker regulations, e.g., the wearing of protective clothing, that may be posted in a factory or chemical laboratory. Therefore, it is crucial to determine the effects of the product on humans who come in contact with the compound.

The risk for exposure to incidental species must be considered for an animal feed additive that will be widely used on farms or ranches. When feeds have been prepared for one species of animal, they may also be available to a different species. Horses, dogs, chickens, and/or cats may all be exposed to spilled cattle feed. Because species-specific toxicity may exist for these drugs, many of which are antibiotics, toxicity testing should be carried out in other species that have a high probability of ingesting feeds. For example, if toxicity testing indicated that horses were more sensitive to the drug than the target species, a distinctive warning label would need to be placed on the food additive container. Drugs available in medicated feed block form should also be tested for incidental consumption by other animals.

The major risk assessment effort connected with products for food-producing animals concerns the determination of the safe concentration of drug residues in edible tissues. The drug sponsor must conduct the standard battery of short-term toxicity tests to determine the general food safety of the drug. These include three or more short-term genetic toxicity tests in two test systems, two subchronic feeding studies (usually in the rat and dog), and a multiple-generation reproduction study and teratology study in the rat. The sponsor thus determines the species most sensitive to the drug and establishes the "NO Effect Level" (NOEL) in mg/kg of body weight for that species.

Complete metabolism studies must be conducted, usually first in the target species, then in an experimental laboratory animal species. The purpose of the laboratory animal study is to demonstrate that the "surrogate" humans used in laboratory testing are exposed to the same metabolites as the human would be upon tissue (meat) ingestion.

Most often it is necessary to use radio-labeled drug, preferably $^{14}C$, to obtain the required level of detection in the various edible tissues of the production animal. Total residue of a chemical in treated animals consists of the unchanged parent compound, unbound free metabolites, and metabolites that are covalently bound to endogenous molecules. Different components of the total residue may have different toxicological potential. Therefore, the sponsor must develop data on the amount, persistence, and chemical nature of the total residue. Any residue consisting of 10 per cent or more of the administered drug is usually considered significant, and chemical identification of the residue is typically required by the FDA.

Conduct of studies in the target species must include consideration of the time required to reach steady-state conditions prior to establishing the drug residue depletion profile. Once steady state is reached, sacrifices at several time intervals after drug administration establish the depletion of the drug from edible tissues and the identification of the target tissue, which is the edible tissue that contains the most drug or drug residue. This tissue will be used to monitor the drug, should a withdrawal period be necessary prior to marketing of the production animal. Further studies are required with the target tissue to establish which residue will be the marker compound for regulatory monitoring purposes.

### Calculating residue concentration

In *General Principles for Evaluating the Safety of Compounds Used in Food-producing Animals* (FDA Guideline 3, July 1994), the procedure is given for calculating the safe residue concentration in an edible tissue which will be consumed by humans. In the following example, assume that the drug in question is a growth modifier proposed for use in broiler chickens and that it will be placed in the feed and administered to the animals for their entire growing period.

**Step 1**. Determine the most sensitive species from the battery of toxicity studies conducted for general food safety. Here it is the dog with a NOEL of 0.5 mg/kg per day in a 6-month conventional toxicology study.

**Step 2**. FDA guidelines require a 1000-fold safety factor based on the duration (6 months or less) of the toxicity study described in Step 1. Accordingly, the Acceptable Daily Intake (ADI) and the Safe Concentration (SC) of total residues are calculated, using a human weight of 60 kg:

$$\text{ADI} = \frac{\text{NOEL}}{\text{Safety Factor}} = \frac{0.5 \text{ mg/kg per day}}{1000} = 0.5 \ \mu\text{g/kg per day}$$

$$\text{SC for muscle} = \frac{\text{ADI} \times \text{Human Weight}}{\text{Daily Tissue Consumption}}$$

$$= \frac{0.5 \ \mu\text{g/kg per day} \times 60 \text{ kg}}{0.5 \text{ kg/day}} = 60 \ \mu\text{g/kg} = 60 \text{ ppb}$$

**Step 3**. Calculate the safe concentration of the active drug. In this example, the drug to be marketed is the potassium salt; therefore, the active drug moiety is 97.5 per cent of the molecular weight, and the safe concentration in muscle tissue is 60/97.5 or 61.5 ppb.

**Step 4**. Determine the safe concentration of residues for other edible tissues. Table 3.1 is based on the FDA's "Consumption Factors" using poultry as an example.

*Table 3.1* Safe concentration of residues

| *Tissue* | *Consumption factor* | *Safe concentration (ppb)* |
|---|---|---|
| Liver | 3 | 184.5 |
| Skin | 2 | 123 |
| Fat | 2 | 123 |

**Step 5**. Establish a validated lower limit of analytical quantitation for the drug marker residue in each edible tissue based on analytical methodology. This example uses 10 ppb $\pm$ standard deviation as the validated detection limit.

**Step 6**. Determine the total residues at selected times after drug administration. See Table 3.2 for an example. The total residues in a target tissue such as liver are depleted below the safe residue concentration for this tissue (184.5 ppb from Step 4) between 6 and 12 h after withdrawal. Interpolation of the unchanged drug to total residue ratio for this time yields a ratio of 0.45. The tolerance for the marker residue in the target tissue is calculated as:

$$\text{Tolerance} = 0.45 \times 184.5 \text{ ppb} = 83 \text{ ppb}$$

**Step 7**. Conduct a trial with the drug for the entire proposed commercial feeding period and analyze the target tissue for residue following withdrawal periods. For instance, the drug is fed for 44 consecutive days, the entire "grow-out" period for

*Table 3.2* Total residues at selected times after drug exposure

| *Time (h)* | *Total residue (ppb)* | *Unchanged drug (ppb)* | *Ratio* |
|---|---|---|---|
| 6 | 297 | 139 | 0.47 |
| 12 | 115 | 49 | 0.42 |
| 24 | 58 | 22 | 0.38 |
| 48 | 34 | 12 | 0.36 |

*Table 3.3* Analysis of residue following drug withdrawal period

| *Time (h)* | *Drug marker levels (ppb)* | *99/95 level (ppb)* |
|---|---|---|
| 6 | 187 | 1013 |
| 12 | 51 | 478 |
| 18 | 49 | 244 |
| 24 | 22 | 134 |
| 36 | <10 | 33 |

broilers. Table 3.3 contains example data for the end of the feeding period. The marker residue tolerance (from Step 6) is 83 ppb. Therefore, 99 per cent of the medicated broiler population, with 95 per cent confidence, would have shown depletion of all residues in all edible tissues below safe concentrations within 36 h following final exposure.

## International considerations in veterinary product development

As in the development of human drugs, a major issue confronting regulatory authorities is the evaluation of biotechnology products. The issues surrounding approval concern the newness of the technology, the difficulty in assessing drugs that have the same structures as endogenously produced compounds, the political climate, and consumer activities that are associated with the use of these products, both for animals and for humans. These issues in the European Union (EU) have led to legislation to ban the sale of biotechnology products for several years.

The concerns of the EU and the US appear to be different, not only for biotechnology products but also for other types of veterinary products used in food-producing animals. For example, the EU has banned the use of steroid analogues in livestock. Whether or not these restrictions remain at their current level or are modified in the future, the regulatory climate in which a drug will be registered should be carefully monitored. Individual country considerations may be very important for certain classes of drugs.

Information about veterinary product development in European countries is available from the European Medicines Evaluation Agency (EMEA) at http://www.eudra.org/en_home.htm.

## Veterinary (off-label) use of human pharmaceuticals

Another potentially controversial issue is the off-label use of human drugs in animals. Human drugs are increasingly prescribed and dispensed to animal owners by veterinarians. Human drugs are used to treat companion animals with diseases for which there are no approved veterinary products. Economic considerations play a large role in the lack of development of specific veterinary products. Human drugs, even those for which there is specific animal information, may not contain any information on the label which suggests a use in animals, and the drug cannot be advertised for such uses, even to veterinarians. The Food and Drug Administration's Compliance Policy Guide 7125.35 (Chapter 25, Veterinary Drugs, July 20, 1992) states that human drugs are considered misbranded if they are used in animals. This policy guide stated, however, that regulatory action will probably not be taken for drugs used in companion animals for which no veterinary version of the drug is approved. In October 1994, the Animal Health Care Reform Bill, an amendment to the Food, Drug and Cosmetic Act, gave veterinarians greater flexibility in administering medications. (See Sec. 608.100 Human-Labeled Drugs Distributed and Used in Animal Medicine (CPG 7125.35) and Animal Medicinal Drug Use Clarification Act of 1994.)

## Relevant CVM guidelines and guidance documents

No. 1 Anticoccidial Guidelines (replaced by Guideline 40)
No. 2 Anthelmintics
No. 3 General Principles for Evaluating the Safety of Compounds Used in Food-Producing Animals 7/94
No. 4 Guidelines for Efficacy Studies for Systemic Sustained Release Sulfonamide Boluses for Cattle
No. 5 Stability Guidelines 12/90
No. 6 Guidelines for Submitting NADA's for Generic Drugs Reviewed by NAS/NRC 10/20/71; rev. 3/19/76
No. 8 Guidelines for Toxicological Investigations (replaced by Guideline Three)
No. 9 Preclearance Guidelines for Production Drugs 10/75
No. 10 Amendment of Section II(G)(1)(b)(4) of the Preclearance Guidelines 10/75
No. 12 Export of New Animal Drugs for Investigational Use Revised 10/92
No. 13 Guidelines for Evaluation of Effectiveness of New Animal Drugs for Use in Free-choice Feeds revision of Medicated Block 1/85
No. 14 Guideline and Format for Reporting the Details of Clinical Trials Using An Investigational New Animal Drug in Food Producing Animals
No. 15 Guideline and Format for Reporting the Details of Clinical Trials Using An Investigational New Animal Drug in Non-food Producing Animals (2277)
No. 16 FOI Summary Guideline 5/85
No. 17 Working Guidelines for Assigning Residue Tolerances replaced by Guideline Three
No. 18 Antibacterial Drugs in Animal Feeds: *Human Health Safety Criteria*
No. 19 Antibacterial Drugs in Animal Feeds: *Animal Health Safety Criteria*
No. 20 Antibacterial Drugs in Animal Feeds: *Antibacterial Effectiveness Criteria*
No. 21 Nutritional Ingredients in Animal Drugs and Feeds rev. 3/93

No. 22 Guideline Labeling of Arecoline Base Drugs Intended for Animal Use
No. 23 Medicated Free Choice Feeds-Manufacturing Control 7/85
No. 24 Guidelines for Drug Combinations for Use in Animals 10/83
No. 25 Guidelines for the Efficacy Evaluation of Equine Anthelmintics 1/79
No. 26 Guidelines for the Preparation of Data to Satisfy the Requirements of Section 512 of the Act Regarding Animal Safety, Effectiveness, Human Food Safety and Environmental Considerations for Minor Use of New Animal Drugs 4/86 (superceded by Guidance 61)
No. 27 New Animal Drug Determinations 7/89
No. 28 Animal Drug Applications Expedited Review Guideline 6/90
No. 29 Guidelines for the Effectiveness Evaluation of Swine Anthelmintics 9/80
No. 30 Guidelines for Anti-infective Bovine Mastitis Product Development 6/85
No. 31 Guidelines for the Evaluation of Bovine Anthelmintics (7/81)
No. 32 Guideline for Threshold Assessment (replaced by Guideline 3)
No. 33 Target Animal Safety Guidelines for New Animal Drugs 6/89
No. 34 Biomass Guideline – Guideline for New Animal Drugs and Food Additives Derived From a Fermentation; Human Food Safety Evaluation (replaced by Guideline Three)
No. 35 Bioequivalence Guidance-Final 1996
No. 36 Guidelines for Efficacy Evaluation of Canine/Feline Anthelmintics 7/85
No. 37 Guidelines for Evaluation of Effectiveness of New Animal Drugs for Use in Poultry Feed for Pigmentation 3/84
No. 38 Guideline for Effectiveness Evaluation of Topical/Otic Animal Drugs 8/84
No. 39 Guideline on the Conduct of Clinical Investigations: Responsibilities of Clinical Investigators and Monitors for Investigational New Animal Drug Studies 10/92 (replaced by Guidance 58)
No. 40 Draft Guideline for the Evaluation of the Efficacy of Anticoccidial Drugs and Anticoccidial Drug Combinations in Poultry 4/92
No. 41 Draft Guideline: Formatting, Assembling, and Submitting New Animal Drug Applications 6/92
No. 42 Series of four guidelines entitled "Animal Drug Manufacturing Guidelines" 1994
No. 43 Draft Guideline for Generic Animal Drug Products Containing Fermentation-derived Drug Substances 10/95
No. 45 Guideline for Uniform Labeling of Drugs for Dairy and Beef Cattle 8/93
No. 48 Guidance for Industry: Submission Documentation for Sterilization Process Validation in Applications for Human and Veterinary Drug Products 11/94
No. 49 Guidance Document For Target Animal Safety And Drug Effectiveness Studies For Anti-microbial Bovine Mastitis Products (Lactating and Non-lactating Cow Products) 4/96
No. 50 Draft Guideline for Target Animal and Human Food Safety, Drug Efficacy, Environmental and Manufacturing Studies for Teat Antiseptic Products 2/93
No. 51 Points to Consider Guideline–Development of a Pharmacokinetic Guideline Enabling Flexible Labeling of Therapeutic Antimicrobials. See Guidance 66 for updated information
No. 52 Guideline for Microbiological Testing of Antimicrobial Drug Residues in Food 1/96

No. 53 Guideline for the Evaluation of the Utility of Food Additives in Diets Fed to Aquatic Animals 5/94
No. 54 Draft Guideline for Utility Studies for Anti-Salmonella Chemical Food Additives in Animal Feeds 6/94
No. 55 Supportive Data for Cat Food Labels Bearing "Reduces Urinary pH" Claims: Guideline in Protocol Development 6/94
No. 56 Protocol Development Guideline for Clinical Effectiveness and Target Animal Safety Trials 11/94
No. 57 Master Files: Guidance for Industry for the Preparation and Submission of Veterinary Master Files 1995
No. 58 Guidance for Industry for Good Target Animal Study Practices: Clinical Investigators and Monitors 5/97
No. 59 Guidance for Industry: Submitting a Notice of Claimed Investigational Exemption in Electronic Format to CVM via E-mail 1/99
No. 61 Guidance For Industry: FDA Approval of New Animal Drugs for Minor Uses and for Minor Species 4/99
No. 62 Guidance for Industry: Consumer-directed Broadcast Advertisements: Draft Guidance 8/97
No. 63 Guidance for Industry: Validation of Analytical Procedures: Definition and Terminology: Draft Guidance 12/97
No. 64 Guidance for Industry: Validation of Analytical Procedures: Methodology: Draft Guidance 12/97
No. 65 Guidance for Industry: Industry-supported Scientific and Educational Activities 11/97
No. 66 Guidance for Industry: Professional Flexible Labeling of Antimicrobial Drugs 8/98
No. 67 Guidance for Industry: Small Entities Compliance Guide for Renderers 2/98
No. 68 Guidance for Industry: Small Entities Compliance Guide for Protein Blenders, Feed Manufacturers, and Distributors 2/98
No. 69 Guidance for Industry: Small Entities Compliance Guide for Feeders of Ruminant Animals with On-farm Feed Mixing Operations 2/98
No. 70 Guidance for Industry: Small Entities Compliance Guide for Feeders of Ruminant Animals without On-farm Feed Mixing Operations 2/98
No. 71 Guidance for Industry: Use of Human Chorionic Gonadotropin (HCG) as a Spawning Aid for Fish 4/98
No. 72 Guidance For Industry: GMPs For Medicated Feed Manufacturers Not Required to Register and be Licensed with FDA 5/98
No. 73 Guidance For Industry: Stability Testing of New Animal Drug Substances and Products: Draft Guidance 7/98
No. 74 Guidance for Industry: Stability Testing For New Dosage Forms of New Animals Drugs: Draft Guidance 7/98
No. 75 Guidance For Industry: Stability Testing: Photostability Testing of New Animal Drug Substances and Products: Draft Guidance 7/98
No. 76 Guidance For Industry: Questions and Answers BSE Feed Regulations 7/98
No. 77 Guidance For Industry: Interpretation Of On-farm Feed Manufacturing And Mixing Operations: Draft Guidance 9/98
No. 78 Evaluation of the Human Health Impact of the Microbial Effects of Anti-

microbial New Animal Drugs Intended for Use in Food-producing Animals 1/99

No. 84 Guidance for Industry: Product Name Placement, Size, and Prominence in Advertising and Promotional Labeling: Draft Guidance 3/99

Chapter 4

# Regulatory aspects and strategy in medical device and bio materials safety evaluation

*Shayne C. Gad*

In the United States (US), according to 201(h) of the Food, Drug and Cosmetic Act, a medical device is defined as an instrument, apparatus, implement, machine, contrivance, implant, *in vitro* reagent, or other similar or related article, including a component, part, or accessory that is:

- recognized in the official National Formulary, or the United States Pharmacopoeia (USP), or any supplement to them, and
- intended for use in the diagnosis of disease or other condition, or in the cure, mitigation, treatment, or prevention of disease, in man or other animals, or intended to affect the structure or any function of the body or man or other animals, and which does not achieve any of its primary intended purposes through chemical action within or on the body of man or other animals, and which is not dependent upon being metabolized for the achievement of any of its principal intended purposes (CDRH, 1992).

Under this definition, devices might be considered as belonging to one of six categories: specialty devices, medical/surgical supplies, imaging systems, other equipment, *in vitro* diagnostics, and health information systems. As a group, these products represented $51 billion in sales in the US in 1995 (Allen, 1996). This was projected to grow to $63 billion in 1998. In 1994, this industry employed 332,000 people in the manufacturing sector (SIC codes 3841–3545).

Specialty devices comprise disposable and long-term single use products which are usually associated with a specific procedure and various types of implantable devices. These are usually technology-intensive and higher priced products which are used only in a specific procedure. Examples include pacemakers, heart valves, knee implants, laparoscopic endoscopes, hip implants, coronary angioplasty balloon catheters, intravenous (IV) pumps and cassettes, vascular grafts, stents, and internal (surgical) staplers. These accounted for $10 billion in sales in 1993 in the US.

Medical/surgical supplies include a broad range of commodity disposables and small reusable devices and instruments. These typically are lower technologically and in price, and enjoy broad use across a range of medical and surgical procedures. Items in this category include sutures, syringes and needles, gloves, IV administration sets, adhesive bandages, IV catheters, surgical packs, caps and gowns, and surgical sponges. These accounted for $9.5 billion in US sales in 1993.

Imaging systems include all types of imaging equipment, contrast agents, and films. Examples include X-ray, angiography, Magnetic Resonance Imaging (MRI), Computerized Axial Tomography (CAT), as well as nuclear and ultrasound systems plus the disposable film and contrast agents used in them. The imaging system category accounted for $9 billion in US sales during 1993.

The other equipment category includes nonimaging equipment systems such as patient monitoring machines, anesthesia machines, critical care respirators, EKG (electrocardiogram) diagnostic systems, and ventilators. In 1993, US sales for this category amounted to $3 billion.

In vitro diagnostics, also sometimes called simply diagnostics, are composed of chemical reagent and instrument systems used to evaluate patients' blood and other body fluids and tissues. Subsegments include clinical chemistry, hematology, microbiology, infectious disease immunology, and other (cancer, etc.) immunodiagnostics. Diagnostics sales during 1995 were $7.9 billion in the US.

The last category, health information systems, encompasses computer hardware and software used for clinical and laboratory purposes. Those computer and software systems used exclusively for patient records and financial data processing are not included in this category or considered medical devices. This category accounted for $500 million in US sales in 1993, but has undoubtedly grown rapidly since then.

## History

As has previously been reviewed by Hutt (1989), the regulation of medical devices has followed a different history than that of drugs. Medical devices go back to at least the Egyptians and Etruscans. Problems with fraudulent devices in the US date back to the late 1700s, though no legislative remedy was attempted until the 1900s. In fact, the legislative history of the 1906 Food and Drug Act contains no references to devices. Devices continued to be regulated under the postal fraud statutes. Such regulation was evidently ineffectual, as fraudulent devices flourished during this period. Starting in 1926, the Food and Drug Administration (FDA) monitored such devices and assisted the US Postal Service in its regulatory actions. Medical devices were covered in the 1938 Act, but only with regard to adulteration and misbranding. Over the intervening years, various committees which examined medical device regulation consistently came to similar conclusions: that the FDA has inadequate authority and resources to regulate the medical device industry. As part of the agreement that resulted in passage of the 1962 amendments, however, all references to medical devices were deleted. The need and demand for increased regulation continued to grow. In 1967, President Lyndon Johnson supported the proposed Medical Device Safety Act, which nevertheless was not well received by Congress. In fact, no legislation pertaining to medical device safety was passed until 1976.

In 1969, at the request of President Richard Nixon, the Department of Health, Education and Welfare (HEW) established a Study Group in Medical Devices, also known as the Cooper Committee, because it was chaired by the Director of the National Heart and Lung Institute, Dr. Theodore Cooper. Its report in 1970 concluded that a different regulatory approach was needed to deal with medical devices. This report initiated the chain of events that culminated in the Medical Device Amendment of 1976. In the interim, the Bureau of Medical Devices and Diagnostic Products was created in 1979. Remarkably, the 1976 Amendment retained the essential provisions of the Cooper Committee Report regarding inventory and classification of all medical devices by class: Class I (general controls), Class II (performance standards), or Class III (premarket approval). These classifications are discussed in greater detail later in this chapter. These remain the essential regulations applicable to medical devices. Both the Drug Price Competition and Patent Restoration Act of 1984 and the Orphan Drug Act of 1983 contained language that made the provisions of the laws applicable to medical devices but did not have provisions unique to medical devices. The recent perceptions, revelations, and controversy surrounding silicone breast implants will probably cause additional changes in the regulation of devices.

As a consequence, 1978 brought guidelines for investigational device exemptions (IDEs, the equivalent of INDAs for drugs). These requirements, as shall be seen later, effectively excluded a wide range of medical devices from regulation by establishing an exemption for those new or modified devices which are equivalent to existing devices. The year 1990 saw the passage of the Safe Medical Devices Act, which made premarketing requirements and postmarketing surveillance more rigorous. The actual current guidelines for testing started with the USP guidance on biocompatibility of plastics. A formal regulatory approach springs from the Tripartite Agreement, which is a joint intergovernmental agreement between the UK, Canada, and the US (with France having joined later). After lengthy consideration, the FDA announced acceptance of International Standards Organization (ISO) 10993 guidelines for testing (ASTM, 1990a,b ; FAO, 1991; MAPI, 1992; O'Grady, 1990; Spizizen, 1992) under the rubric of harmonization. This is the second major trend operative in device regulation: the internationalization of the market place with accompanying efforts to harmonize regulations.

Independent of FDA initiatives, the USP has promulgated test methods and standards for various aspects of establishing the safety of drugs (such as the recent standards for inclusion of volatiles in formulated drug products), which were, in effect, regulations affecting the safety of drugs and devices. Most of the actual current guidelines for the conduct of nonclinical safety evaluations of medical devices have evolved from such quasi-agency actions (such as the USP's 1965 promulgation of biological tests for plastics and ongoing American National Standards Institute (ANSI) standard promulgation).

## Regulatory basis

### *Regulations: general considerations for the US*

#### *Regulations*

The US federal regulations that govern the testing, manufacture, and sale of medical devices are covered in Chapter 1, Title 21 of the Code of Federal Regulations (21 CFR). These comprise nine 6 × 8 inch volumes which stack 8 inches high. This title also covers foods, veterinary products, medical devices, and cosmetics. As these topics will be discussed elsewhere in this book, here we will briefly review those parts of 21 CFR that are applicable to medical devices (Chengelis *et al.*, 1995).

Of most interest to a toxicologist working in this arena would be Chapter 1, Subchapter A (Parts 1–78), which cover general provisions, organization, etc. The

Good Laboratory Practices (GLPS) are codified in 21 CFR 58. The regulations applicable to medical devices are covered in Subchapter H, Parts 800–895 of 21 CFR. As discussed earlier, the term medical device covers a wide variety of products: contact lenses, hearing aids, intrauterine contraceptive devices, syringes, catheters, drip bags, orthopedic prostheses, etc. The current structure of the law was established by the Medical Device Amendment of 1976. Products on the market on the day the amendment was passed were assigned to one of three classes (I, II, or III), based on the recommendation of advisory panels. Medical device classification procedure is described in Part 860. Class I products (the least risk laden) were those for which safety and effectiveness could be reasonably assured by general controls. Such devices are available over the counter to the general public. Class II products were those for which a combination of general controls and performance standards were required to reasonably assure safety and effectiveness. Class II devices are available only with a doctor's prescription. Class III products were those for which general controls and performance standards were inadequate; these were required to go through a premarket approval process. All devices commercially distributed after May 28, 1976 ("preamendment Class III devices") which are not determined to be substantially equivalent to an existing marketed device are automatically categorized as Class III and require the submission of a Premarket Approval (PMA). Please note that these are classifications for regulatory purposes only and are distinct from the classification (HIMA/PHRMA) of product types (e.g., internal versus external) discussed elsewhere in this chapter. Kahan (1995) provides a detailed overview of what comprises general controls, performance standards and such.

As with the subchapter on drugs, much of the subchapter on medical devices in the regulations concerns categorizations and specifics for a wide variety of devices. For a toxicologist involved in new product development, the parts of highest interest are 812 and 814. As with drugs, devices must be shown to be safe and effective when used as intended, and data must be provided to demonstrate such claims. In order to conduct the appropriate clinical research to obtain these data, a sponsor applies to the Agency for an investigational device exemption (IDE), as described in 21 CFR 812. As stated in this section, "an approved investigational device exemption (IDE) permits a device that would otherwise be required to comply with a performance standard or to have premarket approval to be shipped lawfully for the purpose of conducting investigations of that device." Given the broad range of products that fall under the category of medical devices, the toxicological concerns are equally broad; testing requirements to support an IDE are vaguely mentioned in the law, even by FDA standards. In this regard, the law simply requires that the IDE application must include a report of prior investigations which "shall include reports of all prior clinical, animal and laboratory testing." There is no absolute requirement for animal testing, only a requirement that such testing must be reported. There are, of course, standards and conventions to be followed in designing a safety package to support an IDE, and these are discussed in a subsequent section of this chapter.

In order to obtain a license to market a device, a sponsor either submits a 510(k) premarket notification or applies for a PMA, as described in 21 CFR 814. Like an NDA, a PMA application is a very extensive and detailed document that must include, among other things, a summary of nonclinical laboratory studies submitted in the application 921 CFR 814.20(b)(3)(v)(A), as well as a section containing results of the nonclinical laboratory studies with the device, including microbiological, toxicological, immunological, biocompatibility, stress, wear, shelf life, and other laboratory or animal tests as appropriate. As with drugs, these tests must be conducted in compliance with the GLP Regulations. Under the language of the law, a sponsor submits a PMA, which the FDA then "files." The filing of an application means that "FDA has made a threshold determination that the application is sufficiently complete to permit substantive review." Reasons for refusal to file are listed in 814.44(e), and include items such as an application that is not complete and has insufficient justification for the omission(s) present. The agency has 45 days from receipt of an application to notify the sponsor as to whether or not the application has been filed. The FDA has 180 days after filing of a complete PMA (21 CFR 814,40) to send the applicant an approval order, an "approved" letter or a "not approved" letter, or an order denying approval. An "approval order" is self-explanatory and is issued if the agency finds no reason (as listed in 814.45) for denying approval. An "approved" letter 814.44(e) means the application substantially meets requirements, but some specific additional information is needed. A "not approved" letter, 814.45(f), means that the application contains false statements of fact, does not comply with labeling guidelines, or that nonclinical laboratory studies were not conducted according to GLPs, etc. Essentially, an order denying approval means that the sponsor must do substantially more work and must submit a new application for PMA for the device in question. 510(k) premarket approval submissions are less extensive than PMAs, but must still include appropriate preclinical safety data. 510(k)s are supposed to be approved in 90 days.

Actual review and approval times historically have been much longer than the statutory limits. For 1995, the average total review time for Class III products in the US cleared by 510(k) was 579 days (versus 240 or less in the EU) (The Gray Sheet, 1996a). For fiscal year 1996, overall average 510(k) review times (for an expected 5,875 filings) is projected to be 137 days (with low risk exempted devices and refusals to file not being included in the totals or average). Average PMA review times are projected to be 250.days (The Gray Sheet, 1996b). See Chapter 14 for a discussion of general regulatory considerations (such as GLPs) which are applicable to all safety evaluation studies.

### *Regulations versus law*

A note of caution must be inserted here. The law (the document passed by Congress) and the regulations (the documents written by the regulatory authorities to enforce the laws) are separate documents. The sections in the law do not necessarily have numerical correspondence. For example, the regulations on the PMA process is described in 21 CFR 312, but the law describing the requirement for a PMA process is in Section 515 of the FDCA. Because the regulations rather than the laws themselves have a greater impact on toxicological practice, greater emphasis is placed on regulation in this chapter. For a complete review of FDA law, the reader is referred to the monographs by the Food and Drug Law Institute in 1995.

Laws authorize the activities and responsibilities of the various federal agencies. All proposed laws before the US Congress are referred to committees for review and approval. The committees responsible for FDA oversight are summarized on Table 4.1. This table also highlights the fact that authorizations and appropriations (the

*Table 4.1* Congressional committees responsible for FDA oversight

| | |
|---|---|
| *Authorization* | |
| Senate | All public health service agencies are under the jurisdiction of the Labor and Human Resources Committee |
| House | Most public health agencies are under the jurisdiction of the Health and the Environmental Subcommittee of the House Energy and Commerce Committee |
| *Appropriation* | |
| Senate | Unlike most other public health agencies, the FDA is under the jurisdiction of Agriculture, Rural Development, and Related Agencies Subcommittee of the Senate Appropriations Committee |
| House | Under the jurisdiction of the Agriculture, Rural Development, and Related Agencies Subcommittee of the House Appropriations Committee |

funding necessary to execute authorizations) are handled by different committees. Figure 4.1 presents the organization of the Center for Devices and Radiological Health (CDRH). As can has seen by the organizational structure presented in the figure, the categorization of devices for division review purposes is functionally based.

## Organizations regulating device safety in the US

The agency formally charged with overseeing the safety of drugs and devices in the US is the FDA. It is headed by a commissioner who reports to the Secretary of the Department of Health and Human Services (DHHS) and has a tremendous range of responsibilities. Medical devices are overseen by the CDRH, headed by a director. Drugs are overseen primarily by the Center for Drug Evaluation and Research (CDER) (though some therapeutic or Healthcare entities are considered as biologically derived and therefore regulated by the Center for Biologic Evaluation and Research, or CBER). There are also "combination products" (part drug, part device) which may be regulated by either or both CDER/CBER and CDRH, depending on where the expertise is perceived to be in the FDA (CFR, 1992).

Most of the regulatory interaction of a toxicologist involved in assessing the biocompatibility of devices is with the appropriate part of the CDRH, though for combination products the two centers charged with drugs or biologicals may also come into play. Within the CDRH there is a range of groups (called divisions) which focus on specific areas of use for devices (such as general and restorative devices; cardiovascular, respiratory, and neurological devices; ophthalmic devices; reproductive, abdominal, ear, nose, and throat, and radiological devices; and clinical laboratory devices). Within each of these there are engineers, chemists, pharmacologists/toxicologists, statisticians, and clinicians.

There is also at least one nongovernmental body which must review and approve various aspects of devices, setting forth significant "guidance" for the evaluation of safety of devices. This is the USP, and its responsibilities and guidelines are presented later in this chapter.

The other two major regulatory organizations to be considered are the ISO, with ISO 10993 standards (ISO, various dates), and the Japanese Ministry of Health and Welfare (MHW) with its guidelines (MHW, 1995).

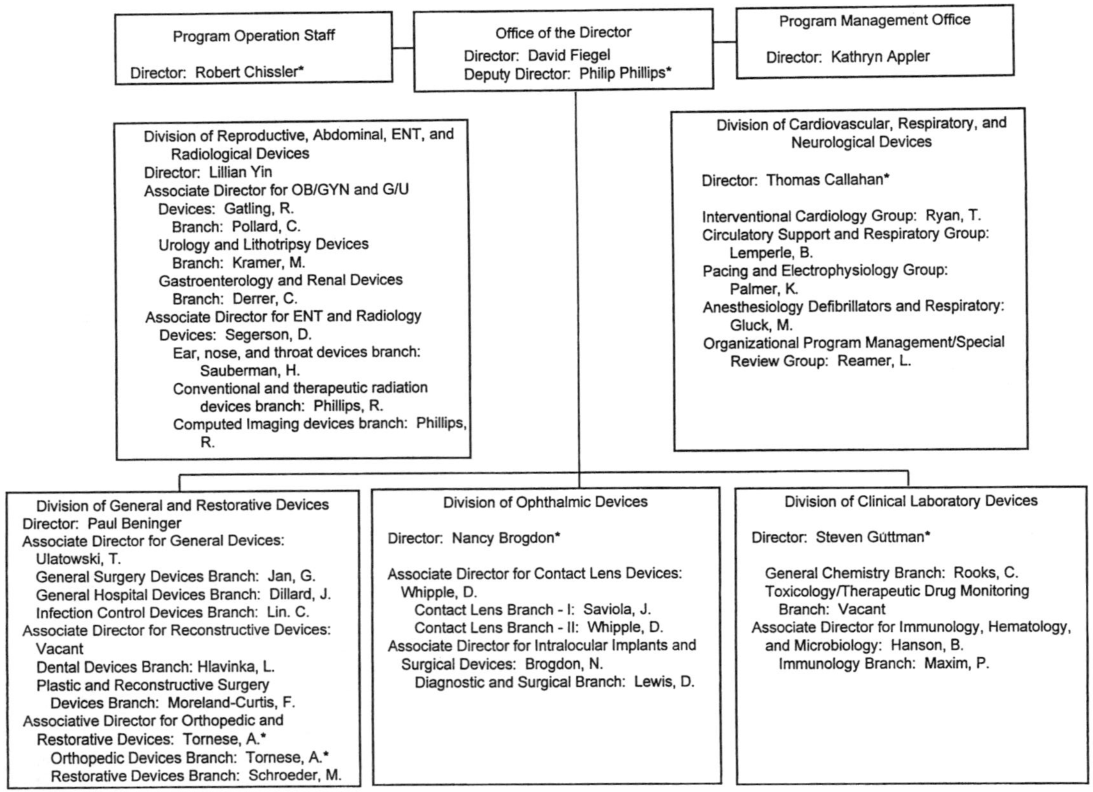


*Figure 4.1* Organization of the Office of Device Evaluation (ODE) for the Center for Devices and Radiological Health (CDRH) of the FDA. Current officials (as of 5/1/1996) are identified by name. ODE evaluates submissions for new device approvals.

## Toxicity testing: medical devices

In a statutory sense, any item promoted for a medical purpose which does not rely on chemical action to achieve its intended effect is a medical device (as discussed earlier). In vitro diagnostic tests are also regulated as medical devices. The regulation of devices under these definitions has had a different history than that of drugs – it has not been as strict and it has evolved at a slower rate. Also, requirements for safety evaluation of devices have not been as strict as those for drugs. The safety concerns are also somewhat different. Toxicologic safety concerns for devices (as opposed to concerns of mechanical safety, such as disintegration of heart valves) are called biocompatibility concerns.

Medical devices are classified as being in three different classes and are regulated accordingly. Class III devices are subject to the greatest degree of regulation and include devices which are implanted in the body, support life, prevent health impairment, or present an unreasonable risk of illness or injury. These are subject to premarketing approval. Class II and Class I devices are subject to lesser control, required only to comply with general controls and performance standards.

There are several governing schemes for dictating what testing must be done on new Class III devices in the general case, with each developed and proposed by a different regulatory organization at different times over the last few years. ICH has attempted to harmonize these requirements so that different (or duplicate) testing would not need to be performed to gain device approval in different national markets. As discussed in the last chapter of this book, there are also specialized testing requirements for some device types such as contact lenses (CDRH, 1995a,b) and tampons (CDRH, 1995c).

As with drugs, all safety testing for devices must be conducted in conformity with GLPs. Table 4.2 presents the existing FDA CDRH requirements for device characterization and testing. The exact nature of the test protocols is based on recommendations by USP, ISO, and others. It should be noted that Class II devices, if new, are also subject to the Tripartite guidelines.

Additional concerns with devices are considerations of their processing after production. Recently, concerns have arisen about the potential for allergies to develop to latex components. Concern has grown not only for natural rubber used in devices with systemic exposure (catheters, stoppers on syringes, and such), but also for such devices which have been washed in fluids used to wash other items which contain latex. It is likely that all devices containing latex will soon have to be labeled as such.

Devices which have systemic exposure need to be sterilized. Radiation and heat can be used for some devices, but others cannot be sterilized in this way. Ethylene oxide or other chemical sterilants must be used, raising concerns that residual sterilants may present problems. At the same time, devices with exposure to the fluid path must be demonstrated to be neither pyrogenic nor hemolytic in their final manufactured form.

1 The selection of material(s) to be used in device manufacture and its toxicological evaluation should initially take into account full characterization of the material, for example, formulation, known and suspected impurities, and processing.
2 The material(s) of manufacture, the final product, and possible leachable chemicals or degradation products should be considered for their relevance to the overall toxicological evaluation of the device.
3 Tests to be utilized in the toxicological evaluation should take into account the bioavailability of the bioactive material, i.e. nature, degree, frequency, duration,

*Table 4.2* FDA device categories and suggested biological testing[a]

| Device categories | | Biological tests | | | | | | | | | | |
|---|---|---|---|---|---|---|---|---|---|---|---|---|
| | | Short-term | | | | | | | | Long-Term | | |
| Body contact duration<br>A—Transient (< 5 minutes)<br>B—Short-term (5 minutes-29 days)<br>C—Long-term (> 30 days) | | Irritation tests | Sensitization assay | Cytoxicity | Acute systemic toxicity | Hemocompatibility/hemolysis | Pyrogenicity (material-mediated) | Implantation tests | Mutagenicity (carcinogenicity) | Subchronic toxicity | Chronic toxicity | Carcinogenesis bioassay |
| Intact surface | A | * | * | * | | | | | | | | |
| | B | * | * | * | | | | | | | | |
| | C | * | * | * | * | | | | | | | |
| External devices | | | | | | | | | | | | |
| Breached or surface comprised | A | * | * | * | * | * | | | | | | |
| | B | * | * | * | * | * | | | | * | | |
| | C | * | * | * | * | * | | * | * | * | * | |
| Intact natural channels | A | * | * | * | * | * | * | * | | | | |
| | B | * | * | * | * | * | * | * | * | * | | |
| | C | * | * | * | * | * | * | * | * | * | * | * |
| Externally communicating devices | | | | | | | | | | | | |
| Blood path indirect | A | * | * | * | * | * | * | * | * | | | |
| | B | * | * | * | * | * | * | * | * | * | | |
| | C | * | * | * | * | * | * | * | * | * | * | * |
| Blood path direct | A | * | * | * | * | * | * | * | * | | | |
| | B | * | * | * | * | * | * | * | * | * | | |
| | C | * | * | * | * | * | * | * | * | * | * | * |
| Bone | A | * | * | * | * | * | * | * | * | | | |
| | B | * | * | * | * | * | * | * | * | * | | |
| | C | * | * | * | * | * | * | * | * | * | * | * |
| Internal devices | | | | | | | | | | | | |
| Tissue and tissue fluids | A | * | * | * | * | * | * | * | * | | | |
| | B | * | * | * | * | * | * | * | * | * | | |
| | C | * | * | * | * | * | * | * | * | * | * | * |
| Blood | A | * | * | * | * | * | * | * | * | | | |
| | B | * | * | * | * | * | * | * | * | * | | |
| | C | * | * | * | * | * | * | * | * | * | * | * |

[a] For these devices with possible leachables or degradation products, e.g., absorbable surfaces, hemostatic agents, etc., testing for pharmacokinetics may be required. Reproductive and developmental toxicity tests may be required for certain materials used for specialized indications. Considerations should be given to long-term biological tests where indicated in the table taking into account the nature and mobility of the ingredients in the materials used to fabricate the device. Differences between sensitization test procedures required by ISO 10993-10 and the MHW guidelines.

and conditions of exposure of the device to the body. This principle may lead to the categorization of devices which would facilitate the selection of appropriate tests.

4 Any *in vitro* or *in vivo* experiments or tests must be conducted according to recognized GLPS followed by evaluation by competent informed persons.
5 Full experimental data, complete to the extent that an independent conclusion could be made, should be available to the reviewing authority, if required.

6 Any change in chemical composition, manufacturing process, physical configuration or intended use of the device must be evaluated with respect to possible changes in toxicological effects and the need for additional toxicity testing.
7 The toxicological evaluation performed in accordance with this guidance should be considered in conjunction with other information from other nonclinical tests, clinical studies, and postmarket experiences for an overall safety assessment.

### Device categories: definitions and examples

Noncontact devices. Devices that do not contact the patient's body directly or indirectly; examples include *in vitro* diagnostic devices.

External devices

- *Intact surfaces*. Devices that contact intact external body surfaces only; examples include electrodes, external prostheses, and monitors of various types.
- *Breached or compromised surfaces*. Devices that contact breached or otherwise compromised external body surfaces; examples include ulcer, burn and granulation tissue dressings or healing devices, and occlusive patches.

Externally communicating devices

- *Intact natural channels*. Devices communicating with intact natural channels; examples include contact lenses, urinary catheters, intravaginal and intraintestinal devices (sigmoidoscopes, colonoscopes, stomach tubes, gastroscopes), endotracheal tubes, and bronchoscopes.
- *Bloodpath, indirect*. Devices that contact the blood path at one point and serve as a conduit for fluid entry into the vascular system; examples include solution administration sets, extension sets, transfer sets, and blood administration sets.
- *Blood path, direct*. Devices that contact recirculating blood; examples include intravenous catheters, temporary pacemaker electrodes, oxygenators, extracorporeal oxygenator tubing and accessories, and dialyzers, dialysis tubing and accessories.

Internal devices

- *Bone*. Devices principally contacting bone; examples include orthopedic pins, plates, replacement joints, bone prostheses and cements.
- *Tissue and tissue fluid*. Devices principally contacting tissue and tissue fluid or mucus membranes where contact is prolonged; examples include pacemakers, drug supply devices, neuromuscular sensors and stimulators, replacement tendons, breast implants, cerebrospinal fluid drains, artificial larynx, vas deferens valves, ligation clips, tubal occlusion devices for female sterilization, and intrauterine devices.
- *Blood*. Devices principally contacting blood; examples include permanent pacemaker electrodes, artificial arteriovenous fistulae, heart valves, vascular grafts, blood monitors, internal drug delivery catheters, and ventricular assist pumps.

Table 4.3 Differences between sensitization test procedures required by ISO 10993-10 and the MHW guidelines

| *ISO 10993-10* | *MHW (1995)* |
|---|---|
| Sample preparation: extraction in polar and/or nonpolar solvents | Two extraction solvents, methanol and acetone, recommended |
| Extraction ratio: extraction ratio is dependent on thickness of device or representative portion | Specific extraction ratios: 10:1 (volume solvent: weight sample) |
| Extract used for testing. If extraction is not possible, the adjuvant and patch test can be utilized | Residue obtained from extraction is redissolved and used for testing. (If residue does not dissolve in DMSO, or a sufficient amount of residue is not obtained, the adjuvant and patch test is recommended). Sufficient amount of residue: 0.1–0.5% (weight residue:weight test material) |

### Biological tests

Also required to properly utilize the tables is a knowledge of the objectives of the specified biological tests. These can be considered as follows:

- *Sensitization assay*. Estimates the potential for sensitization of a test material and/or the extracts of a material using it in an animal and/or human. ISO and MHW procedures are contrasted in Table 4.3.
- *Irritation tests*. Estimates the irritation potential of test materials and their extracts, using appropriate site or implant tissue such as skin and mucous membrane in an animal model and/or human. ISO and MHW procedures are contrasted in Table 4.4; and for eye irritation in Table 4.5.
- *Cytotoxicity*. With the use of cell culture techniques, this test determines the lysis of cells (cell death), the inhibition of cell growth, and other toxic effects on cells caused by test materials and/or extracts from the materials. ISO and MHW procedures are contrasted in Table 4.6.
- *Acute systemic toxicity*. Estimates the harmful effects of either single or multiple exposures to test materials and/or extracts, in an animal model, during a period of less than 24 h. ISO and MHW procedures are contrasted in Table 4.7.
- *Hematocompatibility*. Evaluates any effects of blood contacting materials on hemolysis, thrombosis, plasma-proteins, enzymes, and the formed elements

Table 4.4 Differences in intracutaneous reactivity test procedures required by ISO 10993-10 and the MHW guidelines

| *ISO 10993-10* | *MHW* |
|---|---|
| Number of test animals: three rabbits for 1–2 extracts | Two rabbits for each extract |
| Number of test/control injections per extract: five test and five control injections | Ten test and five control injections |
| Evaluation of responses: quantitative comparison of responses of test and control responses | Qualitative comparison of test and control responses |

Table 4.5 Differences in eye irritation testing procedures outlined in ISO 10993-10 and the MHW guidelines

| *ISO 10993-10* | *MHW (1995)* |
|---|---|
| Time of exposure: 1 s | 30 s |
| Grading scale: classification system for grading ocular lesions | Draize or McDonald–Shadduck scale |

using an animal model. Traditionally, hemolysis, which determines the degree of red blood cell lysis and the separation of hemoglobin caused by test materials and/or extracts from the materials *in vitro*, has been "the" representative test employed. A broader range of primary tests (adding evaluations of thrombosis, coagulation, platelets, and immunology aspects) is currently recommended. ISO and MHW procedures for hemolysis are contrasted in Table 4.8.

- *Pyrogenicity, material mediated.* Evaluates the material mediated pyrogenicity of test materials and/or extracts. ISO and MHW procedures are contrasted in Table 4.9.
- *Implantation tests.* Evaluates the local toxic effects on living tissue, at both the gross level and microscopic level, to a sample material that is surgically implanted into appropriate animal implant site or tissue, e.g. muscle, bone; for 7–90 days. ISO and MHW procedures are contrasted in Table 4.10.
- *Mutagenicity (genotoxicity).* The application of mammalian or nonmammalian cell culture techniques for the determination of gene mutations, changes in chromosome structure and number, and other DNA or gene toxicities caused by test materials and/or extracts from materials. Selected tests representing gene mutation tests (Ames or mouse lymphoma), chromosomal aberration tests (CHO) and DNA effects tests (mouse micronucleous and sister chromatid exchange) should generally be employed. ISO and MHW procedures are contrasted in Table 4.11.
- *Subchronic toxicity.* The determination of harmful effects from multiple exposures to test materials and/or extracts during a period of 1 day to less than 10 per cent of the total life of the test animal (e.g., up to 90 days in rats).

Table 4.6 Differences between cytotoxicity test procedures specified by ISO 10993-5 and the MHW guidelines (MHW, 1995)

| *ISO 10993-10* | *MHW (1995)* |
|---|---|
| Number of cells per dish: 0.5-1 million cells | 40–200 cells/dish |
| Extraction ratio: 60 $cm^2$/20 ml if thickness 80.5 mm; 120 $cm^2$/20 ml if thickness 70.5 mm or 4 g/20 ml | 5 $cm^2$/ml or 1 g/10 ml |
| Exposure period: typically 24–72 h (2 h for filter diffusion test) | 6–7 days |
| Toxicity determination: visual grading and/or quantitative assessments | Quantification of surviving colonies |
| Positive controls: materials providing a reproducible cytotoxic response (e.g., organotin-impregnated polyvinyl chloride) | Segmented polyurethane films containing 0.1% zinc diethyldithiocarbamate and 0.25% zinc dibutyldithiocarbamate |

*Table 4.7* Comparison of grading scales used to score responses of test animals to ASTM and ISO/USP procedures

| *Response* | *ASTM* *Description* | *ISO/USP* |
|---|---|---|
| Normal, no symptoms | Mouse exhibits no adverse physical symptoms after injection | |
| Slight | Mouse exhibits slight but noticeable symptoms of hypokinesis, dyspnea, or abdominal irritation after injection | |
| Moderate | Mouse exhibits definite evidence of abdominal irritation, dyspnea, hypokinesis, ptosis, or diarrhea after injection. (Weight usually drops to between 15 and 17 g) | |
| Marked | Mouse exhibits prostration, cyanosis, tremors, or severe symptoms of abdominal irritation, diarrhea, ptosis, or dyspnea after injection. (Extreme weight loss; weight usually less than 15 g) | |
| Dead, expired | Mouse dies after injection | |
| | Interpretation | Interpretation |
| | The test is considered negative if none of the animals injected with the test article extracts show a significantly greater biological reaction than the animals treated with the control article | The test is considered negative if none of the animals injected with the test article show a significantly greater biological reaction than the animals treated with the control article |
| | If two or more mice show either marked signs of toxicity or die, the test article does not meet the requirements of the test | If two or more mice die, or show signs of toxicity such as convulsions or prostration, or if three or more mice lose more than 2 g of body weight, the test article does not meet the requirements of the test |
| | If any animals treated with a test article shows slight signs of toxicity, and not more than one animal shows marked signs of toxicity or dies, a repeat test using freshly prepared extract should be conducted using groups of ten mice each. A substantial decrease in body weight for all animals in the group, even without other symptoms of toxicity, requires a retest using groups of ten mice each. In the repeat test, the requirements are met if none of the animals injected with the test article shows a substantially greater reaction than that observed in the animals treated with the control article | If any animal treated with a test article shows only slight signs of biological reaction, and not more than one animal shows gross signs of biological reaction or dies, a repeat test should be conducted using groups of ten mice. On the repeat test, all ten animals must not show a significantly greater biological reaction than the animals treated with the control article |

*Table 4.8* Differences in hemolysis test procedures recommended by ISO 10993-4 and the MHW guidelines

| *ISO 10993-4* | *MHW (1995)* |
|---|---|
| Hemolysis can be assessed by any of several validated methods to assay hemoglobin in plasma | Hemolytic index is assessed by measuring hemoglobin at 1, 2, and 4 h by spectrophotometric methods. The hemolysis over this period is expressed as a percentage of the positive control |

*Table 4.9* Comparison of pyrogen test procedures required by ISO 10993-11 and the MHW guidelines

| *ISO 10993-11* | *MHW (1995)* |
|---|---|
| Number of animals: three rabbits required; comparison of febrile response in test animals to baseline temperature for evaluation of pyrogenicity potential | Three rabbits (test) required; comparison to baseline temperature is evaluated as index of pyrogenicity potential |
| Test duration: test measurement intervals: every 30 min for 3 h | Test measurement intervals: every hour for 3 h |
| Evaluation: cut-off for positive febrile response: 0.5°C | Cut-off for positive febrile response: 0.6°C |

*Table 4.10* Differences in ISO 10993-3 and the MHW guidelines for assessing the effects of device or material implantation

| *ISO 10993-3* | *MHW 1995* |
|---|---|
| Time point(s) of assessment: sufficient to achieve steady state(e.g. 2, 4, 6, and 12 weeks) | 7 days and 4 weeks |
| Number of animals: at least three per time period of assessment | At least four per time period |
| Number of samples of evaluation: at least eight per time period for test and control | No minimum number specified |
| Evaluation criteria: comparative evaluation of responses to test and control materials | If more than two of the four test sites in each animal exhibit a significant response compared to control sites, the test is considered positive |

- *Chronic toxicity*. The determination of harmful effects from multiple exposures to test materials and/or extracts during a period of 10 per cent to the total life of the test animal (e.g., over 90 days in rats).
- *Carcinogenesis bioassay*. The determination of the tumorigenic potential of test materials and/or extracts from either single or multiple exposures, over a period of the total life (e.g., 2 years for rat, 18 months for mouse, or 7 years for dog).
- *Pharmacokinetics*. To determine the metabolic processes of absorption, distribution, biotransformation, and elimination of toxic leachables and degradation products of test materials and/or extracts.
- *Reproductive and developmental toxicity*. The evaluation of the potential effects of test materials and/or extracts on fertility, reproductive function, and prenatal and early postnatal development.

*Table 4.11* Differences in genotoxicity testing procedures required by ISO 10993-3 and the MHW guidelines

| *ISO 10993-10* | *MHW 1995* |
|---|---|
| Extraction vehicles: a physiological medium is used and, where appropriate, a solvent (e.g. dimethylsulfoxide) | Recommends methanol and acetone as extracting vehicles |
| Extraction: extract test material and test the extract or dissolve material in solvent and conduct test. The conditions of extraction should maximize the amount of extractable substances, as well as subject the test device or material to the extreme conditions it may be exposed to, without causing significant degradation. Extraction ratio is dependent on thickness of test material | Extract at room temperature at a ratio of 10:1 (solvent: material) and obtain residue (at least 0.1–0.5% (weight of residue/weight of test material)), redissovle in appropriate solvent and test residue |
| | If sufficient residue is unobtainable, extract test material (in ethanol, acetone, or DMSO at 10 g of test material per 20 ml for the Arnes mutagenicity assay, and in cell culture medium at 120 $cm^3$ or 4 g/20 ml for the chromosomal aberration assay), at 37°C for 48 h and test extract. The Ames mutagenicity assay is conducted with a volume of 200 μl per plate |

The tests for leachables such as contaminants, additives, monomers, and degradation products must be conducted by choosing appropriate solvent systems that will yield a maximal extraction of leachable materials to conduct biocompatibility testing. Chapter 3 addresses the issues behind sampling, sample preparation, and solvents.

The effects of sterilization on device materials and potential leachables, as well as toxic by-products, as a consequence of sterilization should be considered. Therefore, testing should be performed on the final sterilized product or representative samples of the final sterilized product. Table 4.12 presents the basis for test selection under the Tripartite Agreement.

## *US pharmacopoeial testing*

The earliest guidance on what testing was to be done on medical devices was that provided in the USP and other pharmacopoeias. Each of the major national pharmacopoeias offers somewhat different guidance. The test selection system for the USP (presented in Table 4.12), which classified plastics as Classes I through VI, is now obsolete and was replaced in usage by the other guidelines presented here. But the actual descriptions of test types, as provided in the USP (and presented in the appropriate chapters later in this book) are still very much operative (USP, 1990).

There are British, European and Japanese pharmacopoeias, of which the latter requires the most attention due to some special requirements still being operative if product approval is desirable.

*Table 4.12* Classification of plastics (USP XXII)[a]

| *Plastic classes*[b] | | | | | | *Tests to be conducted* | | | |
|---|---|---|---|---|---|---|---|---|---|
| *I* | *II* | *III* | *IV* | *V* | *VI* | *Test material* | *Animal* | *Dose* | *Procedures*[c] |
| X | x | x | X | x | X | Extract of sample in sodium chloride inspection | Mouse | 50 ml/kg | A (iv) |
| X | x | x | X | x | x | | Rabbit | 0.2 ml/animal at each of ten sites | B |
| X | x | x | X | x | | Extract of sample in 1 in 20 | Mouse | 50 ml/kg | A (iv) |
| X | x | x | X | x | | Solution of alcohol in sodium chloride injection | Rabbit | 0.2 ml/animal at each of ten sites | |
| | x | | X | x | | Extract of sample in polyethylene glycol 400 | Mouse | 10 g/kg | A (ip) |
| | | x | X | | | | Rabbit | 0.2 ml/animal at each of ten sites | |
| | | x | X | x | x | Extract of sample in vegetable oil | Mouse | 50 ml/kg | A (ip) |
| | | | X | x | x | | Rabbit | 0.2 ml/animal at each of ten sites | B |
| | | x | | | x | Implant strips of sample | Rabbit | Four strips/animal | C |

[a] Tests required for each class are indicated by "x" in appropriate rows.

[b] A (ip), systemic injection test (intraperitioneal); A(iv), systemic injection test (intravenous); B, intracutaneous; C, implantation test (intramuscular implantation)

[c] The table lists the biological tests that might be applied in evaluating the safety of medical devices and/or polymers. This does not imply that all the tests listed under each category will be necessary or relevant in all cases. Tests for devices made of metals, ceramics, biological materials, etc., are not included here but are under consideration. Categorization of medical devices is based on body contact and contact duration.

### *ISO testing requirements*

The European Economic Community (EEC) has adopted a new set of testing guidelines for medical devices under the aegis of ISO (ISO, 1990; The Gray Sheet, 1992). The ISO 10993 guidelines for testing provide a unified basis for international medical device biocompatibility evaluation, both in terms of test selection (as presented in Tables 4.13 and 4.14) and test design and interpretation (Table 4.15). In 1996, the US FDA also announced that it would adhere to ISO 10993 standards for device biocompatibility evaluation.

This international standard specifies methods of biological testing of medical and dental materials and devices and their evaluation with regard to their biocompatibility. Because of the many materials and devices used in these areas, the standard offers a guide for biological testing.

### *MHW requirements*

The Japanese ISO test selection guidelines vary from those of FDA and ISO and are summarized in Table 4.16.

Actual test performance standards also very, as shown in Tables 4.3–4.10.

Committees dealing with materials and devices must decide on tests and test series relevant to the respective materials and devices, It is the responsibility of the product

*Table 4.13* ISO initial evaluation tests

| Device categories<br>Body contact duration<br>A—Limited exposure<br>B—prolonged or repeated exposure<br>C—permanent contact | | Biological tests: Cytotoxicity | Sensitization | Irritation or Intracutaneous | Acute systemic toxicity | Subchronic toxicity | Mutagenicity | Pyrogenicity | Implantation | Hemocompatability |
|---|---|---|---|---|---|---|---|---|---|---|
| **Surface devices** | | | | | | | | | | |
| Skin | A | x | x | x | | | | | | |
| | B | x | x | x | | | | | | |
| | C | x | x | x | | | | | | |
| Mucous membranes | A | x | x | x | | | | | | |
| | B | x | x | x | | | | | | |
| | C | x | x | x | | x | x | | | |
| Breached surface | A | x | x | x | | | | | | |
| | B | x | x | x | | | | | | |
| | C | x | x | x | | x | x | | | |
| **Externally communicating** | | | | | | | | | | |
| Blood path indirect | A | x | x | x | | | | x | | x |
| | B | x | x | x | | | | x | | x |
| | C | x | x | | x | x | x | x | | x |
| Tissue/bone communicating | A | x | x | x | | | | | | |
| | B | x | x | | | | x | | x | |
| | C | x | x | | | | x | | x | |
| **Internal devices** | | | | | | | | | | |
| Circulating blood | A | x | x | x | x | | | x | | x |
| | B | x | x | x | x | | x | x | | x |
| | C | x | x | x | x | x | x | x | | x |
| **Implant devices** | | | | | | | | | | |
| Bone/tissue | A | x | x | x | | | | | | |
| | B | x | x | | | | x | | x | |
| | C | x | x | | | | x | | x | |
| Blood | A | x | x | x | x | | | x | x | x |
| | B | x | x | x | x | | x | x | x | x |
| | C | x | x | x | x | x | x | x | x | x |

committees to select adequate test methods for products. The standard contains animal tests, but tries to reduce those tests to the justifiable minimum. Relevant international and national regulations must be observed when animals are used.

ISO 10993 is based on existing national and international specifications, regulations, and standards wherever possible. It is open to regular review whenever new research work is presented to improve the state of scientific knowledge. Tables 4.4 and 4.5 provide the test matrices under ISO 10993. Subsequently, specific guidance on individual test designs, conduct and interpretation has been provided as Subparts 2–11 of ISO-10993 (Table 4.15).

*Table 4.14* ISO special evaluation tests

| Device categories<br>Body contact duration<br>A—Limited exposure<br>B—prolonged or repeated exposure<br>C—permanent contact (time limits to added) | | Biological tests: Chronic toxicity | Carcinogenicity | Reproductive/developmental | Degradation |
|---|---|---|---|---|---|
| **Surface devices** | | | | | |
| Skin | A | | | | |
| | B | | | | |
| | C | | | | |
| Mucous membranes | A | | | | |
| | B | | | | |
| | C | | | | |
| Breached surface | A | | | | |
| | B | | | | |
| | C | | | | |
| **Externally communicating** | | | | | |
| Blood path indirect | A | | | | |
| | B | | | | |
| | C | x | x | | |
| Tissue/bone communicating | A | | | | |
| | B | | | | |
| | C | | x | | |
| **Internal devices** | | | | | |
| Circulating blood | A | | | | |
| | B | | | | |
| | C | x | x | | |
| Bone/tissue | A | | | | |
| | B | | | | |
| | C | x | x | | |
| Blood | A | | | | |
| | B | | | | |
| | C | x | x | | |

## *Medical Devices Directive guideline*

The EEC had adopted its own guidance, Council Directive 93/42/EEC (1993), for manufacturers of medical devices (The Gray Sheet, 1992; European Committees on Standardization, 1991). According to the preamble of the Medical Devices Directive (MDD), the classification rules are "based on the vulnerability of the human body taking account of the potential risks associated with the technical design and manufacture of the devices." The classification rules are presented in Annex IX of the MDD. Implementation of the MDD requires that the classification rules are applied in accordance with the intended purpose, or most-critical specified use, of a device. Should more than one rule apply, the strictest takes precedence.

Except for the special rules 13–18, which will probably not be combined with the active device rules, the classification rules contained in Annex IX distinguish between

*Table 4.15* ISO/ANSI/AAMI standards

| | *ISO designations* | *Year issued* |
|---|---|---|
| Guidance on selection of tests | 10993-1 | 1992 |
| Animal welfare requirements | 10993-2 | 1992 |
| Tests for genotoxicity, carcinogenicity and reproductive toxicity | 10993-3 | 1992 |
| Selection of tests for interactions with blood | 10993-4 | 1992 |
| Tests for cytotoxicity: *in vitro* methods | 10993-5 | 1992 |
| Tests for local effects after implantation | 10993-6 | 1994 |
| Ethylene oxide sterilization residuals | 10993-7 | 1995 |
| Guidance for reference materials | 10993-8 | 1998 |
| Degradation of materials related to biological testing | 10993-9 | 1994 |
| Tests for irritation and sensitization | 10993-10 | 1995 |
| Tests for systemic toxicity | 10993-11 | 1994 |
| Sample preparation and reference materials | 10993-12 | 1998 |
| Identification and quantification of degradation products from polymers | 10993-13 | 2000 |
| Identification and quantification of degradation products from ceramics | 10993-14 | 2000 |
| Identification and quantification of degradation products from metals and alloys | 10993-15 | 2000 |
| Toxicokinetic study design for degradation products and leachables | 10993-16 | 2000 |

two categories of device: noninvasive and invasive, Both these categories are further divided into four rules. Among noninvasive devices, distinction is made among:

- Devices for channeling or storing substances for infusion, administration, or introduction into the body
- Devices for the biological or chemical modification of liquids for infusion
- Devices that come into contact with injured skin
- All other noninvasive devices

Invasive devices are divided into those that are introduced into natural body orifices and those that are surgically introduced into the body. Classification criteria of the surgically introduced devices include:

- Duration (transient, short-term, or long-term use)
- Interaction (biological, chemical, or ionizing radiation)
- Location (heart, central circulatory, or central nervous systems)

Four additional rules are stated for active medical devices, and within these a distinction is made among:

- Devices used for therapy
- Devices for diagnosis
- Devices for the administration and/or removal of substances to and from the body
- All other active devices

Classification criteria for medical devices are based on potential hazards, taking into account:

- Nature, density, and site of energy application

*Table 4.16* Japanese MHW test selection guidelines

| Device Categories | | Initial Evaluation | | | | | | Supplemental Evaluation | | | | | |
|---|---|---|---|---|---|---|---|---|---|---|---|---|---|
| | Body contact | Contact duration | Cytotoxicity | Sensitization | Irritation or intracutaneous | Systemic toxicity (acute) | Subchronic toxicity | Genotoxicity | Pyrogen | Implantation | Hemocompatibility | Chronic Toxicity | Carcinogenicity |
| Surface Devices | Skin | A | X | X | X | | | | | | | | |
| | | B | X | X | X | | | | | | | | |
| | | C | X | X | X | | | | | | | | |
| | Mucosal membrane | A | X | X | X | | | | | | | | |
| | | B | X | X | X | | | | | | | | |
| | | C | X | X | X | | X | X | | | | | |
| | Breached/ compromised surface | A | X | X | X | | | | | | | | |
| | | B | X | X | X | | | | | | | | |
| | | C | X | X | X | | X | X | | | | | |
| External Communicating Devices | Blood path indirect | A | X | X | X | X | | | X | | X | | |
| | | B | X | X | X | X | | | X | | X | | |
| | | C | X | X | | X | X | X | X | | X | X | X |
| | Tissue/bone dentin communicating | A | X | X | X | | | | | | | | |
| | | B | X | X | | | | X | | X | | | |
| | | C | X | X | | | | X | | X | | | X |
| | Circulating blood | A | X | X | X | X | | | X | | X | | |
| | | B | X | X | X | X | | X | X | | X | | |
| | | C | X | X | X | X | X | X | X | | X | X | X |
| Implant Devices | Bone/tissue | A | X | X | X | | | | | | | | |
| | | B | X | X | | | | X | | X | | | |
| | | C | X | X | | | | X | | X | | X | X |
| | Blood | A | X | X | X | X | | | X | X | X | | |
| | | B | X | X | X | X | | X | X | X | X | | |
| | | C | X | X | X | X | X | X | X | X | X | X | X |

A = Temporary contact (24 hours) B = Short and medium term contact (24 hours -29 days) C = Long term contact (>30 days)

- Substance involved
- Part of the body concerned
- Mode of application or immediate danger to the patient with respect to cardiac performance, respiration, and/or central nervous system

Reduction of nonactive characteristics: by combining active and nonactive characteristics, the following assumptions regarding active medical devices were made:

- Only those nonactive characteristics that would lead to a higher class were considered. Therefore, all Class I characteristics of rules 1–8 were omitted.
- Implantable and long-term surgically invasive devices (rule 8) that are also active should be covered by the Active Implantable Medical Device (AIMD) Directive 90/385/EEC (although definitions of "implantable" in the AIMD and MDD are slightly different).
- Connection to another active medical device, as described in rules 2 and 5, will not change the class of an active medical device.
- Energy supplied in the form of ionizing radiation (rules 6 and 7) is sufficiently covered under rule 10.

- Because surgically invasive active devices are not intended to be wholly or mainly absorbed by, or chemically changed in, the body (rules 6 and 7), these characteristics were omitted.
- Devices "intended to administer medicines" (rules 6 and 7) are sufficiently covered under rule 11.

## CE marking of devices

After June 13, 1998, all medical products distributed in Europe have had to bear the CE mark. ISO 9000 certification supplements and supports an assessment of conformity to the MDD, which must be performed by a certification body appointed by the EU member states (Haindl, 1997). To qualify for the CE mark, manufacturers of Class IIa, IIb, and III devices must be certified by a notified body to Annex II, V, or VI of the MDD (also known as 93/42/EEC) and comply with the essential requirements of the directive. Manufacturers of active implantables and IVDs have separate directives to contend with. When auditing for compliance, the notified body will check a number of items in addition to a manufacturer's QA system, including technical files, sterility assurance measures, subcontracting procedures, recall and vigilance systems, and declarations of conformity. Depending on the classification and certification route, some devices will also require an EC-type examination or a design review by the notified body.

Manufacturers of Class I products, who require minimal interaction with a notified body, appear to be the clear winners in this scheme, but even they must deal with a number of vague or confusing requirements (see Table 4.17). Simply classifying their products according to the dictates of 93/42/EEC, Annex IX, can be a tricky affair, and faulty classification can lead to bigger problems. The simplified flowcharts in Figure 4.2 should help manufacturers determine whether their products qualify as Class I devices. For more difficult products, manufacturers may need to refer to a consultant or obtain a suitable software program.

Classification is based on the intended and declared use of a product, not solely on its salient features. The Class I designation usually – but not always – excludes sterile

*Table 4.17* Which products are class I?

The classification of a product refers to its intended use. The following is a simplified listing of Class I products:

- Noninvasive (and nonactive) devices that do not modify the biological or chemical composition of blood or liquids intended for infusion; store blood, body liquids, or tissues for administration; or connect to an active medical device
- Dressings intended only as a mechanical barrier or for absorption of exudates
- Invasive products for use in natural body orifices and stomas for no longer than 1 h or in the oral or nasal cavity or ear canal for up to 30 days
- Surgical invasive products if they are reusable instruments and not intended for continuous use of more than 1 h
- Active devices that administer neither energy nor substances to the body nor are made for diagnosis

Class I products cannot:

- Incorporate medicinal products (drugs) or animal tissue
- Be intended for contraception or the prevention of sexually transmitted diseases

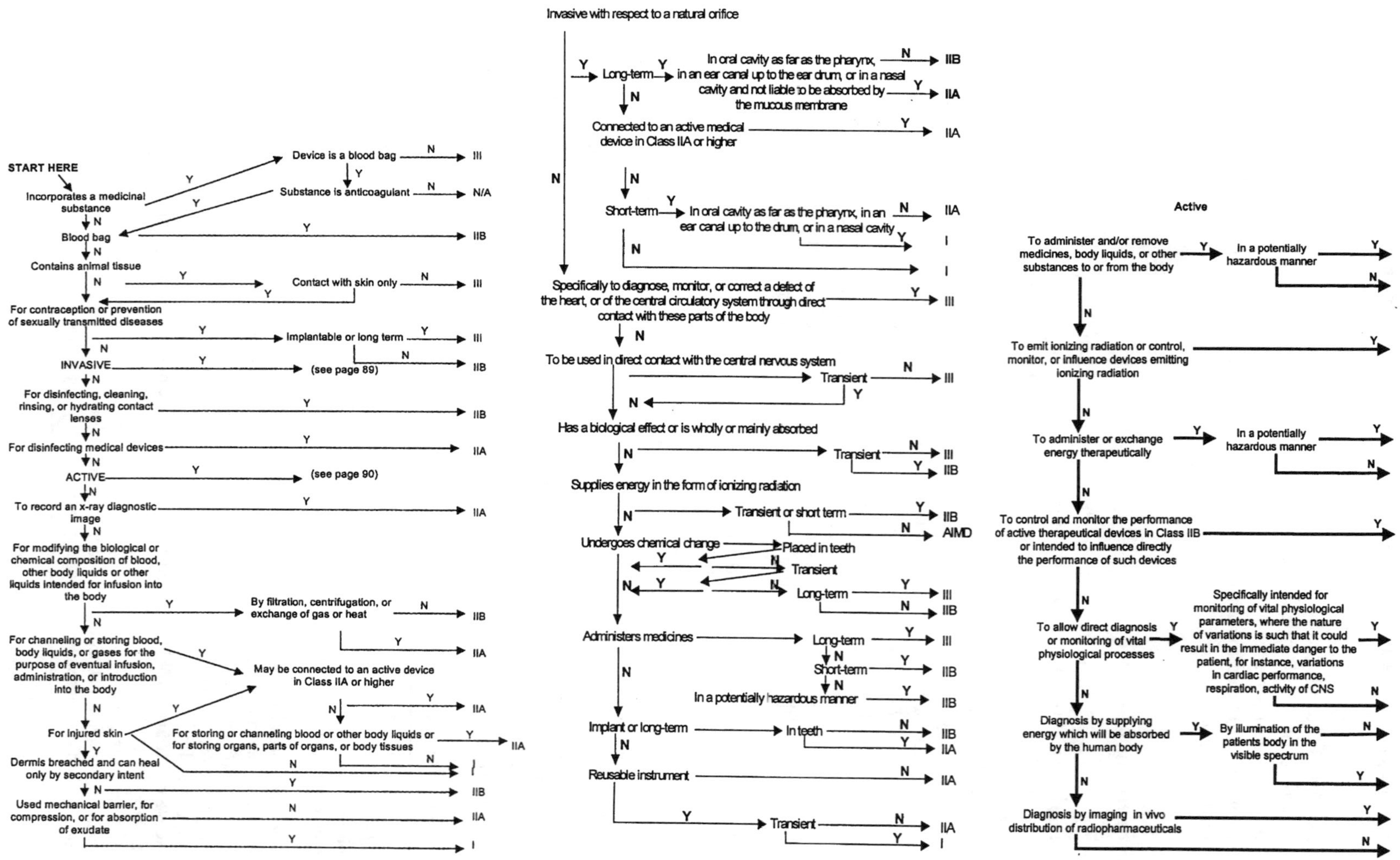


*Figure 4.2.* Medical device classification flowchart.

products and measuring devices that measure physiological parameters or require a high degree of accuracy. So, for example, a reusable scalpel is Class I, but a sterile scalpel is Class IIa; a scalpel blade for the reusable device is Class I, but if it is supplied sterile, it is Class IIa. A stethoscope, a simple graduated syringe (not for injection pumps), and a measuring spoon for administering an expectorant are not considered measuring devices, although a hand-driven blood-pressure gage and a digital thermometer are.

All of the classification rules are included in the directive, but they are not easy to understand. An EC working group has drawn up a separate paper known as MEDDEV 10/93 (Final Draft Guidelines on Medical Device Classification, 1993) to explain the rules and provide some practical guidelines. For example, the directive stipulates that reusable surgical instruments belong in the Class I designation as long as they are not intended for more than an hour of continuous use. According to this definition, items such as scissors and tweezers, even if they are used in a 6-h operation, are still considered Class I devices because they are not used continuously during that time.

Even if a Class I product is supplied sterile, the manufacturer must issue a self-declaration of conformity. In this case, the manufacturer need only certify the QC system governing those aspects of manufacture concerned with securing and maintaining sterile conditions. If the device is packaged and sterilized by a company that works with a certified process, then the manufacturer must only validate the process for the particular device and submit the results to a notified body. The manufacturer still needs certification by a notified body with regard to the performance aspects relating to sterility and measurement function; the notified body will also want to inspect the manufacturer's facility. Nonetheless, the procedure is far less complicated than a full production audit.

All manufacturers applying for CE marking privileges – including manufacturers of Class I devices – must prepare the proper technical documentation; appoint a "responsible person" within the EEC; design product labels and labeling according to 93/42/EEC, Annex I, paragraph 13; and sign a declaration of conformity. The technical dossier should not pose a major problem for manufacturers familiar with device master files. A list of required dossier contents is given in Table 4.18. For biological material testing, Europe uses the ISO 10993 (EN 30993) protocols, but test results according to the Tripartite Agreement (or USP XXIII) are accepted. Every electrical device must also be proven to comply with the EMC requirements defined in the MDD; suppliers of preassembled electrical components may have the appropriate test results already available. Reformatting an existing device master file is not necessary, only creating an index that cross-references the essential requirements of the directives with the device file contents. The master file is a controlled document, as defined in ISO 9000, and manufacturers would do well to regard it as highly confidential.

The technical dossier is closely linked to the responsible person, a representative in the EEC governed by European law and authorized by the manufacturer to oversee routine regulatory affairs. Specifically, the responsible person must ensure compliance with the European vigilance system, which covers both postmarket surveillance and adverse-incident reporting. For example, if a patient were injured by a device, or if a patient would have been injured had the caregiver not intervened, the responsible person would have to investigate the incident together with the device's manufacturer and file a report with the competent authorities. Moreover, the European authorities must be able to obtain the master file in case of trouble; therefore, the manufacturer must either store the file or its abbreviated form with the responsible person or draw up

*Table 4.18* Contents of a device master file

| |
|---|
| 1. EC declaration of conformity and classification according to Annex IX of the MDD |
| 2. Name and address of the manufacturer's European responsible person |
| 3. Product description, including: |
| All variants |
| Intended clinical use |
| Indications/contraindications |
| Operating instructions/instructions for use |
| Warnings/precautions |
| Photographs highlighting the product |
| Photographs highlighting the usage |
| Brochures, advertising, catalog sheets, marketing claims (if available) |
| 4. Product specifications including: |
| Parts list, list of components |
| Specifications of materials used, including data sheets |
| List of standards applied |
| Details of substance(s) used (in the event of drug–device combination) |
| QA specifications (QC specs, in-process controls, etc.), etc. |
| Labeling, accompanying documents, package inserts (DIN EN 289, prEN 980) |
| Instruction for use (prEN 1041) |
| Service manual |
| 5. Product verification, including: |
| Testing data and reports, functionality studies, wet lab or benchtop testing |
| Materials certificates/reports on biological tests |
| EMC testing and certificates |
| Validation of the packaging/aging studies |
| Compatibility studies (connection to other devices) |
| Risk Analysis (DIN EN 1441). |
| Clinical Experience. |
| List of requirements (Annex I) indicating cross-reference with documentation |

a contractual agreement that gives the agent the right to access the master file without delay if required by the authorities. The agent must be available all year, as the time frame for notification could be as short as 10 days. Ideally, the responsible person should be familiar with the national regulation in all member states.

The simplest way to maintain a European address would be to appoint a distributor as the responsible person, although this course is not without potential problems. The selected distributor does not need certification as long as the manufacturer's name and CE mark are on the product labeling. The name of the responsible person must also appear on the label, package insert, or outer packaging, even if the product is sold by a completely different distributor in another country. There is no official rule or proposal regarding how many responsible persons a manufacturer should have, but each one must appear on the labeling; therefore, appointing more than one is of limited use. The responsible person should be selected with great care; device master files (Table 4.18) must be made available to the responsible person in the event of patient injury or near injury, and many distributors are potential competitors. Class I devices, by nature, will rarely lead to patient injury, but manufacturers should still consider labeling issues when choosing a representative. It is easy to change distributors, but changing the responsible person means changing all the product labeling. As an alternative, manufacturers can contract with a professional agency to serve as a representative completely independent from any distribution network.

The issue of labeling is itself a source of contention. Not all countries have decided yet whether they will insist on having their own language on device labels. Many countries have rather imprecise rules, dictating that their national language must appear only if necessary. Manufacturers can reduce potential trouble by using the pictograms and symbols defined in the harmonized European standard EN 980. For instructions of use, manufacturers are advised to use all 12 languages, used in the European Economic Area. The requirements for labeling are presented in Annex I, paragraph 13, of the MDD; some devices may be subject to additional requirements outlined in product standards.

Class I products fall under the jurisdiction of local authorities, but who serves as those authorities may differ from country to country. In Germany, for example, there are no clear-cut regulations that define the competence of the local authorities, except in the case of danger to the patient. European product liability laws more or less give the consumer the right to sue anybody in the trade chain. Normally, claims would be filed against the manufacturer, but it is possible that there will be claims against a responsible person. This is a rather new legal situation, and the rules will be determined by court decisions. It is hoped that Class I products will not instigate many court actions, but clearly, even manufacturers of Class I devices will have a host of new concerns under the CE marking scheme.

## Road map to testing

Determining what testing is required for the development and approval of a new medical device can be a complex issue. This is even more the case after the issue of when to perform necessary tests is factored in. Post approval, one must determine what ongoing testing is required to ensure continued safety of the product. Understanding the complexities requires careful consideration of some key concepts.

### *Key concepts*

There are ten major categories to consider in evaluating and establishing the safety of a medical device, and in so doing defining what testing must be performed. These are presented as a list in Table 4.19, and are considered in detail in this chapter.

*Table 4.19* Key concepts in medical device safety assessment

| | |
|---|---|
| 1. | Condition of use |
| 2. | Materials/components/products |
| 3. | Chemical and physical property considerations |
| 4. | Factors of influence |
| 5. | Prior knowledge |
| 6. | Types and uses of tests |
| 7. | Reasonable man |
| 8. | Qualification versus process control |
| 9. | Tiers of concern: consumer/healthcare provider/manufacturer |
| 10. | Sterilization |

### Condition of use

The starting point for evaluating the safety of a near (or potential) medical device (or material for use in devices) must be understanding both how it is intended to be used (which governs the type of contact it will have with the end use consumers, i.e., patients, and therefore the areas of potential risk and the applicable regulations) and how it is likely to be used (or misused).

Intended use starts with developing an objective statement of what purpose the device is to serve, and therefore how it is to be in contact with the patient (skin/body surface, only body cavity, indirectly with a fluid path (such as and most commonly the blood stream) within the body) and for how long there is to be patient contact. The categories for type of contact are drawn from the nation's regulatory guidelines presented earlier, but actual devices may fit into several categories. It is also important to know more details of the contact (such as what body cavity contact is with – mouth, nasal, vagina, anal, etc.).

The duration of exposure in use should also be established. This should be the cumulative duration for any patient, and not just the single time/use duration. (That is, if a device is to be used 5 min a day each day for a week, it should be considered to have an approximately 35-min cumulative patient exposure). For most devices, the intended use for any one patient is a single time. But if the device (say a glove) is used by a Healthcare provider, over the course of a day, cumulative exposure can be extensive.

One must also consider unintended uses or expected abuses of the device. People use devices in ways that are not planned, such as using elastic bandages to cover wounds. This is especially (but not exclusively) the case in the care of children and the elderly, who are more likely to be susceptible to adverse effects. It should also be kept in mind that, though many electronic devices (disposable gloves and syringes, for example) are intended to be single-use disposables, in poorer cultures this frequently may not be the case. One cannot guard against every – or even most unusual – device usage, but should exercise some consideration of what the most likely misuses are.

### Materials, components, and products

What is actually sold for use by or on the end use consumer (the patient) is what is regulated by the various government agencies. This is the product, which must be evaluated for biocompatibility in conformance with applicable guidelines, in the form or forms (sterilized or unsterilized) that it is intended to be sold.

But products are frequently composed of components. Simple examples are a disposable syringe (needle, barrel, plunger, lubricant, and stopper) or a surgical prep set (scrub, disinfectant, razor, etc.). Changing a component can significantly alter the biocompatibility of a product. And certain components, by the nature of both their composition and exposure to patients, are more likely to present biocompatibility problems. An example is the common disposable plastic syringe, of which billions are used each year. For the syringe, the most likely problem component is the stopper – the flexible piece at the end of the plunger. The stopper is most commonly made of natural rubber, and has direct contact with fluids entering the body (and frequently a fluid path). This is the most common problem component for a syringe.

Components, of course, are manufactured from or composed of materials. Materials (polymers, elastomers, steel, etc.) are the fundamental starting point for development of a device, and are very frequently not produced by the device (or component) manufacturer, but are rather provided by an outside vendor.

Almost all biocompatibility problems (the exceptions being due to sterility, sterilization and cleanliness) for devices are due to the materials used in a device. Table 4.20 provides a concise list of material-based considerations for safety of a device.

## *Chemical and physical property considerations*

Engineers involved in the design and development of new medical devices are primarily concerned with the physical properties essential for the proper functioning of the device. Accordingly, the most important aspects of materials being used in device construction are its physical properties. Toxicologists and others responsible for device safety (the subject of this volume) are primarily concerned with the chemical nature and properties of materials used in devices, but must also be conversant with physical property considerations. For that reason, this section presents a primer on the chemical nature and chemical and physical properties of materials used in devices.

The majority of components of medical devices are constituted from a small number of categories of materials. The important categories of materials in device formulation and construction are the following: water, stainless steels, polymers, elastomers, sili-

*Table 4.20* Raw material characterization

**Chemical characterization**
List materials
List potential extractables or leachables from materials
Physiochemical tests, USP
  Water and isopropanol extracts
International pharmacopoeial tests
Infrared analysis
Document polymer identification
Chromatographic characterization
  Molecular weight distribution
  Additive and/or extract analysis
Trace metals
Specific gravity
Moisture content

**Physical characterization**
Hardness
Surface characterization
Color, opacity or haze
Strength properties
  Tensile/elongation
  Flexure
  Compression
Thermal analysis
Viscosity, melting point, refractive index

**Biological characterization**
[nothing listed]

cones and natural fibers (cotton and wood pulp, primarily). Each of these needs to be considered in turn. Additionally, biological derived materials are seeing increased utilization in devices (Kambric *et al.*, 1986), but are of such diverse nature that is not currently possible to adequately overview them in this volume.

These are interactions which occur between material used in a device and the organism ("host") that it has contact with. Device materials having systemic contact with a host ("biomaterials") may be degraded by the host by a number of chemical means which should be kept in mind when considering the use of any of the materials described in this section. These pathways of chemical degradation of biomaterials include:

- Hydrolysis (acid, base, neutral aqueous media)
- Oxidation (corrosion, chain cleavage)
- Thermolysis
- Photooxidation
- Specific enzyme-catalyzed hydrolysis or oxidation
- Attack by complex media (culture media, serum, blood, gastric juices, urinary fluids, phagocyte-containing fluid, etc.)
- Chain cleavage due to mechanical fracture

### Water

Water, in one way or another, is involved in the production of virtually every medical device. For many devices (particularly diagnostics), it is also incorporated into the device. Yet water tends to be invisible to many considering device biocompatibility and safety.

Water's greatest uses are in cleaning and rinsing devices and their components. Purified water is obtained by distillation, ion-exchange treatment, reverse osmosis, or other suitable processes. Such water is prepared from source material complying with the regulations of the US Environmental Protection Agency (USEPA) for drinking water. Water can be a problem in device safety and biocompatibility generally due to the presence of things in it which do not comply with USEPA and other regulations (such as microorganisms, pesticides, organics, other pyrogens, and heavy metals), as well as contaminants from other devices or components which may have been previously rinsed in the water (such as latex from gloves or stoppers). Water is usually a source of problems in production of devices, and not in the development stage.

### Metals

A variety of metals sees significant use in medical devices, though with the exception of stainless steel their use in patient contacting situations is largely limited to implants. The uses are a reflection of the properties of the various metals, as summarized in Gad (1997).

### Factors influencing test selection

Actual decisions as to what testing is to be done are based on a complex set of reasons, some of which are particular to the company involved and some of which are generally applicable. The author labels these reasons as "factors of influence," and believes that they can be summarized as belonging to the seven categories presented in Table 4.21. The first of these factors, regulatory requirements, was extensively covered in the first chapter. The others are discussed briefly here.

### *Perceptions*

It should be kept in mind that what people believe/perceive is as important as what is real. What materials are used, how a device is designed and what testing is done are significantly influenced by current public and Healthcare provider beliefs. Concerns about and memories of silicones, latex, toxic shock syndrome, etc., may dictate more extensive testing than regulations. Beliefs can also influence device acceptance, such as the case of IntraUterine Devices (IUDs) as contraceptive devices after the publicity around the Dalkon Shield.

### *Hazard identification*

The most fundamental requirement in testing is to quickly identify (or eliminate the possibility of) any significant hazards, their services and how to eliminate or minimize them if they are present. Many of the tests used for medical devices are really designed to act as sensitive screens for hazards. They purposely maximize the potential to get a positive response (that is, they are very sensitive). Such tests share a number of common characteristics and do not establish the relevance of such findings of hazard to real life device use.

### *Risk assessment*

The process of taking the results of toxicity and biocompatibility tests of literature findings and all other sources of information and then relating them to actual device use in the marketplace is risk assessment. The need to be able to perform a meaningful and convincing risk assessment may require the conduct of additional tests which allow for the quantification of risk (which screens usually do not). It should be pointed out at this point that all substances (even water and green apples) are toxic at some dose. The real life hazard is when the exposure that harm may occur at is within the realm of likely exposure. Table 4.22 addresses the point of relative toxicity. ISO (1996) has issued specific guidelines addressing risk and hazard assessment for devices and biomaterials.

*Table 4.21* Factors of influence on safety test selection

| |
|---|
| Regulatory requirements |
| Perceptions |
| Hazard identification |
| Risk assessment |
| Animal welfare concerns |
| Claims |
| Time and economics |

*Table 4.22* Classification of chemical hazards[a]

| *Commonly used term* | *Routes of administration* | | | |
|---|---|---|---|---|
| | *Single oral dose, rats $LD^{50}$* | *Inhalation 4 h vapor exposure mortality 1/6–4/6 rats (ppm)* | *Single application to skin of rabbits $LD^{50}$* | *Probable lethal dose for man* |
| Extremely toxic | 1 mg or less/kg | 10 | 5 mg or less/kg | A taste, a drop, one grain |
| Highly toxic | 1–50 mg/kg[b] | 10–100 | 5–43 mg/kg | One teaspoonful (4 ml) |
| Moderately toxic | 500-500mg/kg | 100–1000 | 44–340 mg/kg | 1 ounce (30 g) |
| Slightly toxic | 0.5–5 g/kg | 1000–10,000 | 0.35–2.81 g/kg | 1 pint (250 g) |
| Practically g/kg quart or nontoxic | 5–15 g/kg | 10,000–100,000 | 22.6 or more g/kg | $>$ 1 quart or $>$ 1 l |

[a] From Deichemann and Gerard, 1969; Gad and Chengelis (1998).

[b] By law, those materials with oral $LD_{50}$s of 50 mg/kg or less in rats are classified as Class B poisons and must be labeled "Poison." Class A poisons are defined not by testing, but rather by inclusion on a regulatorily mandated list (CFR 173, Section 173.326). "S 173.326 Poison A. (a) For the purpose of Parts 170–189 of this subchapter extremely dangerous poison. Class A. are poisonous gases or liquids of such nature that a very small amount of the gas, or vapor of the liquid, mixed with air is dangerous to life. This class includes the following: (1) Bromactone. (2) Cyanogen. (3) Cyanogen chloride containing less than 019 per cent water. (4) Diphosgene. (5) Ethyldichlorarsine. (6) Hydrocyanic acid (see Note 1 of this paragraph). (7) [Reserved]. (8) Methyldichlorarsine. (9) [Reserved]. (10) Nitrogen peroxide (tetroxide). (11) [Reserved]. (12) Phosgene (diphosgene). (13) Nitrogen tetroxide-nitric oxide mixtures containing up to 33.2 per cent weight nitric oxide. Note 1: diluted solutions of hydrocyanic acid of not exceeding 5 per cent strength are classed as poisonous articles. Class B (see S 173-343)." (b) Poisonous gases or liquids, Class A. as defined in paragraph (a) of this section, except as provided in S 173.331, must not be offered for transportation by rail express. [239 FR 18753, Dec. 29, 1964. Redesignated at 32 FR 5606, Apr. 5, 1967, and amended by Amdt. 173-94, 41 FR 16081, Apr. 15, 1976; Amdt. 173-94A, 41 FR 40883, Sept. 20, 1976].

### *Claims*

Claims are what is said in labeling and advertising, and may be either of a positive (therapeutic or beneficial) or negative (lack of an adverse effect) nature. The positive or efficacy claims are not usually the direct concern of the toxicologist though it must be kept in mind that such claims both must be proved and can easily exceed the limits of the statutory definition of a device, turning the product into a drug or combination product.

Negative claims such as "nonirritating" or "hypoallergenic" also must be proved, and are generally the responsibility of the product safety professional to provide proof of. There are special tests for such claims.

### *Time and economics*

The final factors of influence or arbitrator of test conduct and timing are the requirements of the marketplace, the resources of the organization and the economic worth of the product.

Plans for filings with regulatory agencies and for market launches are typically set before actual testing (or final stage development) is undertaken, as the need to be in the marketplace in a certain timeframe is critical. Such timing and economic issues are beyond the scope of this volume, but must be considered.

### *Prior knowledge*

The appropriate starting place for the safety assessment of any new chemical entity, particularly a potential new material for a medical device, is to first determine what is already known about the material and whether there are any close structural or pharmacological analogues (pharmacological analogues being agents with assumed similar pharmacological mechanisms). Such a determination requires complete access to the available literature. In using this information, one must keep in mind that there is both an initial requirement to build a data file or base, and a need to update such a store on a regular basis. Updating a database requires not merely adding to what is already there, but also discarding out-of-date (i.e., now known to be incorrect) information and reviewing the entire structure for connections and organization.

The first step in any new literature review is to obtain as much of the following information as possible:

1 Correct chemical identity including molecular formula, Chemical Abstracts Service (CAS) Registry number, common synonyms, trade names, and a structural diagram. Gosselin *et al.* (1984); Ash and Ash (1994, 1995) are excellent sources of information on existing commercial products, their components, and uses.
2 Chemical composition (if a mixture) and major impurities.
3 Production and use information.
4 Chemical and physical properties (physical state, vapor pressure, pH, solubility, chemical reactivity, etc.)
5 Any structurally related chemical substances that are already on the market or in production.
6 Known or presumed biological properties.

Collection of the above information is not only important for hazard assessment (high vapor pressure would indicate high inhalation potential, just as high and low pH would indicate high irritation potential), but the prior identification of all intended use and exposure patterns may provide leads to alternative information sources; for example, drugs to be used as antineoplastics or antibiotics may already have extensive toxicology data obtainable from government or private sources. A great deal of the existing toxicity information (particularly information on acute toxicity) is not available in the published or electronic literature because of concerns about the proprietary nature of this information and the widespread opinion that it does not have enough intrinsic scholarly value to merit publication. This unavailability is unfortunate, as it leads to a lot of replication of effort and expenditure of resources that could be better used elsewhere. It also means that an experienced toxicologist must use an informal search of the unpublished literature of his colleagues as a supplement to searches of the published and electronic literature.

There are now numerous published texts that should be considered for use in literature-reviewing activities. An alphabetic listing of 23 of the more commonly used sources for safety assessment data is provided in Table 4.23. Obviously, this is not a complete listing and consists of only the general multipurpose texts that have a wider range of applicability for toxicology. Texts dealing with specialized classes of agents (e.g., disinfectants) or with specific target organ toxicity (neurotoxins and teratogens) are generally beyond the scope of this text. Parker (1988) should be consulted

*Table 4.23* Published information sources for safety assessment

| *Title* | *Author, date* |
|---|---|
| Annual Report on Carcinogens | National Toxicology Program, various |
| *Burger's Medicinal Chemistry* | Wolff (1996) |
| *Carcinogenically Active Chemicals* | Lewis (1991) |
| *Catalog of Teratogenic Agents* | Shepard, 1988 |
| *Chemical Hazards of the Workplace* | Proctor and Hughes (1978) |
| *Chemically Induced Birth Defects* | Schardein (1985) |
| *Clinical Toxicology of Commercial Products* | Gosselin *et al.* (1984) |
| *Contact Dermatitis* | Cronin (1980) |
| Criteria Documents | NIOSH (various) |
| Current Intelligence Bulletins | NIOSH (various) |
| *Dangerous Properties of Industrial Materials* | Sax (1985) |
| *Documentation of the Threshold Limit Values for Substances in Workroom Air* | ACGIH (1986) |
| *Handbook of Toxic and Hazardous Chemicals* | Sittig (1985) |
| Hygienic Guide Series | AIHA (1980) |
| *Hamilton and Hardy's Industrial Toxicology* | Finkel (1983) |
| *Medical Toxicology* | Ellenhorn *et al.* (1997) |
| *Merck Index* | Budavari (1989) |
| NIOSH/OSHA *Occupational Health Guidelines for Chemical Hazards* | Mackison (1981) |
| *Patty's Industrial Hygiene and Toxicology* | Clayton and Clayton (1981) |
| *Physician's Desk Reference* | Barnhart (1991) |
| *Registry of Toxic Effects of Chemical Substances (RTECS)* | NIOSH (1984) |
| *Casarett and Doull's Toxicology: The Basic Science of Poisons* | Klassen (1996) |
| *Toxicology of the Eye* | Grant (1993) |

for details on the use of these texts. Parker (1988), Sidhu *et al.* (1989); Wexler *et al.* (2000) should be consulted for more extensive listings of the literature and computerized databases.

### *Miscellaneous reference sources*

There are some excellent published information sources covering some specific classes of chemicals, for example, heavy metals, plastics, resins, or petroleum hydrocarbons. The National Academy of Science series *Medical and Biologic Effects of Environment Pollutants* covers 10–15 substances considered to be environmental pollutants. *CRC Critical Reviews in Toxicology* is a well-known scientific journal that over the years has compiled over 20 volumes of extensive literature reviews of a wide variety of chemical substances. A photocopy of this journal's topical index will prevent one from overlooking information that may be contained in this important source. Trade organizations such as the Fragrance Industry Manufacturers Association and the Chemical Manufacturers Association have extensive toxicology databases from their research programs that are readily available to toxicologists of member companies. Texts that deal with specific target organ toxicity – neurotoxicity, hepatotoxicity, or hematotoxicity – often contain detailed information on a wide range of chemical structures. Published information sources like the *Target of Organ Toxicity* series (Taylor&Francis, now halfway through revision) or few examples of the types of publications that often contain important information on many industrial chemicals may be useful either directly or by analogy. Upon discovery that the material one is evaluating may possess target organ toxicity, a cursory review of these types of texts is warranted.

In the last decade, for many toxicologists the on-line literature search has changed from an occasional, sporadic activity to a semicontinuous need. Usually, nontoxicology-related search capabilities are already in place in many companies. Therefore, all that is needed is to expand the information source to include some of the databases that cover the types of toxicology information one desires. However, if no capabilities exist within an organization one can approach a university, consultant, or a private contract laboratory and utilize their on-line system at a reasonable rate. It is even possible to access most of these sources from home using a personal computer. The major available on-line databases are as follows:

A. *National Library of Medicine*. The National Library of Medicine (NLM) information retrieval service contains the well-known and frequently used Medline, Toxline, and Cancerlit databases. Databases commonly used by toxicologists for acute data in the NLM service are the following:

1 Toxicology Information Online (Toxline) is a bibliographic database covering the pharmacological, biochemical, physiological, environmental, and toxicological effects of drugs and other chemicals. It contains approximately 1.7 million citations, most of which are complete with abstract, index terms, and CAS Registry numbers. Toxline citations have publication dates of 1981 to the present. Older information is on Toxline 65 (pre-1965–1980).
2 Medical Information Online (Medline) is a database containing approximately 7 million references to biomedical journal articles published since 1966. These arti-

cles, usually with an English abstract, are from over 3000 journals. Coverage of previous years (back to 1966) is provided by back files, searchable online, that total some 3.5 million references.

3 Toxicology Data Network (Toxnet) is a computerized network of toxicologically oriented data banks. Toxnet offers a sophisticated search and retrieval package that accesses the following three subfiles:

- Hazardous Substances Data Bank (HSDB) is a scientifically reviewed and edited data bank containing toxicological information enhanced with additional data related to the environment, emergency situations, and regulatory issues. Data are derived from a variety of sources including government documents and special reports. This database contains records for over 4100 chemical substances.
- Toxicology Data Bank (TDB) is a peer-reviewed databank focusing on toxicological and pharmacological data, environmental and occupational information, manufacturing and use data, and chemical and physical properties. References have been extracted from a selected list of standard source documents.
- Chemical Carcinogenesis Research Information System (CCRIS) is a National Cancer Institute-sponsored database derived from both short- and long-term bioassays on 2379 chemical substances. Studies cover carcinogenicity, mutagenicity, promotion, and cocarcinogenicity.

4 Registry of Toxic Effects of Chemical Substances (RTECS) is the NLM's online version of the National Institute for Occupational Safety and Health's (NIOSH) annual compilation of substances with toxic activity. The original collection of data was derived from the 1971 Toxic Substances Lists. RTECS data contain threshold limit values, aquatic toxicity ratings, air standards, National Toxicology Program carcinogenesis bioassay information, and toxicological/carcinogenic review information. The National Institute for Occupational Safety and Health is responsible for the file content in RTECS, and for providing quarterly updates to NLM: RTECS currently covers toxicity data on more than 106,000 substances.

*B. The Merck Index* is now available online for up-to-the minute access to new chemical entities.

### Search procedure

As mentioned in the Introduction, chemical composition and identification information should already have been obtained before the chemical is to be searched. With most information retrieval systems this is a relatively straightforward procedure. Citations on a given subject may be retrieved by entering the desired free text terms as they appear in titles, keywords, and abstracts of articles. The search is then initiated by entering the chemical CAS number and/or synonyms. If you are only interested in a specific target organ effect – for instance, carcinogenicity – or specific publication years, searches can be limited to a finite number of abstracts before requesting the printout.

Often it is unnecessary to request a full printout (author, title, abstract). You may choose to review just the author and title listing before selecting out the abstracts of interest. In the long run, this approach may save you computer time, especially if the number of citations being searched is large.

Once you have reviewed the abstracts, the last step is to request photocopies of the articles of interest. Extreme caution should be used in making any final health hazard determination based solely on an abstract or nonprimary literature source.

### Monitoring published literature and other research in progress

Although there are a few other publications offering similar services, the *Life Sciences* edition of *Current Contents* is the publication most widely used by toxicologists for monitoring the published literature. *Current Contents* monitors over 1180 major journals and provides a weekly listing by title and author. Selecting out those journals you wish to monitor is one means of selectively monitoring the major toxicology journals.

Another mechanism for monitoring research in progress is by reviewing abstracts presented at the annual meetings of professional societies such as the Society of Toxicology, Teratology Society, Environmental Mutagen Society, and American College of Toxicology. These societies usually have their abstracts prepared in printed form; for example, the current *Toxicologist* contains over 1700 abstracts presented at the annual meeting. Copies of the titles and authors of these abstracts are usually listed in the societies respective journals, which, in many cases, would be reproduced and could be reviewed through *Current Contents*.

### Other testing considerations

Safety assessment tests used for medical devices can generally be considered as either hazard identification/screens or special studies uniquely designed for specific problems or types of devices (Tables 4.24 and 4.25). The remainder of this chapter will largely look at how each of the significant types of such tests are performed and interpreted. Earlier in this chapter the author summarized the common varieties of available biocompatibility tests and their objectives, as well as where in the text they are considered in detail.

#### *Reasonable man*

The reasonable man is a concept in law which, though not universally applicable, still provides guidance as to what one can expect from those that use devices (and what, therefore, the limits are on uses for which the manufacturer of the device should be considered responsible for ensuring safety). The standard of reasonableness is obviously open to interpretation, but does provide a conceptual basis for determining what uses one must ensure a device is safe for (and for one must supply a precautionary label) and which it is not. The "test" employed in a legal sense is one of foreseeability, i.e., would a reasonable man in the defendant's position foresee a measurable risk to the plaintiff? (Madden, 1992).

#### *Qualifications versus process control*

Most of this book addresses testing from the point of view of what is done to quality a product – to get it access to the marketplace. Such testing is done, at a minimum, to

*Table 4.24* Product and process validation[a]

**Environmental control**
Environmental monitoring program
Microorganism identification
Viable and nonviable particulate analysis

**Manufacturing process control** – initial qualification and ongoing control
Raw material characterization (compare effects of process on characteristics determined in phase I)
  Infrared analysis
  Cytotoxicity
  Physiochemical tests (USP, JP, etc.)
Other materials characterization tests
  Bioburden testing
  Process water system validation
Purified water monograph tests, USP
Water for injection monograph tests, USP
  Endotoxin concentrations (LAL testing)
  Quality device cleaning processes
  Package qualifications

**Sterility**
Bioburden testing and organism identification
Biological indicator studies (sport count, D-value)
Sterilization cycle development
Sterilization cycle validation, plan for periodic revalidation
Dose determination studies (AAMI) plan for quarterly dose audits
Sterility tests
EO dissipation curve studies and assessment of user exposure levels (AAMI/ISO)
Package validation

**Finished product qualification** – single use or reusable
Physical testing for function and performance stability
Chemical residues
Testing for bacterial endotoxins
  In vitro, limulus amebocyte lysate (LAL)
  In vivo, rabbit pyrogen tests
Biocompatibility
  Cytotoxicity test
  Hemocompatibility test
Special material and device tests
  Chemistry tests
  Microbiology tests
  Toxicology tests
Nonviable particulate analysis
Label claim (instructions) for reusable devices
  Decontamination
  Cleaning
  Disinfection/sterilization
Other product specific testing
Shelf life stability qualification
  Accelerated aging studies
  Real time aging

[a] This is a series of qualification studies to demonstrate that manufacturing process controls are sufficient for preproduction quality assurance requirements and product specifications. Testing is performed to verify the effectiveness of such control and to evaluate the biological effects of processing aids added during manufacture.

*Table 4.25* Routine testing[a]

**Release testing**
Endotoxin concentration
  Limulus amebocyte lysate (LAL), USP
Pyrogenicity
  Rabbit test, USP
Safety test, USP
  Infusion/transfusion assemblies
Sterility testing
Microbial limit test, USP
Cytotoxicity, USP/ISO
Materials characterization

**Periodic audit testing**
Endotoxin concentration
  Limulus amebocyte lysate (LAL), USP
Pyrogenicity
  Rabbit test, USP
Cytotoxicity
  Raw materials
  Finished products
In vitro hemolysis test for blood contract products
EO residual testing
Materials characterization
Physical testing
Particulate testing
Bioburden testing

[a] Release testing involves what is performed routinely to satisfy GMP and ISO requirements for finished product testing prior to the release of product for distribution. In addition, Phase IV includes testing that may be incorporated into the manufacturer's quality assurance audit program by conducting periodic raw material and finished product testing in order to document that materials and product conform to specifications.

meet specific regulatory requirements which one can determine by consulting the appropriate guidelines. However, biocompatibility testing does not end once a product is approved for the marketplace. Rather, some form of testing must be conducted on an ongoing basis to ensure that the lots of product that enter the marketplace over time continue to be safe. The testing to be done to ensure such is generally specified in the Device Master File (DMF), but what tests are done and with what frequency is left to the judgment of the manufacturer (who is, however, charged in the GMP with conducting an adequate program of periodic testing to ensure the continued quality and safety of the product). Such testing is usually derived from the results of qualification testing and product and (manufacturing) process validation studies (Table 4.24). Careful consideration (and statistical analysis of these and the variables that are involved in the manufacturing process generally identify which biocompatibility tests best serve to identify when the product is not as it should be, due to either the process not being in control (or there having been a series of small incremental changes which in summation have altered the process) or when changes in vendor supplied materials have occurred. A statistical analysis of the data will also clarify sampling strategies and required frequency of testing. This will lead to specification of a routine testing program for lot release, most commonly utilizing approaches shown in Table 4.25. Changes in materials or processes involved in the manufacture of an approved device may require filing of a 510 (K) and additional testing. The FDA has provided a flow chart (Figure 4.3) as guidance in this decision.

The DMFs on plant manufacturing SOPs (standard operating procedures) need to specify what happens when a lot fails routine or release testing. It is sometimes wise to have a conditional two-tier test scheme – an inexpensive but somewhat sensitive screening test (such as cytotoxicity) which is performed on some specified regular basis, and a second, more specific (and expensive) test which is conducted in those cases where a lot fails the screening test.

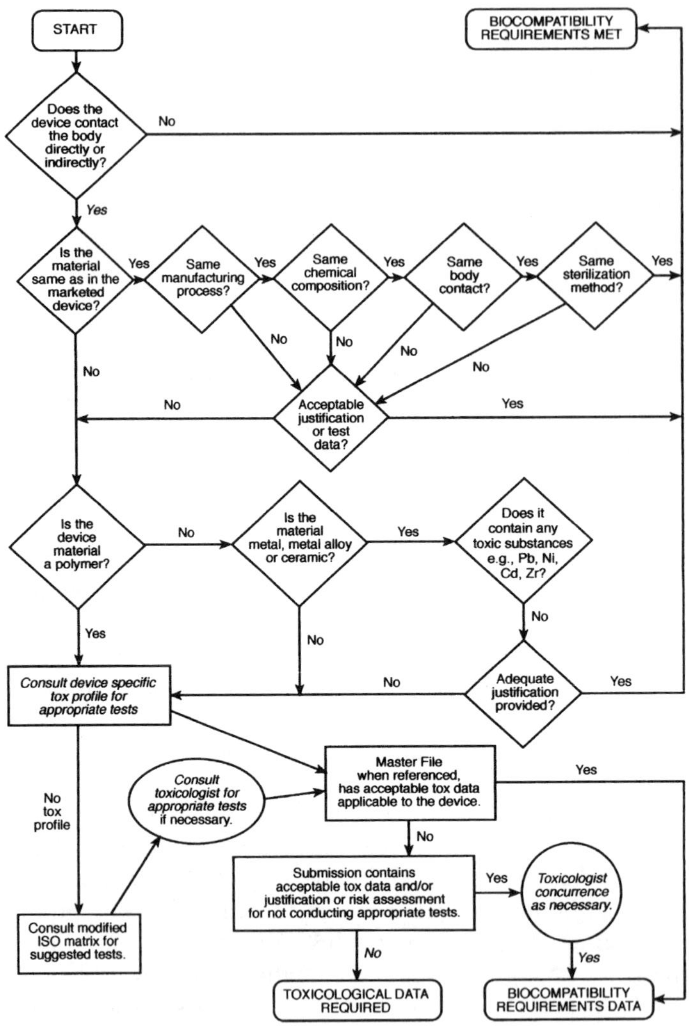


*Figure 4.3* Biocompatibility flow chart for the selection of toxicity tests for 510(k)S (FDA, 1995).

### *Tiers of concern: consumers, Healthcare providers and manufacturing employees*

This book focuses primarily on the tests done to meet regulatory requirements for new product approval. Such requirements are intended to ensure the safety on the end use consumer of devices, the patients. Knowing what the intended use and claims are for a product, it is generally easy to identify what routes, duration, and extent of "exposure" or "dosing" will be.

However, patients represent only the final tier of those who will be exposed to a device. There are (at least) two other tiers that we must consider. The others are the healthcare providers and those involved in actually producing and packaging the devices. Healthcare providers include nurses, doctors, laboratory technicians, pharmacists, and public health workers. Though they do not use the devices on a daily basis, they will handle and apply or administer the products. As such, they will have different routes and durations of exposure which must be considered and evaluated for safety. Dermal exposure in particular is likely to be more extensive.

Likewise, those involved in manufacturing and packaging the product will have significantly different exposures. For these individuals, we must also be concerned with exposure to materials used in device construction and formulation. Here, the potential for inhalation exposure is most likely.

### *Sterilization and cleanliness*

It goes without saying that microbial contamination of devices must be controlled and that appropriate steps must be taken to sterilize products and materials. The subject is addressed later in this volume in some detail. It should also be remembered, however, that the means of sterilization (ethylene oxide, radiation, chemical sterilization or steam) may both affect device quality and, in some cases, carry their own biocompatibility concerns. Here, residuals are the issue.

And finally, it must be stressed that cleanliness, in the sense of exclusion of foreign matter (even seemingly innocuous things like lint and dust) is essential. If such foreign materials should gain entry to the body, they can trigger dangerous immune modulated responses (Turco and Davis, 1973). The FDA has specifically considered the problem of particles in medical devices from the perspective of physiological effects and provided guidance on the issue (Marlowe, 1980).

## ***Risk assessment***

The reality is that not all materials used on devices are entirely safe. Generally, if one looks long enough at small enough quantities, some type of risk can be associated with every material. Risk can be defined as the possibility of harm or loss. Health risk, of course, is the possibility of an adverse effect on one's health. Risk is sometimes quantified by multiplying the severity of an event times the probability the event will occur, so that

$$\text{Risk} = \text{severity} \times \text{probability}$$

While this equation appears useful in theory, in practice it is difficult to apply to the biological safety of medical devices. The process known as health-based risk assessment attempts to provide an alternative strategy for placing health risks in perspective (Stark, 1998).

## Standards and guidances

A paradigm for the risk assessment process has been detailed in a publication prepared by the US National Academy of Sciences (Hayes, 1994). Although devised primarily for cancer risk assessment, many of the provisions also apply to the assessment of other health effects. The major components of the paradigm are (1) hazard identification, (2) dosage-response assessment, (3) exposure assessment, and (4) risk characterization (Ecobichon, 1992).

The general approach to risk assessment was adapted to medical devices via the draft CEN standard *Risk Analysis*, published in 1993,[1] and more recently via the ISO standard, *ISO 14538 – Method for the establishment of Allowable Limits for Residues in Medical Devices Using Health-Based Risk Assessment*, published in 1996.[2] At the present time, the FDA is also working to develop a health-based risk assessment protocol adapted to medical devices. Informally called the Medical Device Paradigm, the document is not yet generally available (Brown and Stratmeyer, 1997).[3]

Some manufacturers may object that regulators are once again attempting to impose a "drug model" on medical devices. However, we shall see in the following pages that judicious application of these risk assessment principles can provide a justification for using materials that carry with them some element of risk, and that may, under traditional biocompatibility testing regimes, be difficult to evaluate or be deemed unsuitable for medical device applications.

## Method

**Hazard identification**. The first step in the risk assessment process is to identify the possible hazards that may be presented by a material. This is accomplished by determining whether a compound, an extract of the material, or the material itself produces adverse effects, and by identifying the nature of those effects. Adverse effects are identified either through a review of the literature or through actual biological safety testing.

**Dose-response assessment**. The second step is to determine the dose response of the material – that is, what is the highest weight or concentration of the material that will not cause an effect? This upper limit is called the *allowable limit*. There are numerous sources in the literature of data from which to determine allowable limits; some will be more applicable than others, and some may require correction factors.

**Exposure assessment**. The third step is to determine the exposure assessment by quantifying the *available dose* of the chemical residues that will be received by the patient. This is readily done by estimating the number of devices to which a patient is likely to be exposed in a sequential period of use (for instance, during a hospital stay) or over a lifetime. For example, a patient might be exposed to 100 skin staples following a surgical procedure, or to two heart valves in a lifetime; thus, the amount of residue available on 100 skin staples or two heart valves would be determined.

**Risk characterization**. Characterizing the risk constitutes the final step of the process. The allowable limit is compared with the estimated exposure: if the allowable limit is greater than the estimated exposure by a comfortable safety margin, the likelihood of an adverse event occurring in an exposed population is small, and the material may be used.

### *Case studies*

We can best get a sense of how these standards work by looking at some actual medical case studies that illustrate the risk assessment process (Stark, 1997).

**Nitinol implant.** Nitinol is an unusual alloy of nickel and titanium that features the useful property of "shape memory." A nitinol part can be given a particular shape at a high temperature, then cooled to a low temperature and compressed into some other shape; the compressed part will subsequently deploy to its original shape at a predetermined transition temperature. This feature is particularly beneficial for vascular implant applications in which the shape of the device in its compressed state eases the insertion process. The nitinol deploys as it is warmed by the surrounding tissue., expanding to take on the desired shape of a stent, filter, or other device. The transition temperature depends on the alloy's relative concentrations of nickel and titanium: a typical nickel concentration of 55–60 per cent is used in medical devices, since this gives a transition temperature at approximately the temperature of the body (37°C).

*Hazard identification.* One concern with using nitinol in implant applications is the potential release of nickel into the body. Although nickel is a dietary requirement, it is also highly toxic – known to cause dermatitis, cancer subsequent to inhalation, and acute pneumonitis from inhalation of nickel carbonyl, and to exert a toxic effect on cellular reproduction. It is a known sensitizer, with approximately 5 per cent of the domestic population allergic to this common metal, probably through exposure from costume jewelry and clothing snaps. *The biocompatibility question at hand is whether or not in vivo corrosion of nitinol releases unsafe levels of nickel.*

*Dose-response assessment.* A search of the world medical literature revealed that the recommended safe level of exposure to nickel in intravenous fluids is a maximum of 35 μg/day (Stark, 1996). This value can be taken as an allowable limit of nickel exposure for a 70 kg(154 lb) adult.

The intravenous fluid data are based on subjects that are comparable to the patients who will be receiving nitinol implants. The data are for humans (not animals), for ill patients (not healthy workers or volunteers), and for similar routes of exposure (intravenous fluid and tissue contact). For these reasons, no safety correction factor need be applied to the allowable limit of exposure.

*Exposure assessment.* The available dose of nickel from nitinol implants can be estimated from data found in the literature. In one study, dental arch wires of nitinol were extracted in artificial saliva, and the concentration of nickel measured in the supernatant. Corrosion reached a peak at day 7, then declined steadily thereafter. The average rate of corrosion under these conditions was 12.8 μg/day per $cm^2$ over the first 28 days.

*Risk characterization.*A comparison of the available dose with the allowable limit for intravenous fluid levels shows that there is approximately a three-fold safety margin, assuming that the implanted device is a full 1 $cm^2$ in surface area. (Devices with less surface area will contribute even less to the nickel concentration and have an even larger safety margin.) Considering the high quality of the data, a three-fold safety margin is sufficient to justify using nitinol in vascular implants.

**Wound-dressing formulation.** Today's wound dressings are highly engineered products, designed to maintain the moisture content and osmotic balance of the wound bed so as to promote optimum conditions for wound healing. Complex constructions of hydrocolloids and superabsorbers, these dressings are sometimes used in direct tissue contact over full-thickness wounds that penetrate the skin layers.

*Hazard identification.* There have been reports in the literature of patients succumbing to cardiac arrest from potassium overload, with the wound dressing as one of the important contributors of excess potassium in the bloodstream. The effects of potassium on cardiac function are well characterized. Normal serum levels for potassium are 3.8–4 mEq/l. As the potassium concentration rises to 5–7 mEq/l, a patient can undergo cardiac arrest and die. *The biocompatibility issue to be explored is whether or not a wound-dressing formulation might release dangerous levels of potassium if used on full-thickness wounds.*

*Dose-response assessment.* An increase of approximately 1 mEq/l of potassium is unlikely to provoke mild adverse events in most patients. Assuming that the average person's blood volume is 5 l, a one-time dose of 5 mEq of potassium may begin to cause adverse reactions. This value can be considered to be the allowable limit of potassium for most patients.

*Exposure assessment.* Let us suppose that each dressing contains 2.5 g of potassium bicarbonate. Since the molecular weight of potassium bicarbonate is 100 g/mole, each dressing contains 0.025 mole of sodium bicarbonate, or 0.025 mEq of potassium ion. If a patient were to use four dressings in a day, the available dose of potassium would be 0.1 mEq/day.

*Risk characterization.* Comparing the available dose of potassium (0.1 mEq) to the allowable limit (5 mEq) shows that there is a 50-fold safety margin. Considering that patients may be small in size, may have kidney impairment, or may receive potassium from additional sources such as intravenous fluids, this safety margin is too small, and so the dressing should be reformulated.

**Perchloroethylene solvent**. A manufacturer of metal fabricated parts uses perchloroethylene to clean the finished pieces. Perchloroethylene has many advantages as a cleaner and degreaser: it is highly volatile, does not damage the ozone layer, and is very effective as a precision cleaning solvent. The most common use of perchloroethylene is in the dry cleaning industry, but it is also commonly used in the electronics industry to clean circuit boards.

*Hazard identification.* The downside of perchloroethylene is that it is highly toxic, with a material safety data sheet several pages in length listing adverse effects ranging from dizziness to death. Biocompatibility testing on solvent-cleaned parts would be meaningless; the solvent concentration on the part is so small that any effects of the solvent would be masked by the natural biological process of the test animals. *The biocompatibility question that must be answered is whether or not sufficient residual perchloroethylene remains on the cleaned metal parts to pose a health hazard.*

*Dose-response assessment.* Threshold limit values (TLVs) are values that indicate the maximum level of a chemical that a healthy worker could take in on a daily basis over the course of his or her work life without experiencing any adverse effects. The TLV for perchloroethylene is 50 ppm/day (50 ml of perchloroethylene per $10^3$ l of air) by inhalation. The average person inhales 12,960 l of air per day, making this equivalent to 650 ml of perchloroethylene per day. Since the vapor density of perchloroethylene is 5.76 g/l, the TLV is equal to 3.7 g of perchloroethylene per day by inhalation.

Because TLVs for inhalation – as opposed to direct tissue exposure – are determined based on healthy individuals (not ill patients), we will divide the TLV by an uncertainly factor of 100, i.e., ten to account for a different route of exposure and ten to account for healthy-to-ill persons. By this method, we obtain an allowable perchloroethylene limit of 37 mg/day.

*Exposure assessment.* To calculate an available dose of perchloroethylene, we need some additional information. In this case, the manufacturer brought a number of cleaned metal pieces into equilibrium within a closed jar, then analyzed the headspace above the pieces by using a high-pressure liquid chromatography to determine the concentration of perchloroethylene released. The concentration of perchloroethylene was undetectable by high-performance liquid chromatography. Since the limits of this analytical method are 2 ppb, this value was taken as the concentration of perchloroethylene in the headspace. Taking the weight of the metal pieces, the number of pieces tested, and the volume of the headspace, it was calculated that the amount of perchloroethylene per single piece was a maximum of 1.0 ng/piece. If we suppose that a patient might be exposed to a maximum of 50 pieces over a lifetime, then the maximum available dose of perchloroethylene from the pieces would be 50 ng.

*Risk characterization.* A comparison of the available dose (50 ng) to the allowable limit (37 mg/day) indicates an ample safety margin.

**Ligature material**. A manufacturer purchases commercial black fishing line to use as a ligature in a circumcision kit. Because the ligature is not "medical grade," a cytotoxicity test is routinely conducted as an incoming inspection test. It was assumed that a negative cytotoxicity test would be associated with an acceptable incidence of skin irritation.

*Hazard identification.* A newly received lot of the fishing line failed the cytotoxicity test. The extraction ratio of this material– of indeterminate surface area – was 0.2 g/ml, with a 0.1 ml aliquot of sample extract being applied to a culture dish. Thus, 0.2 g/ml × 0.1 ml = 0.02 g represents a toxic dose of fishing line.

*Dose-response assessment.* A titration curve was obtained on the sample extract. If the sample was diluted 1:2, the test was still positive; however, if the sample was diluted 1:4, the test was negative. Thus, 0.02 g/4 = 0.005 g of fishing line, the maximum dose that is not cytotoxic. This value was called the allowable limit of fishing line.

*Exposure assessment.* Each circumcision kit contained about 12 inches of line, but only about 4 inches of material was ever in contact with the patient. Since an 8-yard line was determined to weigh 5 g, the available dose of fishing line was calculated to be 5 g/ 288 inches × 4 inches = 0.07 g.

*Risk characterization.* A comparison of the available dose (0.07 g) with the allowable limit (0.005 g) convinced the manufacturer to reject the lot of fishing line.

### *Sources of data*

Data for calculating the allowable limit of exposure to a material can come from many sources, most of them promulgated by industrial and environmental hygienists and related agencies (Hayes, 1994).

TLVs are time-weighed average concentrations of airborne substances. They are designed as guides to protect the health and well-being of workers repeatedly exposed to a substance during their entire working lifetime (7–8 h/day, 40 h/week). TLVs are published annually by the American Conference of Governmental Industrial Hygienists (ACGIH).[4] Biological Exposure Indices (BEIs) are also published annually by ACGIH. These are the maximum acceptable concentrations of a substance at which a worker's health and well-being will not be compromised.

Other published guides include Workplace Environmental Exposure Levels (WEELs), from the American Industrial Hygiene Association;[5] Recommended Exposure Limits (RELs), from the US National Institute for Occupational Safety and Health;[6] and Permissible Exposure Limits (PELs), from the US Occupational Safety and Health Administration.[7] In the US, PELs have the force of law.

Another important limit measurement, Short-Term Exposure Limits (STELs), are defined as the maximum concentration of a substance to which workers can be exposed for a period of up to 15 min continuously, provided that no more than four excursions per day are permitted, and with at least 1 h between exposure periods. The STEL allows for short-term exposures during which workers will not suffer from irritation, chronic or irreversible tissue damage, or narcosis of sufficient degree to increase the likelihood of injury, impair self-rescue, or materially reduce work efficiency. Some substances are given a "ceiling" – an airborne concentration that should not be exceeded even momentarily. Examples of substances having ceilings are certain irritants whose short-term effects are so undesirable that they override consideration of long-term hazards.

### *Uncertainty factors*

An uncertainty factor is a correction that is made to the value used to calculate an allowable limit. It is based on the uncertainty that exists in the applicability of the data to actual exposure conditions. Typically, uncertainty factors range in value from 1 to 10. For example, a correction factor of 10 might be applied for data obtained in animals rather than humans, or to allow for a different route of exposure. In other words, for every property of available data that is different from the actual application, a correction factor of between 1 and 10 is applied. If our first example had been of a small amount of data obtained in animals by a different route of exposure, an uncertainty factor of 1000 might be applied.

### *Safety margins*

A safety margin is the difference or ratio between the allowable limit (after correction by the uncertainty factor) and the available dose. How large does a safety margin need to be? Generally, a safety margin of 100 × or more is desirable, but this can depend on the security of the risk under consideration, the type of product, the business risk to the company, and the potential benefits of product use.

## References

ACGIH. *Documentation of the Threshold Limit Values for Substances in Workroom Air*, 5th ed. Cincinnati, OH: American Conference of Governmental Industrial Hygienists, 1986.

AIHA *Hygienic Guide Series*, Vols. I and II. Akron, OH: American Industrial Hygiene Association, 1980.

Allen A. *Medical Device Industry Fact Book*, 3rd ed. Santa Monica, CA: Canon Communications, Inc., 1996.

Ash M., Ash I. *Cosmetic and Personal Care Additives*. Electronic Handbook. Brookfield, VT: Gower, 1994.

Ash M, Ash I. *Food Additives*. Electronic Handbook. Brookfield, VT: Gower, 1995.

ASTM. *1990 Annual Book of ASTM Standards, Vol. 13.01, Medical Devices*. Philadelphia, PA: ASTM, 1990a.

ASTM. Standardization in Europe: a success story. *ASTM Standardization News*, 1990b;38.

Barnhart ER. *Physician's Desk Reference*. Oradell, NJ: Medical Economics Company, 1991.

Brown RP, Stratmeyer M. Proposed approach for the biological evaluation of medical device materials. In: *Proceedings of the Medical Design and Manufacturing East 97 Conference and Exposition*. Santa Monica, CA: Canon Communications, 1997. pp. 205–18.

Budavari S. *The Merck Index*, 11th ed. Rahway, NJ: Merck and Company, Inc., 1989.

CDRH. *Regulatory Requirements for Medical Devices: A Workshop Manual*. Washington, DC: Center for Device and Radiological Health, HHS Publication FDA 92-4165, August 1992.

CDRH. *Premarket Notification (510(k)) Guidance Document for Contact Lens Car Products*. Washington, DC: Center for Device and Radiological Health, Food and Drug Administration, 1995a.

CDRH. *Testing Guidelines for Class III Soft (Hydrophilic) Contact Lens Solutions*. Washington, DC: Center for Device and Radiological Health, Food and Drug Administration, 1995b.

CDRH. *Draft Guidance for the Content of Premarket Notifications for Menstrual Tampons*. Washington, DC: Center for Device and Radiological Health, Food and Drug Administration, 1995c.

CFR. FDA's policy statement concerning cooperative manufacturing arrangements for licensed biologics. *Fed Reg* 1992;57:55544.

Chengelis CP, Holson JF,Gad SC. *Regulatory Toxicology*. New York: Raven Press, 1995.

Clayton DG, Clayton FE. *Patty's Industrial Hygiene and Toxicology*, 3rd revised rd., Vols. 2A–C. New York: John Wiley & Sons, 1981.

Council Directive 93/42/EEC of 14 June 1993 Concerning Medical Devices, *Offic J Eur Commun* 1993;36:1.

Cronin E. *Contact Dermatitis*. Edinburgh: Churchill Livingston, 1980.

Deichemann W, Gerard H. *Toxicology of Drugs and Chemicals*. New York: Academic Press, 1969.

Ecobichon DJ. *The Basis of Toxicology Testing*. Boca Raton, FL: CRC Press, 1992.

Ellenhorn MJ, Schonwald S, Ordog G, Wasserberger J. *Medical Toxicology*. 2nd ed. New York: Elsevier, 1997.

FAO. *Report of the FAO/WHO Conference on Food Standards, Chemicals in Food and Food Trade (in cooperation with GATT)*, Vol. 1, Rome, March 18–27, 1991.

Final Draft Guidelines on Medical Device Classification, MEDDEV 10/93, Brussels, European Commission, October 1993.

Finkel AJ. *Hamilton and Hardy's Industrial Toxicology*, 4th ed. Boston, MA: John Wright PSG Inc., 1983.

Gad SC. *Safety Evaluation of Medical Devices*. New York: Marcel Dekker, 1997.

Gad SC, Chengelis CP. *Acute Toxicology*. La Jolla, CA: Academic Press, 1998.

Gosselin RE, Smith RP, Hodge HC. *Clinical Toxicology of Commercial Products*, 5th ed. Baltimore, MD: Williams & Wilkins, 1984.

Grant WM. *Toxicology of the* Eye, 4th ed. Springfield, IL: Charles C. Thomas, 1993.

The Gray Sheet. EC 'Medical Devices' Directive slated for adoption in mid-1993, EC Commission official says: CEN estimate development of 92 standards for Directive, M-D-D-1 reports. *The Gray Sheet* October 12, 1992.

The Gray Sheet. European Union Class III device approvals average 240 days or less, HIMA survey says: study release intended to bolster support for FDA reform legislation. *The Gray Sheet*, February 26, 1996a:7–8.

The Gray Sheet. FDA 510(k) average review time for fiscal 1996 projected to be on par with FY 95 figure of 137 days: PMA average review time expected to drop to 250 days, *The Gray Sheet*, March 25, 1996b.

Haindl H. CE marking via self-declaration. *Med DeviceDiagn Industry* 1997:86–90.
Hayes AW.. *Principles and Methods of Toxicology*, 3rd ed. New York: Raven Press, 1994. pp. 26–58.
Hutt PB. A history of government regulation and misbranding of medical devices. *Food Drug Cosm Law J* 1989;44(2):99–117.
ISO (1990). *ISO 9000 International Standards for Quality Management, Vision 2000 – A strategy for International Standards' Implementation in the Quality Arena During the 1990's* (2nd ed., compendium), EEC, Brussels.
ISO. *Biological Evaluation of Medical Devices*. ISO 10993, Parts 1–12. Geneva: International Organization for Standardization, various dates.
ISO. *Risk and Hazard Assessment of Medical Devices*. ISO 14538, ISO, Brussels, 1996.
Kahan JS. *Medical Devices – Obtaining FDA Market Clearance*. Watham, MA: Parexel, 1995.
Kambric HE, Muraboyoshi S, Nose Y. Biomaterials in artificial organs. *Chem Eng News* 1986;(April 14):30–48.
Klassen CD. *Casarett and Doull's Toxicology*. New York: McGraw Hill, 1996.
Lewis RJ. *Carcinogenically Active Chemicals*. New York: Van Nostrand Reinhold, 1991.
Mackison F. *Occupational Health Guidelines for Chemical Hazards*. Washington, DC: Department of Health and Human Services, National Institute for Occupational Health and Safety/Occupational Safety and Health Administration (NIOSH)/Department of Labor (OSHA) DHHS no. 81-123, Government Printing Office, 1981.
Madden MS. *Toxic Torts Handbook*. Boca Raton, FL: Lewis Publishers, 1992.
MAPI. The European community's new approach to regulation of product standards and quality assurance (ISO 9000): what it means for US manufacturers. *MAPI Economic Report ET-218*, January, 1992.
Marlowe DE. *Particles in Medical Devices*. Springfield, VA: US Food and Drug Administration. National Technology Information Service (PB81-131625), US Department of Commerce, 1980.
MHW. *Japanese Guidelines for Basic Biological Tests for Medical Devices and Materials*, notification no. 99. Tokyo: Pharmaceutical Affairs Bureau, Ministry of Health and Welfare, July 27, 1995.
NIOSH. *Registry of Toxic Effects of Chemical Substances*, 11th ed., Vols. 1–3. Washington, DC: National Institute for Occupational Safety and Health, Department of Health and Human Services DHHS No. 83-107, 1983 and RTECS Supplement DHHS 84-101, 1984.
O'Grady J. Interview with Charles M. Ludolph, *ASTM Standardization News*, 26, February, 1990.
Parker, C.M. (1988). Available toxicology information sources and their use, *Product Safety Evaluation Handbook* (S.C. Gad, Ed.). Marcel Dekker, New York, pp. 23–41.
Proctor NH, Hughes JP. *Chemical Hazards of the Workplace*. Philadelphia, PA: J.B. Lippincott, 1978.
Sax NI. *Dangerous Properties of Industrial Materials*, 6th ed. New York: Van Nostrand Reinhold, 1985.
Schardein JL. *Chemically Induced Birth Defects*. New York: Marcel Dekker, 1985.
Shepard TH. *Catalog of Teratogenic Agents*, 9th ed. Baltimore, MD: Johns Hopkins University Press, 1998.
Sidhu KS, Stewart TM, Netton EW. Information sources and support networks in toxicology. *J Am Coll Toxicol* 1989;8:1011–1026.
Sittig M. *Handbook of Toxic and Hazardous Chemicals*, 2nd ed. Park Ridge, NJ: Noyes Publications, 1985.
Stark NJ. Literature in review: biological safety of Parylene C. *Med Plas Biomat* 1996;3(2):30–35.
Stark NJ. Case studies: using the world literature to reduce biocompatability testing. *Proceedings of the Medical Design and Manufacturing East 97 Conference and Exposition*. Santa Monica, CA: Canon Communications, 1997. pp. 205-1–205-7.
Stark NJ. Conducting health-based risk assessment of medical materials. *Med Plastics Biomater* 1998;Sept./Oct.:18—25.
Turco S, Davis NM. Clinical significance of particular matter: a review of the literature. *Hosp Pharm* 1973;8:137–140.

USP. Biological tests-plastics. *The United States Pharmacopoeia*, XXI revision. Rockville, MD: United States Pharmacopoeial Convention, Inc., 1990. pp 1235–1238.

Wexler P, Hakkinen PJ, Kennedy G, Stoss FW. *Information Resources in Toxicology*, 3rd ed. San Diego, CA: Academic Press, 2000.

Wolff ME. *Burger's Medicinal Chemistry*. New York: John Wiley and Sons, 1996.

## Further reading

AAMI. *AAMI Standards: Vol. 4, Biological Evaluation of Medical Devices*. Arlington, VA: AAMI, 1996.

AAMI. *AAMI Standards: Reduced Devices – Risk Management – Part 1: Applications*. AAMI/ISO 14971-1. Arlington, VA: AAMI, 1998.

APHIS. Animal and Plant Health Inspection Service, United States Department of Agriculture, *Fed Reg* 1989;54(168): 36112–36163.

European Committee for Standardization. *CEN Annual Report 1991*. Brussels, 1991.

FDLI. *Compilation of Food and Drug Laws*, Volumes I and II. Washington, DC: Food and Drug Law Institute, 1995.

Food and Drug Administration (FDA). Good laboratory practice regulations: final rule. *Fed Reg* 1987;52(172).

Food and Drug Administration (FDA). *EPA Bluebook Memorandum #G95: Use of International Standard ISO-10993, Biological Evaluation of Medical Devices Part I: Evaluation and Testing*. Washington, DC: Food and Drug Administration, 1995.

Fries RC. *Medical Device Quality Assurance and Regulatory Compliance*. New York: Marcel Dekker, Inc., 1999.

Gad SC. *In Vitro Toxicology*, 2nd ed. Philadelphia, PA: Taylor and Francis, 1999.

Gad SC, Taulbee S. *Handbook of Data Recording, Maintenance and Management for the Biomedical Sciences*. Boca Raton, FL: CRC Press, 1996.

Goering PL, Galloway WD. Toxicology of medical device material. *Fund Appl Toxicol* 1989;13:193–195.

Heller MA. *Guide to Medical Device Regulation*, Volumes 1 and 2. Washington, DC: Thompson Publishing Group, 1999.

ICH. *S2B Guidelines on Genotoxicity: A Standard Battery for Genotoxicity Testing of Pharmaceuticals*. Geneva: International Conference on Harmonization, July 1997.

Kapp R. Available toxicology information sources and their use. In: Gad SC, editor. *Product Safety Evaluation Handbook*, 2nd ed. New York: Marcel Dekker, 1999. pp. 23–41.

Lang LA. A review of latex hypersensitivity. *Toxic Subst Mech* 1996;15:1–11.

McAnultz PA. Biocompatibility testing of medical devices for worldwide marketing. *Eur Pharm Contractor* 1999;8:94–100.

National Toxicology Program. *Annual Plan for Fiscal Year 1985*. Washington, DC: Department of Health and Human Services, NTP-85-0055. Government Printing Office, 1985a.

National Toxicology Program. *Fourth Annual Report on Carcinogens*. Washington, DC: Department of Health and Human Services, 1985b. PB 85-134633.

National Toxicology Program. *Review of Current DHHS, DOE, and EPA Research Related to Toxicology*. Washington, DC: Department of Health and Human Services, NTP-85-056. Government Printing Office, 1990.

National Institute for Occupational Safety and Health (NIOSH). *NIOSH Criteria for a Recommended Standard for Occupational Exposure*. Cincinnati, OH: Department of Health, Education and Welfare.

NIOSH. *Current Intelligence Bulletins*. Cincinnati, OH: National Institute for Occupational Safety and Health, Department of Health, Education and Welfare.

Rechen E, Barth DJ, Marlowe D, Kroger L. FDA use of international standards in the premarket review process. *Biomed Instrum Technol* 1998;Sept./Oct.:518–526.

Reeum AF. *Handbook of Biomaterials Evaluation*. Philadelphia, PA: Taylor and Francis, 1999.

Regulatory Affairs Focus. European update. *Regulat Affairs Focus* 1996;1(4):8.

Spizizen G. The ISO 9000 standards: creating a level playing field for international quality. *Natl Prod Rev* Summer, 1992, pp. 32–38.

Von Recum AF.*Handbook of Biomaterials Evaluation,* 2nd ed. Philadelphia, PA: Taylor and Francis, 1999.

The Wilkerson Group. *Forces Reshaping the Performance and Contribution of the US Medical Device Industry*. Washington, DC: Health Industry Manufacturers Association, 1995.

Chapter 5

# Food additives and nutrition supplements

*Vasilios H. Frankos and Joseph V. Rodricks*

The modern age of food safety began about a century ago, when Harvey Wiley of the Department of Agriculture took up the study of food adulteration. Wiley, who had gained notoriety because of his contribution to the commercialization of sucrose, led a determined effort to seek out and put an end to the use of substances in food that could deceive or harm consumers. Wiley and his so-called poison squad (whose members sometimes used themselves as test subjects to identify harmful substances) were also deeply involved in efforts to enact legislation that would place regulatory authority in the hands of the federal government. The Pure Food and Drugs Act of 1906 that resulted from the efforts of Wiley and other activists of the Progressive Era was the first national law that dealt explicitly with poisonous substances. The law made illegal any food found to be adulterated, which meant that it contained an "added poisonous or...deleterious ingredient which may render [the food] injurious to health".

Under this early law the government had the burden of coming forward with evidence of adulteration. The "may render" phrase turned out to be significant, because it suggested that the government was not required to demonstrate conclusively that people would be harmed from an adulterant, but only that there was a reasonable possibility that harm might occur. This interpretation was affirmed in a 1914 Supreme Court decision, and gradually gave rise to the view that results from studies in experimental animals could be used to identify substances possibly harmful to humans.

The development of the tools of experimental toxicology in the second quarter of this century also gave rise to concerns about harmful properties of chemicals that had not been recognized when the 1906 law was enacted, and the traditional notion of a "poison" was expanded to include a range of adverse effects that could not be easily recognized except through the use of animal studies. These growing concerns arising in toxicology laboratories, combined with the results from the ever-sharpening and more sensitive instruments of the analytical chemists, led in the 1950s to a series of legislative efforts to amend the federal food safety laws. These various amendments, which are discussed in detail in the relevant sections of this chapter, called for investigation of food additive safety and for assurance that such additives would be safe prior to their introduction into the food supply. In particular, the 1958 Food Additives Amendment to the Food, Drug and Cosmetic Act (FDCA) of 1938 placed upon manufacturers the burden of showing that food additives are reasonably free of harm; the federal government, represented by the Food and Drug Administration (FDA), was placed in the position of setting forth the types of toxicity and chemistry studies needed to assess the safety of food additives and of specifying the criteria to be used in judging safety. To meet these demands, FDA toxicologists, assisted by representatives of academic institutions and industry, developed during the 1950s and 1960s the first sets of modern toxicology protocols. These early protocols did much to shape current testing programs in many areas of regulation, and their various offspring are central to most of the topics of this book.

## Composition of the diet and the role of food additives

The human diet is an exceedingly complex mixture of chemicals, most of natural origin. In addition to the dietary constituents that are the sources of human nutrition, the staggering array of plant and animal products consumed as food contain untold numbers of naturally occurring compounds of great structural diversity. Some of these impart desirable aesthetic properties into foods, such as flavor and color, but most are present simply because they play roles in the lives of the plants and animals used as human foods. Based on rough extrapolation from the numbers of natural food constituents that have been identified, it is reasonable to estimate that there are tens and perhaps hundreds of thousands of such substances in the human diet.

The processes of food and beverage preparation – primarily cooking, smoking, and fermentation – bring about many chemical changes and introduce compounds not found in raw products. Human beings manipulate foods in other ways to achieve certain technical effects. Food preservation using various salts is an ancient practice, and the addition of other substances, some natural, some synthetic, to color, emulsify, flavor, preserve, sweeten, and otherwise alter physical characteristics is hardly of recent origin.

Additional chemical substances are introduced, usually in very small amounts, as by-products of agriculture, food processing, and packaging. Among these are crop-use pesticides, drugs used in food-animal production, and substances that migrate from food contact surfaces and packaging. Finally, the diet is not entirely protected from unwanted contaminants of both natural (bacterial and fungal metabolites) and industrial origin. No other environmental medium to which human beings are directly exposed is as chemically complex as the diet (Rodricks, 2001).

The study of diet and health is an immense scientific enterprise; in the past decade especially, enormous advances have been made in understanding how the diet as a whole and even individual dietary constituents both enhance and impair health. This chapter is not concerned with these broad issues of diet and health, but rather will focus on the specific and narrow subject of the safety of food additives. Although many toxicologists are beginning to turn their attention to the larger issues of diet and health, legal and regulatory requirements remain heavily focused on additives, and the toxicologist has a heavy burden to meet in satisfying those requirements. The objectives of this chapter are to summarize these requirements and to provide some guidance on how they are to be met. The concluding section is devoted to some of the emerging issues pertaining to the role of diet in human health that are now beginning to create new demands on toxicologists.

A discussion of the important legal and regulatory distinctions among the various substances that are added to foods and beverages is offered in the next section, followed by a presentation of toxicity testing requirements as set forth in what has come to be called the FDA's "Redbook". Because of its increasing importance, some attention is devoted to the issue of risks to the nonhuman environment. This chapter also contains a review of issues of data evaluation, risk assessment, and safety criteria. Food additive petitions are then discussed. A later section deals with substances Generally Recognized as Safe (GRAS) which, while added to food, are not food additives as these terms are defined in law. The final section deals with a variety of emerging issues.

## Regulatory definitions

### General legal requirements

The FDCA recognizes and distinguishes among three broad categories of food constituents and contaminants (Roberts, 1981):

1 Substances intentionally added to food, both directly and indirectly
2 Substances that are natural components of food
3 Substances that may contaminate food.

The law also imposes substantially different regulatory and technical requirements upon these three classes. In particular, substances that are directly or indirectly added to food (of which there are several subgroups) can be legally introduced only if they have been shown to meet certain safety criteria. The FDA has the principle authority to enforce the FDCA and to ensure that added substances meet those safety criteria. Manufacturers have the burden of showing that the criteria are met. A major component of meeting this burden is the development of appropriate toxicity data.

### Substances intentionally added to food

The FDCA defines a food additive as:

> ...any substance the *intended* use of which results or may reasonably be expected to result, *directly* or *indirectly*, in its becoming a component of or otherwise affecting the characteristics of any food (including any substance intended for use in producing, manufacturing, packing, processing, preparing, treating, packaging, transporting, or holding food; and including any source of radiation intended for any such use...) (FDCA, 21 Code of Federal Regulations, emphases added).

It is clear from this language that intention is a key criterion by which food additive status is to be judged. It is also clear that the intended use of a substance is not limited to its direct addition to food, but includes any use that "results or may reasonably be expected to result" in the introduction of some substance into food that would not otherwise be present. The important distinctions among so-called direct and indirect additives will be further discussed below.

The FDCA creates some further distinctions. In particular, the law goes on to exclude several groups of intentionally added substances from the legal category of food additives. Thus, the FDCA recognizes as distinct:

1 GRAS
2 Color additives

3 Pesticide chemicals in or on raw agricultural products
4 Drugs in food-producing animals

Although these four groups of substances might be considered either direct or indirect "additives" in the popular and even technical use of the term, they are not "food additives" in the legal or regulatory sense.

### *Dietary supplements*

The FDA traditionally considered dietary supplements to be composed only of essential nutrients, such as vitamins, minerals, and proteins. The Nutrition Labeling and Education Act of 1990 added "herbs, or similar nutritional substances", to the term "dietary supplement". Through the Dietary Supplements Health and Education Act of 1994 (DSHEA), Congress expanded the meaning of the term "dietary supplements" beyond essential nutrients to include such substances as ginseng, garlic, fish oils, psyllium, enzymes, glandulars, and mixtures of these.

The DSHEA established a formal definition of "dietary supplement" using several criteria. A dietary supplement:

- Is a product (other than tobacco) that is intended to supplement the diet that bears or contains one or more of the following dietary ingredients: a vitamin, a mineral, an herb or other botanical, an amino acid, a dietary substance for use by man to supplement the diet by increasing the total daily intake, or a concentrate, metabolite, constituent, extract, or combinations of these ingredients.
- Is intended for ingestion in pill, capsule, tablet, or liquid form.
- Is not represented for use as a conventional food or as the sole item of a meal or diet.
- Is labeled as a "dietary supplement".
- Includes products such as an approved new drug, certified antibiotic, or licensed biologic that was marketed as a dietary supplement or food before approval, certification, or license (unless the Secretary of Health and Human Services (HSS) waives this provision).

For decades, the FDA regulated dietary supplements as foods, in most circumstances, to ensure that they were safe and wholesome, and that their labeling was truthful and not misleading. An important facet of ensuring safety was the FDA's evaluation of the safety of all new ingredients, including those used in dietary supplements, under the 1958 Food Additive Amendments to the Federal Food, Drug, and Cosmetic Act (FD&C Act). However, with passage of the DSHEA, Congress amended the FD&C Act to include several provisions that apply only to dietary supplements and dietary ingredients of dietary supplements. As a result of these provisions, dietary ingredients used in dietary supplements are no longer subject to the premarket safety evaluations required of other new food ingredients or for new uses of old food ingredients. They must, however, meet the requirements of other safety provisions.

Signed by President Clinton on October 25, 1994, the DSHEA acknowledges that millions of consumers believe dietary supplements may help to augment daily diets and provide health benefits. Congress's intent in enacting the DSHEA was to meet the concerns of consumers and manufacturers to help ensure that safe and appropriately labeled products remain available to those who want to use them. In the findings associated with the DSHEA, Congress stated that there may be a positive relationship between sound dietary practice and good health, and that, although further scientific research is needed, there may be a connection between dietary supplement use, reduced health-care expenses, and disease prevention.

### GRAS substances

When Congress enacted the 1958 food additive amendments to the FDCA, certain food ingredients that had long been in use were exempted from the premarket testing and approval processes introduced for food additives. An ingredient in use prior to January 1, 1958 could be classified as GRAS based on a demonstration that it had a common use in food. Substances could also be classified as GRAS through "scientific evaluation procedures". The principal criterion for GRAS status is documentation that a substance is "generally recognized, among experts qualified by scientific training and experience to evaluate its safety, as having been adequately shown through scientific procedures (…or experience based on common use in food) to be safe under the conditions of its intended use".

Congress developed the GRAS concept because it did not consider it either wise or practical to subject substances such as baking soda, salt, pepper, vinegar, etc., to the new requirements it had just imposed upon food additives. Additional comment on the evaluation of GRAS substances is provided in a later section of this chapter.

### Color additives

Substances used to impart color to food are subject to a variety of legal requirements (e.g., batch-by-batch certification by the FDA for synthetic colors) not found in the food additive regulations. In general, however, the safety criteria for color additives are identical to those used for food additives. Although color additives are not explicitly considered in this chapter, the testing and evaluation procedures described can be considered applicable to them.

### Pesticides and animal drugs

These "added" substances are dealt with elsewhere in this volume and will not be further discussed here.

### Direct and indirect food additives

At the present time there are thousands of directly and indirectly added food substances regulated as "food additives" in the strict, legal sense; most are indirect food additives. These substances have been the subject of *food additive*petitions (discussed further below) submitted to the FDA since 1958. These food additive petitions contain all information pertaining to safety and were found adequate by the Agency to meet its criteria for approval. The conditions of approval for these food additives, and therefore the allowed uses, are spelled out in specific regulations, codified in Sections 170–199,

Volume 21 of the Code of Federal Regulations. Table 5.1 lists some representative examples of currently approved direct and indirect food additives.

## Legal burdens for proof of safety

Under the law, manufacturers or users ("petitioners") must satisfy the FDA's safety criteria prior to the marketing of a food additive. The FDA's role is to specify the safety criteria and the type and quantity of the data necessary to satisfy these criteria, although petitioners certainly have a major role in decisions regarding the types of data appropriate in specific cases. In addition to information on additive chemistry, manufacturing, and purity, the FDA requires two major types of data. First, information has to be provided on expected intake of the additive by humans exposed to it because of its proposed uses. The second type of information concerns the toxicity of the additive, as uncovered in various experimental studies. As will be shown in detail later in this chapter, the FDA's safety criteria are based on the need to ensure an adequate margin of safety between the expected level of human intake and the exposure levels that create adverse health effects.

The DSHEA Act amends the adulteration and safety provisions of the FD&C Act. Under DSHEA a dietary supplement is adulterated if it or one of its ingredients presents "a significant or unreasonable risk of illness or injury" when used as directed on the label, or under normal conditions of use (if there are no directions). A dietary supplement that contains a new dietary ingredient (i.e., an ingredient not marketed for dietary supplement use in the US prior to October 15, 1994) may be adulterated when there is inadequate information to provide reasonable assurance that the ingredient will not present a significant or unreasonable risk of illness or injury. The Secretary of HHS may also declare that a dietary supplement or dietary ingredient poses an imminent hazard to public health or safety. However, like any other foods, it is a manufacturer's responsibility to ensure that its products are safe and properly labeled prior to marketing. The FDA has, however, the legal burden of showing that a supplement may present a health risk before it can act.

*Table 5.1* Selected examples of direct and indirect food additives regulated by the FDA[a]

| *Major categories* | *Examples* |
|---|---|
| *Direct food additives* | |
| Food preservatives | Butylated hydroxytoluene (BHT), butylated hydrozyanisole (BHA), nitrates, nitrites |
| Anticaking agents | Silicon dioxide, calcium silicate |
| Coating agents, films | Courmarone-indene resin, polyacrylamide |
| Gums, chewing gum bases | Arabinogalactan, carrageenan |
| Special dietary, nutritional agents | Nicotinamide-ascorbic complex, kelp |
| Nonnutritive sweeteners | Aspartame |
| *Indirect food additives* | |
| Adhesives and components of coatings | Acrylate ester copolymer coating |
| Paper and paperboard components | Acrylamide–acrylic acid resins |
| Polymers | *n*-Butyl methacrylate |
| Adjuvants, production acids and sanitizers | Hydrogen peroxide |

[a] Title 21, Code of Federal Regulations, Parts 170–199.

Several distinctions between the FDA's requirements for food additives and those for human drugs should be mentioned. Under the FDCA, for example, drug manufacturers are required to go beyond animal ("preclinical") testing and conduct a variety of clinical trials in human subjects. No such legal requirement exists for food additives, although, as will be seen, some petitioners undertake such studies, particularly for high-intake, direct additives. A second difference between human drugs and food additives is that the FDA considers both the health benefits (efficacy) of a drug and its potential risks when deciding on its approvability. No such balancing is permitted for food additives. Although a food additive must be shown to produce a desired technical effect (an emulsifier must emulsify), its approval depends entirely upon the demonstration of its safety – the FDCA is a "risk only" statute with respect to food additives. A discussion of recent trends regarding health benefit claims for foods appears later in this chapter. A final difference between a food additive and a human drug is that, under its use conditions, the latter is expected to produce biological effects, whereas a food additive should be without any measurable effect.

### *The Delany amendment*

Substances that display carcinogenic properties in animal tests are not permitted to be intentionally added to food in any amount. The Delaney Amendment to the FDCA, incorporated in 1958, puts the matter as follows:

> No additive shall be deemed safe if it is found to induce cancer when ingested by man or animal or, if it is found after tests which are appropriate for the evaluation of the safety of food additives, to induce cancer in man or animal.

Although guidance on the enforcement of this controversial amendment is quite clear with respect to substances directly added to food, there are important circumstances in which ambiguities arise, particularly in connection with indirect food additives and manufacturing by-products. A later section of this chapter concerns such situations and demonstrates how the use of quantitative risk assessment resolves these ambiguities.

## Guide to safety assessment ("the Redbook")

### *Safety evaluation principles*

The FDA has described its currently preferred approach to safety assessment of food additives in a publication entitled *Toxicological Principles for the Safety Assessment of Direct Food Additives and Color Additives Used in Food.* This document is commonly known as "the Redbook" because of the color of its cover when it was originally published in 1982 (USFDA, 1982). This FDA publication sets forth practices that have evolved over the years based on knowledge of toxicological properties associated with certain types of chemical compounds. In 1993, the FDA proposed revisions to the Redbook as part of its attempts to harmonize its toxicity testing guidelines with those published by other agencies, countries and international organizations. The document, known as "Redbook II", is available only as a draft at this time (USFDA, 1993).

The overall approach to safety assessment presented in the Redbook has not changed in the recent revisions and is still organized around four basic premises (USFDA, 1982 , 1993). First, the Agency presumes that some toxicological information is necessary for every food additive. The Agency's second premise is that the amount of safety data required for a particular food additive is dictated by what is called a *level of concern*. The level of concern is based on the magnitude of potential human intake of an additive and its molecular structure.

The fourth premise noted by the FDA is that the initial evaluation of testing requirements by the Agency can be adjusted when toxicological data suggest that a significant or unexpected adverse effect is found to be associated with ingestion of a particular additive.

## Levels of concern

### Direct food additives

The Redbook uses the concept of "concern" as a fundamental part of the safety assessment for direct food additives. This concept of concern is used as a cost-effective means for gathering the necessary safety information and as a dictate for how much testing is necessary in particular cases. On a relative basis, the level of concern is a predicted measure of the extent to which the use of a particular additive may present a potential health hazard. All initial estimates of concern levels are subject to revision as data are collected.

The levels of concern for various anticipated levels of intake for direct food additives, as given in the 1982 Redbook, are presented in Table 5.2. As shown in Table 5.2, a compound is first assigned a level of expected toxicity based on its molecular structure. These levels are designated by category: A (low toxicity), B (moderate toxicity), or C (high toxicity). The structure category assignment procedures described in the Redbook follow a decision tree format. In moving through the decision tree, the petitioner is asked questions concerning the additive's chemical structure, the number and amounts of unidentified components in the food additive, and its predicted metabolites. If fewer than 90 per cent of the components of the additive have been structurally characterized, the additive is automatically placed into the high toxicity category C. Three tables are referred to in the Redbook's decision tree (Tables A, B, and C); they contain examples of structures corresponding to the three categories.

Examples of compounds that are assigned to category A (low toxic potential) include: simple aliphatic and noncyclic hydrocarbons; monocyclic hydrocarbons; fats; fatty acids; simple aliphatic and noncyclic (saturate) monofunctional alcohol; ketones; aldehydes; acids; esters; ethers; and normal human metabolites of carbohydrates and lipids. Examples of category B (moderate toxic potential) compounds include nonconjugated olefins (excluding unsaturated fatty acids and fats); inorganic salts of iron, copper, zinc, tin, amino acids; polypeptides; and proteins. Examples of compounds assigned to category C (toxicological potency is likely to be high) are the most varied – more that 50 are listed in the Redbook as category C compounds, including organic halides; amides and imines; conjugated alkenes; polynuclear aromatic hydrocarbons; and compounds with nitro groups, *N*-nitroso groups, azide groups, and purine groups.

Once a compound has been categorized based on structure, a level of concern is

*Table 5.2* Levels of concern for direct food additives of specified structure category at different concentrations in the diet (intake)

| *Expected toxicity based on structure* | *Anticipated human exposure (ppm in diet)* | *Concern level* | | |
|---|---|---|---|---|
| | | *I (least)* | *II* | *III (most)* |
| A. Low | 1.0 | | | + |
| | 0.05 | | + | |
| | < 0.05 | + | | |
| B. Moderate | 0.5 | | | + |
| | 0.25 | | + | |
| | < 0.025 | + | | |
| C. High | 0.25 | | | + |
| | 0.0125 | | + | |
| | < 0.0125 | + | | |

derived based on the anticipated human intake (see Table 5.2). These levels are designated Concern Levels I, II, and III.

Once a level of concern has been determined for a food additive, the Redbook lists groups of studies that are then required, as a minimum, to support its safety assessment. A compound assigned a Concern Level of I requires only minimal toxicological data and would initially be tested in a short-term feeding study (at least 28 days in duration) in a rodent species and in short-term tests for carcinogenic potential. Assignment to Concern Level II indicates a need for additional toxicity studies to allow observation of most toxic endpoints, other than late developing histopahological changes. A compound in this category is tested in a 90-day feeding study in a rodent species, a 90-day feeding study in a nonrodent species, a multigeneration reproduction study with a developmental toxicity phase in rodent species, and a battery of short-term tests for carcinogenic potential. A compound assigned to Concern Level III is required to undergo the most extensive testing. In addition to the studies required for a Concern Level II substance, it is necessary to conduct carcinogenicity studies in two rodent species, a chronic feeding study of at least 1 year in duration in a rodent species, a multigeneration reproduction study with teratology phase in a rodent species, a nonrodent long-term feeding study, and short term tests for carcinogenic potential.

As noted, test requirements may be modified as data are acquired. Close communication with Agency scientists is needed to avoid unnecessary testing and to ensure that important studies are not overlooked.

## *Indirect food additives*

### THRESHOLD OF REGULATION EXEMPTION

Under 21 CFR §170.39, the FDA states that if it can be demonstrated that a substance used in a food-contact article that may be expected to migrate into food results in a dietary concentration of that substance at or below 0.5 parts per billion), corresponding to dietary exposure levels at or below 1.5 μg/person per day, the FDA will consider that substance to present no health or safety concerns. Consequently, this substance will be exempt from regulation as a food additive because it becomes a component of food at levels that are below the Threshold of Regulation ("TOR"). Known carcinogens are not excempt.

This regulatory option requires that the information on which the TOR exemption claim is based be submitted to the FDA for review. If the FDA concurs with the TOR analysis, the substance will be added to the list of approved TOR exemptions that is maintained by the Agency and is publicly available.

### INDIRECT FOOD ADDITIVE PETITION

Notifications for food contact substances must contain sufficient scientific information to demonstrate that the substance that is the subject of the notification is safe for the intended use (21 U.S.C. 348(h)(1)). Because the safety standard is the same for all food additives, whether subject to the petition process or the Premarket Notification (PMN) process, information in a PMN should be comparable to that recommended for inclusion in a food additive petition.

Section 309 of the Food and Drug Administration Modernization Act of 1997 (FDAMA), amends Section 409 of the Federal Food, Drug, and Cosmetic Act (FFDCA) (21 U.S.C. 348) to establish a PMN procedure as the primary method by which the FDA regulates food additives that are food contact substances. A food contact substance is any substance that is intended for use as a component of materials used in manufacturing, packing, packaging, transporting, or holding food if the use is not intended to have any technical effect in the food.

If a PMN submission cannot be justified it will be necessary to prepare a formal indirect food additive petition for the additive. Regardless of whether a PMN or petition is submitted the following information is required:

- The name and all pertinent information characterizing the additive including, where it is available, its chemical identity and composition. This includes any impurities in the additive that may migrate into the food. Specific information might include (as available) chemical name, common or trade names, Chemical Abstract Service ("CAS") Registry Number, chemical formulas, structures and molecular weights, composition (manufacturing processes, process reagents, chemical equations, impurities), physical and chemical properties/specifications of the additive, and analytical methodologies used to characterize the substance.
- A statement of the condition of the proposed use of the additive including all directions, recommendations, and suggestions proposed for its use.
- All relevant data bearing on the physical or other technical effects the additive is intended to produce, and the quantity of the additive required to produce such effects. Also, information should be furnished that demonstrates the minimum level required to accomplish the intended technical effect.
- A description of the practical methods for determining the quantity of the additive in or on food, and any substance formed in or on food, because of its use. The methods must be specific, precise, accurate, and reliable and be able to be tested by collaborative studies in typical analytical laboratories.
- Migration (extraction) data. Complete requirements, including extraction methodologies, are found in the FDA guidance document entitled, *Recommendations for Chemistry Data for Indirect Food Additive Petitions* (June 1995) and

*Guidance for Industry: Preparation of Premarket Notifications for Food Contact Substances: Chemistry Recommendations* (USFDA, 1999a).

- Full reports of investigations made with respect to the safety of the additive, both published and unpublished.
- Evaluation of the safety of consumption of residues/extractables from the additive including determination of an Acceptable Daily Intake ("ADI") for the additive itself, calculations of the Estimated Daily Intake ("EDI") of the additive in the total diet, and a comparison of the EDI to the ADI.
- Environmental assessment: evaluation of the environmental impact of the production, use, and disposal of the additive and its components.

The EDI is calculated using methods outlined in the FDA's *Recommendations for Chemistry Data for Indirect Food Additive Petitions* and FDA's *Guidance for Industry: Preparation of Premarket Notifications for Food Contact Substances: Chemistry Recommendations* (USFDA 1999a). The EDI is based on a calculation of the amount of additive that could potentially migrate from the food-contact material into various foods, and a subsequent calculation of the amount of those foods that would be consumed by a person each day. Other uses of the additive will be added to the calculated EDI to estimate the Cumulative EDI (CEDI). The FDA uses the CEDI to assign a "level of concern" to the compound that is the subject of the petition. The level of concern determines the extent of toxicological testing needed for approval.

The type and amount of toxicological testing (both animal and *in vitro* studies) necessary for approval is specified in the FDA guidance entitled, *Preparation of Premarket Notifications for Food Contact Substances: Toxicology Recommendations* (USFDA, 1999b). The FDA uses the concept of "concern level" in determining how much testing is necessary. For indirect additives, the level of concern is based solely on anticipated human exposure.

The Agency recommends that the following toxicology studies be performed to assess the safety of a food contact substance (and its constituents if appropriate) with the indicated CEDIs:

1 CEDIs $< 0.5$ ppb ( $< 1.5$ μg/day)

- No toxicity studies are recommended for a food contact substance or constituent with an estimated CEDI less than 0.5 ppb.
- However, available information on the potential carcinogenicity of such a substance should be discussed in a Comprehensive Toxicoogical Profile (CTP) (e.g., carcinogenicity studies, genetic toxicity studies, and structural similarity to known mutagens or carcinogens). For a carcinogenic constituent of a food contact substance, the CTP should contain an estimate of the potential human risk from the constituent due to the proposed use of the food contact substance.

2 CEDIs $> 0.5$ and $< 50$ ppb ( $> 1.5$ to $< 150$ μg/day)

- The potential carcinogenicity of food contact substances and their constituents should be evaluated using genetic toxicity tests. The recommended genetic toxicity tests include: (1) a test for gene mutations in bacteria and (2) an *in vitro* test with cytogenetic evaluation of chromosomal damage using mammalian cells *or* an *in vitro* mouse lymphoma tk$^{\pm}$ assay. The Agency prefers the mouse lymphoma tk$^{\pm}$ assay because this assay measures heritable genetic damage in living cells and is capable of detecting chemicals that induce either gene mutations or chromosomal aberrations, including genetic events associated with carcinogenesis. In performing the mouse lymphoma tk$^{\pm}$ assay, either the soft agar or the microwell method should be used.

- Other available information on the potential carcinogenicity of these substances should be discussed in CTPs (e.g., carcinogenicity studies, genetic toxicity studies, and structural similarity to known mutagens and carcinogens. For a carcinogenic constituent of a food contact substance, the CTP should estimate the potential human risk from the constituent due to the proposed use of the food contact substance.

3 CEDIs $> 50$ ppb and $< 1$ ppm ( $>$ 150 to $<$ 3000 μg/day)

- The potential carcinogenicity of food contact substances and/or their constituents with estimated CEDIs greater than 50 ppb but less than 1 ppm should be evaluated using genetic toxicity tests. The recommended genetic toxicity tests include: (1) a test for gene mutations in bacteria; (2) an *in vitro* test with cytogenetic evaluation of chromosomal damage using mammalian cells *or* an *in vitro* mouse lymphoma tk$^{\pm}$ assay (the mouse lymphoma assay is preferred); and, (3) an *in vivo* test for chromosomal damage using rodent hematopoietic cells. The Agency prefers the mouse lymphoma tk $\pm$ assay because this assay measures heritable genetic damage in living cells and is capable of detecting chemicals that induce either gene mutations or chromosomal aberrations, including genetic events associated with carcinogenesis. In performing the mouse lymphoma tk $\pm$ assay, either the soft agar or the microwell method should be used.
- Other available information on the potential carcinogenicity of these substances should be discussed in CTPs (e.g., carcinogenicity studies, genetic toxicity studies, and structural similarity to known mutagens or carcinogens). For a carcinogenic constituent of a food contact substance, the CTP should estimate the potential human risk from the constituent due to the proposed use of the food contact substance.
- The potential toxicity of a food contact substance and its constituents should be evaluated by two subchronic oral toxicity tests, one in a rodent species and one in a non-rodent species. The studies should provide an adequate basis for determining an ADI for the food contact substance or its constituents in the indicated range of CEDIs. In addition, the results of these studies will help determine whether longer-term or specialized toxicity tests (e.g., metabolism studies, teratogenicity studies, reproductive toxicity studies, neurotoxicity studies, immunotoxicity studies) are needed to assess the safety of these substances.

4 CEDIs $> 1$ ppm ( $>$ 3000 μg/day)

When the estimated CEDI of a food contact substance or a constituent is expected to be greater than 1 ppm, the Agency requires that a food additive petition be submitted for the food contact substance.

ESTIMATING CEDIS

Estimates of indirect additive intake are usually derived by extraction studies with food-simulating solvents as described in the *Recommendations for Chemistry Data for Indirect Food Additives Petitions* (USFDA, 1988) and the FDA's *Guidance for Industry: Preparation of Premarket Notifications for Food Contact Substances: Chemistry Recommendations* (USFDA, 1999a). The description that follows outlines the general approach to estimation of additive intake taken by the FDA, but the Agency may follow different procedures in specific cases when it considers the general approach inapplicable.

The design of the extraction experiments is discussed in detail in the FDA guidelines (USFDA, 1988, 1999) and includes consideration of the type of extraction vessel used, the concentration of the sample used in the extraction study, the thickness and surface area of the sample extracted, the volume of extracting solvent, the conditions of the extraction (the food simulant used), the time and temperature of the extraction, and the characterization of the substances extracted.

When designing the extraction studies, efforts should be made to mimic the use of the indirect additive. Previously, the FDA recommended four food simulants for extraction experiments:

1 Distilled water for nonacid food (pH above 5.0)
2 3 per cent acetic acid for foods (pH 5.0 or below)
3 8 per cent or 50 per cent ethanol for foods containing alcohol
4 Heptane for fatty foods

Several studies have shown, however, that these systems can sometimes underpredict or overpredict actual migration (USFDA, 1988). As a result, the guidelines have been revised and now recommend that 8 per cent ethanol be used to simulate extraction into both aqueous an acidic foods, that 8 per cent or 50 per cent ethanol be used for alcoholic foods, and that food oils (such as corn oil) be used to simulate extraction into fatty foods. The guidelines also suggest that no single solvent will effectively simulate a food oil for all polymers that contact food. Therefore, the Agency lists certain specific polymers and the fatty-food simulants that it considers appropriate:

1 Polyolefins and ethylene-vinyl acetate copolymers require 95 per cent ethanol
2 Rigid polyvinyl chloride requires 50 per cent ethanol
3 Polystyrene and rubber-modified polystyrene require 50 per cent ethanol.

The migration data gathered using the FDA guidelines are intended to provide estimates of the highest level of migration to foods that might result from use of the new additive. Once suitable extraction data have been gathered, the values are used to calculate exposure to the additive, an estimate that depends not only on the extent of migration into food but also on the fraction of a person's diet that is likely to contact materials containing the additive. The FDA uses the term *Consumption Factor* (CF) to describe that portion of the diet likely to contact specific packaging material. The FDA defines the CF as the ratio of the weight of food containing a specific packaging material to the weight of all goods packaged. Examples of CF values used by the Agency for different packaging categories are shown in Table 5.3. The minimum CF used by the Agency is 0.05.

*Table 5.3* Consumption factors (CF)

| *Package category* | *CF* | *Package category* | *CF* |
|---|---|---|---|
| *A. General* | | | |
| Glass | 0.1 | Adhesives | 0.14 |
| Metal – polymer coated | 0.17 | Retort pouch | 0.05 |
| Metal – uncoated | 0.03 | Microwave susceptor | 0.01 |
| Paper – polymer coated | 0.2 | | |
| Paper – uncoated and clay coated | 0.1 | | |
| Polymer | 0.4 | | |
| *B. Polymer* | | | |
| Polyolefins | 0.35 | PVC | 0.1 |
| LDPE | 0.12 | Rigid | 0.05 |
| LLDPE | 0.06 | Semirigid | 0.05 |
| HDPE | 0.13 | Polyester | 0.05 |
| PP | 0.04 | Cellophane | 0.01 |
| Polystyrene | 0.1 | Nylon | 0.02 |
| Impact | 0.04 | Acrylics, phenolics, etc. | 0.15 |
| Non-impact | 0.06 | EVA | 0.02 |
| | | All others[a] | 0.05 |

[a] As discussed in the text, a minimum CF of 0.05 will be used initially for all exposure estimates.

Before the CF values can be used with the data on migration derived from extraction experiments to derive estimates of probable intake, information must be available on the nature of the food that will likely contact the packaging material. To account for the variable nature or the foods contacting each packaging material, food-type distribution factors(s) have been estimated by the Agency for each type of packaging material; they indicate the fraction of the food contacting each material that is aqueous, acidic alcoholic, and fatty. These values assume that any new additive used in a particular packaging material is added to all of the products that use that packaging material (Table 5.4).

The $f_{\mathrm{T}}$ values are then used along with the CF values and the migration data to estimate the expected concentration $[M]$ of the new additive in food that contacts the packaging material, as follows:

$$[M] = (f_{\text{aqueous and acidic}})(\text{migration value for 8\% ethanol}) + (f_{\text{alcohol}}) \times (\text{migration value for 50\% ethanol}) + (f_{\text{fatty}})(\text{migration value for food oil})$$

The EDI for the additive is estimated as the product of concentration $[M]$, the CF, and the total weight of food consumed by an individual per day (3,000 g/person per day):

$$\text{EDI (mg/person per day)} = 3{,}000 \text{ g/person per day} \times [M] \times \text{CF}$$

The EDI is used together with information exposure from other uses of the indirect additive to establish the CEDI used to establish the level of toxicological testing that will be required for approval.

Although reliable data can be gathered on the concentration of a direct food additive in a particular food, it is sometimes difficult to estimate the pattern of use of that food

*Table 5.4* Food-type distribution factors ($f_T$)

| *Package category* | *Food-type distribution ($f_T$)* | | | |
|---|---|---|---|---|
| | *Aqueous*[a] | *Acidic*[a] | *Alcoholic* | *Fatty* |
| *A. General* | | | | |
| Glass | 0.08 | 0.36 | 0.47 | 0.09 |
| Metal – polymer coated | 0.16 | 0.35 | 0.40 | 0.09 |
| Metal – uncoated | 0.54 | 0.25 | 0.01[b] | 0.20 |
| Paper – polymer coated | 0.55 | 0.04 | 0.01[b] | 0.40 |
| Paper – uncoated and clay coated | 0.57 | 0.01[b] | 0.01[b] | 0.41 |
| Polymer | 0.49 | 0.16 | 0.01[b] | 0.34 |
| *B. Polymer* | | | | |
| Polyolefins | 0.67 | 0.01[b] | 0.01[b] | 0.31 |
| Polystyrene | 0.67 | 0.01[b] | 0.01[b] | 0.31 |
| Impact | 0.85 | 0.01[b] | 0.04 | 0.10 |
| Nonimpact | 0.51 | 0.01 | 0.01 | 0.47 |
| Acrylics, phenolics, etc. | 0.17 | 0.40 | 0.31 | 0.12 |
| PVC | 0.01[b] | 0.23 | 0.27 | 0.49 |
| Acrylonitrile, ionomers, PVDC | 0.01[b] | 0.01[b] | 0.01[b] | 0.97 |
| Polycarbonates | 0.97 | 0.01[b] | 0.01[b] | 0.01[b] |
| Polyesters | 0.01[b] | 0.97 | 0.01[b] | 0.01[b] |
| Polyamides (nylons) | 0.10 | 0.10 | 0.05 | 0.75 |
| EVA | 0.30 | 0.28 | 0.28 | 0.14 |
| Wax | 0.47 | 0.01[b] | 0.01[b] | 0.51 |
| Cellophane | 0.05 | 0.01[b] | 0.01[b] | 0.93 |

[a] For 10% ethanol as the food simulant for aqueous and acidic foods, the food-type distribution factors should be summed.

among consumers. There are a number of reasons for the difficulty. Not to be underestimated is the complexity of American dietary patterns that vary over time by region, subculture, age, and even sex. There are three databases currently available that attempt to provide reasonable estimates of food consumption (Pao *et al.*, 1982, Pennington, 1983, USDA, 1997).

The most recent survey (USDA, 1997) uses the results of the 1994–1996 USDA Continuing Survey of Food Intakes by Individuals (CSFII), the latest in an ongoing series of USDA surveys[1]. This represents the most appropriate data to use in assessing potential exposure to food additives by Americans. The CSFII was conducted using a stratified, multistage area probability sample of the non-institutionalized US population. Dietary intake data for individuals of all ages were collected by 24-h recall in interviews conducted on each of two non-consecutive days.

The CSFII dataset includes 2-day food intake data for over 15,000 individuals. Demographic information (e.g., age, sex, race, region, pregnancy status, nursing status) is available for each individual, allowing exposure analysis for a wide variety of subpopulations. Body-weight data for individuals may be used to calculate exposure per unit body weight.

CSFII results include intake data for over 5,000 foods, coded in a hierarchical structure to provide for easy aggregation and disaggregation of data. In addition, commodity-based "recipes" have been developed for each of the over 5,000 CSFII food codes, allowing assessment of potential exposure to additives in specific food ingredients. Intake means or distributions may be calculated per meal or other short time period for acute exposure estimates, or over both survey days for chronic exposure estimates.

The data presented by Pao *et al.* (1982) report average daily intakes and average intakes per meal for commonly eaten foods. The data were gathered through the Nationwide Food Consumption Survey conducted by the US Department of Agriculture (USDA) in the continental US from April 1977 to March 1978, and are an update of information gathered in 1965 (Pao and Burk, 1975). The data presented in the document are based on responses from 38,000 individuals in 16 different sex and age groups. The tables presented in the document for 200 different foods list the number and proportion of individuals who reported using each food, the number of occasions during a 3-day period on which the individual ate the food, and the quantity of food eaten per occasion per day. The data for quantity consumed per day and per eating occasion are presented as averages with a standard deviation and as percentiles (5th, 25th, 50th, 75th, 90th, 95th, and 99th percentiles). The authors point out that the approach of presenting averages for actual users of a food item (as is done in their report) is more appropriate than that of presenting averages for whole population groups (per capita figures), because the latter includes both users and nonusers. One of the weaknesses of the report is that simultaneous ingestion of more than one food item was assumed not to occur; therefore, the data may underestimate the ingestion of an additive that might be present in more than one food.

Another publicly available source of food intake data is a paper by Pennington (1983). This paper is an update of the lists of food commonly eaten and the average intakes of those foods. The FDA uses this survey in conjunction with random food product analytical testing, to determine background exposures to food contaminants. The objective of the Pennington paper was to develop information on representative US diets using approximately 200 different foods for eight age–sex groups. The diets developed were based on data from the USDA's Nationwide Food Consumption Survey and the Second National Health and Nutrition Examination Survey (NHANES II), which was conducted by the National Center for Heal Statistics in 1976–1980. For each age–sex group, the total daily food intake reported was approximately 90 per cent or more of the weight of the foods actually consumed. This is, therefore, a "market basket" approach to estimating daily food intake, and presents average ingestion from an average meal. No data are available for the 90th percentile consumer.

In addition to using these two publicly available sources, a company may elect to collect its own data through a consumer–market survey. This exercise may yield more reliable and current data.

In the Redbook the FDA specifies that the safety assessment for a new food additive is to be "based upon the probable consumption of the additive". Such data can be derived from the data provided by the Pao *et al.* (1982) and Pennington (1983) papers. The FDA generally focuses on 90th percentile consumption data when comparing intake of a food additive to the derived acceptable daily intake for that additive (discussed further below). In some cases where there is exceptionally high intake in a particular subgroup, the FDA may use 95th or even 99th percentile data.

## Toxicology: strategies and protocols

As discussed earlier, the extent and types of toxicological studies required to support the safety of either direct or indirect food additives are dependent on both the EDI and the potential of demonstrated toxicity of the substance under consideration. The types of experimental studies performed also depend on the nature of the toxicity investigated.

Before describing the studies, however, it is important to review some of the proposed changes for toxicology testing that have been discussed in Redbook II. These proposed changes include:

*Short-term genetic toxicity studies*: a modified battery is recommended including *Salmonella typhimurium* reverse mutation assay, *in vitro* mutagenicity in mammalian cells, and *in vivo* cytogenetics.

*Short-term and subchronic feeding studies in rodents and nonrodents*: the proposed guideline adds screening for neurotoxicity and immunotoxicity and recommends that rodents be single-caged. Further, complete histopathology is not recommended for *all* animals on study.

*Carcinogenicity studies with rodents*: proposed changes included recommendations for single-caging of animals, bioassays beginning with at least 50 animals/sex per group, testing to be terminated at 104 weeks, and histopathology to be performed on all animals in the study.

*Reproduction and developmental toxicity studies*: two generations, one litter per generation, are recommended as a minimum reproductive toxicity assessment. If toxicity is identified, the study should then be expanded. Animals should also be screened for neurotoxicity and immunotoxicity. The assessment of male reproductive effects is also expanded.

The comments that follow pertain to both the 1982 Redbook requirements and those proposed in the recent draft revision.

### Acute oral toxicity studies

Studies of acute oral toxicity are used to provide information on the type of toxicity that is associated with high doses of the test compound (e.g., neurotoxicity, cardiotoxicity) and to identify target organs. The results of such tests can be useful to define dose ranges for longer-term toxicity studies but are not usually used for setting doses in studies of greater than 14 days duration. In general, high-dose acute gavage studies are of very limited usefulness in setting doses for long-term dietary studies.

In the past, acute oral toxicity studies and determination of $LD_{50}$ values were commonplace. Recently, the Agency has focused not on the number of animals that die at a given dose, but rather have proposed that the tests examine toxic effects on organ systems and the recovery of the animals from the administration of high doses of the test compound (Kokoski *et al.*, 1990).

### Short-term feeding studies

Short-term studies favored by the FDA are usually from 7 to 28 days in duration, with the animals exposed repeatedly to the chemical in their diets. This type of study is required for Concern Level I compounds and is useful for assessing the toxic characteristics of a new compound; it is often used as a range-finding study for subchronic and chronic studies. Important information on the target organs for future studies can be collected from these studies. Most protocols are designed for 28 days in duration and involve multiple dose groups, with the animals observed daily for overt signs of toxicity. Gross necropsies are performed on all animals, including those that die during the course of the study.

### *Subchronic feeding studies*

Subchronic feeding studies also examine the toxicity of a compound after repeated dosing for periods of 90 days up to as long as 1 year. This type of study is required for Concern Level II compounds and provides more detailed information on the toxicity associated with the test compound, including target organs and potency. By extending the dosing period over several months, any effects attributable to accumulation of the chemical in tissues can be examined. Studies should allow for determination of a "no observable adverse effect" level. These studies do not provide information on the carcinogenic potential of a compound; however, if the food additive of interest has been assigned a Level III concern, the subchronic studies serve to provide dose information for chronic studies. In the case of substances in Concern Levels I and II, data from subchronic tests are often used for the ultimate determination of safety (see earlier discussion) (Kokoski *et al.*, 1990; USFDA, 1988).

Subchronic toxicity studies usually involve at least three treatment groups and one control group. The tests are most often carried out in rodents and dogs. The group sizes are generally 20 rodents/sex per group or four dogs/sex per group. Blood and urine sampling is performed periodically throughout the studies for determination of insidious toxicity and to aid in target organ identification. At sacrifice, detailed gross necropsies and histopathology are performed on all test and control animals. The test are designed to mimic human exposure and may involve administration in the diet through drinking water, in tablets, or by gavage (USFDA, 1982).

### *Reproductive and developmental toxicity studies*

Reproductive and developmental toxicity testing is designed to determine the effects of a substance on a variety of endpoints, including gonadal function, estrous cycles, mating behavior, conception, parturition, lactation, weaning, and growth and development of the offspring. The mechanisms of any effects elicited are rarely apparent from the results of such testing; however, the data do provide information on the effects of the chemical on neonatal morbidity and mortality and on the teratogenic potential of the test substance. The studies are required for compounds of Concern Levels II and III, are usually conducted with rats, and involve dosing of both male and female animals.

Detailed descriptions of the study design of a multigeneration reproduction and developmental toxicity study with a teratogenicity phase are provided in the Redbook. Briefly, the food additive is administered using a method to mimic human ingestion (diet or drinking water). Three test levels and a control group are included for parental animals of both generations (P and $F_0$). The animals are commonly treated before mating, during pregnancy, and through weaning of the $F_1$ offspring. Selected $F_1$

offspring are treated during their growth into adulthood, mating, production of an $F_2$ generation, and until the $F_2$ generation is 21 days old. For each generation, two successive litters ($F_{1a}$, $F_{1b}$ and $F_{2a}$, $F_{2b}$) are examined to provide greater confidence in the reliability of the data. Groups usually include 20 males and 20 females. At least 10 days should separate the weaning of the first litter and mating for the second litter.

In a teratogenicity phase of any multigeneration study, the test substance must be administered during in-utero development. Multiple dose groups are also included as well as a control. The dams are sacrificed 1 day before parturition. The uterus is removed and examined for embryonic or fetal deaths, live fetuses, and any evidence of malformations of skeletal or soft tissues. Ovaries are examined for the number of corpora lutea. Live fetuses are weighed, sexed, and examined for external abnormalities. A selected number of fetuses are examined for soft tissue malformations, usually by random selection of one-third of the group. The remaining two-thirds of the fetuses are examined for skeletal defects.

### *Chronic toxicity/carcinogenicity studies*

Chronic toxicity and carcinogenicity studies are required for a Concern Level III food additive and are often combined into the study. The objective of the study is to determine not only the toxic effects due to prolonged, repeated exposure to the compound, but also to determine its carcinogenic potential. These studies are lifetime studies in rodents and are generally 104 weeks in duration. For food additives that are categorized at Concern Level III, the FDA recommends that such studies be done in two rodent species by the oral route of administration.

The studies are usually designed to include several satellite groups for interim sacrifices at 3 and 6 months and at 1 year to determine the compound-related effects that are not due to aging. The Redbook also recommends periodic observation of the animals for signs, onset and progression of toxic effects, hematological and organ function tests, and clinical examinations for neurological and ocular changes.

It is important to realize that the power of a carcinogenesis bioassay that uses groups of only a few dozen animals is limited for determining carcinogenic activity. Therefore, definitive evidence of carcinogenicity is difficult to establish from the results of a single study. Factors such as histological diagnosis, sensitivity of the bioassay, and variability in background tumor incidence all contribute to the problem. Other correlative information is often used in the evaluation of the "weight of the evidence" of carcinogenic potential of a particular chemical. Such correlative data include results from short-term genotoxicity testing, structure-activity relationships, evidence of dose–response relationships, the number of strains and/or species tested, pharmacokinetic handling or metabolism of the compound, and the degree of the tumor response (incidence, site, type). Because the Delaney Amendment prohibits the use of carcinogenic food additives, the interpretation of carcinogenicity test results has an exceedingly important impact on the safety assessment process. It is this situation that leads the FDA to use a "weight-of-the-evidence" approach to interpret carcinogenicity bioassays and to assess the safety of food additives.

### *Human data (clinical studies)*

Under the FDCA, clinical data are required to gain approval of a drug for human use. There is no such requirement for food additives. Instead, the safety assessment process for food additives can rest solely on the results from experimental animal studies. In cases where human data are available, however, the data may be incorporated into the safety assessment of the food additive.

In cases where human intake is expected to be relatively large, some manufacturers choose to conduct some form of human investigation after a thorough set of preclinical studies has been completed and evaluated. There are, however, cases where clinical studies for food additives may be required, e.g., macroingredients such as noncaloric fat substitutes. There are several issues relating to the safety assessment of fat substitutes that are best resolved not through animal testing but through human studies (Munro, 1990). First, it may be asked whether noncaloric fat substitutes reduce total energy intake from the diet when consumed chronically, Second, does long-term consumption of such products lead to reduced micronutrient or essential fatty acid status? These types of questions may not be adequately addressed in animal models because high intake of macroingredients in rodents has been shown to induce alterations in normal physiology, leading to spurious toxicological effects of no consequence to humans (Munro, 1990). Therefore, clinical studies are more useful for answering such questions.

Attention is currently focused on a class of food and food additive that may be regulated in the future, and for which the FDA will impose a requirement for human studies to support any claims made. These substances have been called nutraceuticals. In this case, the food additive is put into the food product to produce some health benefit. The Nutrition Labeling and Education Act of 1990 proposed four claims that the FDA has determined can be supported by scientific data. These include, increased fat intake is associated with increased risk of heart disease; increase fat intake is associated with increased risk of cancer; increased sodium intake is associated with hypertension; and decreased calcium intake is associated with osteoporosis. If any other relationships between a food product or constituent and a health effect were to be considered by the FDA, clinical studies would most likely be required in the petition because clinical studies would be most appropriate to answer questions about the validity of the claim.

### Environmental effects of additives

Because of the requirements of the National Environmental Policy Act (NEPA), the FDA is required to assess the environmental implications of its regulatory decisions (21 CFR Part 25, April 26, 1985). Petitioners are therefore required to prepare and environmental assessment before the FDA will approve a food additive petition. The issues addressed in testing programs designed to comply with the FDA's environmental assessment requirement include the intended use; physical–chemical properties; degree of metabolism following use; environmental fate in air, water, and soil; predicted environmental concentrations; potential toxicological effects on aquatic and terrestrial species; and environmental implications of manufacturing and ultimate disposal by the consumer.

The FDA has no standardized requirements for environmental studies. Needed studies are instead determined by evaluating the potential environmental exposure and toxicity information available for the additive. One of the first considerations is the route of its introduction into the environment. Introduction into the environment can occur anytime during the life cycle of the food additive during its manufacture, use, or disposal. Some ingested food additives may, for example, enter the environment through sewage. Chemicals used to produce food additives also may be added to waste water streams at manufacturing or processing plants. Other routes of introduction for food additives include solid waste disposal in landfills, composing of foods, and incineration of solid wastes.

Once the routes of introduction into the environment have been identified, data on the levels of introduction, rates of incorporation into soil, and environmental fate are collected to predict the final concentration of the additive in the relevant environmental media. The level of introduction of substance into the environment is derived by combining information on the rate of consumption of that substance, the amount existing in food, and the metabolism of the substance (Harrass *et al.*, 1990). The EDI value can be used in conjunction with waste water engineering data to estimate the amount of material received by a typical waste water treatment facility (Harrass *et al.*, 1990).

When possible, processes that affect the transport and transformation of food additives are used when estimating the environmental concentration. If, for example, an ingested food additive is resistant to biodegradation and is strongly adsorbed to sewage sludge, most of an ingested dose of that additive would be present in sewage sludge that is then added to soil. Useful data for this assessment include chemical stability (hydrolysis, photolysis), biodegradability, and mobility in waste media (water solubility, soil sorption, volatility).

Once the amount of substance released into the environment has been estimated, the environmental assessment involves examination of available data on toxicity to animals, plants, and other organisms at the ecosystem level in each environmental compartment (air, freshwater, estuarine, marine, and terrestrial ecosystems). The toxicity data base is then compared with the level of environmental exposure to arrive at an assessment of risk.

### *Human risk assessment*

The toxicology data enable the assigning of Acceptable Daily Intake (ADI) levels for human consumption of food additives. The procedure is generally as follows:

1 The study revealing the most sensitive indicator of toxicity is identified.
2 The highest No-Observable-Effect Level (NOEL) from the study is identified.
3 The NOEL is divided by a safety factor, the magnitude of which is a function of several considerations, to be discussed below.
4 The result of these determinations is the ADI. It is assumed that individuals can experience a daily intake of an additive at levels up to its ADI for their full lifetimes without significant risk.

This approach, introduced by the FDA in the early 1950s and widely used through out the world since, is based on the principle that the toxic effects of any substance (with the possible exception of carcinogens, see below) can be avoided if intakes do not exceed the threshold of toxicity (Frankos, 1985; Rodricks, 2001). The experimental NOEL represents the threshold of effect applicable to experimental animals. The human population is likely to contain individuals more sensitive than laboratory animals; moreover, the variability in toxic response among members of the human population may be substantial. It has thus become common practice to use safety factors to compensate for the uncertainties associated with these possibilities.

Several aspects of the NOEL/safety factor approach should be noted:

1 If the NOEL derives from a chronic toxicity study, the typical safety factor is 100 (10 for each of the two major sources of variability).
2 If the NOEL derives from a subchronic toxicity study, and a chronic ADI is desired, an additional factor of 10 is introduced.
3 If the NOEL derives from a developmental/reproductive toxicity study revealing a Type I effect, a factor of 1,000 may be used.
4 The magnitudes of the 'standard' safety factors can be altered in specific situations if data are available to suggest human sensitivities or variabilities are larger or smaller than is suggested by the factors of 10. (Data from human clinical studies, particularly concerning metabolic profiles, may provide the basis for such determinations. They may also yield information concerning the most appropriate animal species, which may not be the most sensitive.)
5 Although probably highly protective, the ADI is not considered to be a sharp dividing line between safe and unsafe intakes. Its use cannot guarantee that every individual in a large population is absolutely not at risk, nor can it be assumed that exposures greater than the ADI, especially if they occur only infrequently, create a significant risk.
6 The ADI approach is not used for carcinogens (see below).

## *EDI*

As described in a previous section, petitioners are required to supply information that can be used to provide a reliable estimate for the daily human intake of the additive, the EDI. As noted earlier, such information is used to guide decisions about the extent of toxicity evaluation needed to support a safety determination. More refined and reliable estimates may be needed when the final safety decision is being confronted.

The FDA typically seeks to ensure that the EDI for the 90th percentile consumer of foods or beverages in which the additive will be present falls below the ADI. Thus, for each dietary item that may contain the additive, data on the additive's maximum concentration and on human consumption rates for the food item, including that for the 90th percentile consumer, must be presented. If the EDI does not exceed the ADI, the additive is approvable. Petitioners frequently initially seek very limited uses of new additives, so that the EDI is only a small fraction of the ADI. When experience with such limited use of the additive has accumulated and reveals no significant difficulties, the petitioner may seek additional uses such that the EDI will increase.

Note that regulations that issue following the approval of a food additive petition will specify the maximum allowable level of the additive – the tolerance – in each item of the diet for which approval has been granted. Tolerances are based on the ADI because human consumption of all items of the diet that can contain the additive, at the 90th percentile consumption rate, will yield an EDI that is less than the ADI when the additive is present at or below the tolerance.

The use of the 90th percentile consumption rate is discretionary on the part of the Agency, and the FDA may use other (higher) consumption rates where necessary to ensure adequate health protection.

### *Carcinogens and risk assessment*

As noted previously, the Delaney Amendment is unambiguous with respect to the intentional and direct addition of carcinogens to food: it prohibits the establishment of tolerances for such substances. Moreover, it appears that the FDCA allows the FDA very little flexibility with respect to the interpretation of animal carcinogenicity data in regard to their relevance to humans. Certainly the Agency can decide whether experimental design and conduct are appropriate and whether the additive causes an increased incidence of tumors in the laboratory animals; however once these decisions are made and it is concluded that the additive is an animal carcinogen, the Agency apparently has no freedom to inquire further into the relevance of the test results to humans. Many scientists question the wisdom of the Delaney requirement, and some legal experts have argued that the FDA does have the authority to make such inquires and to act upon them, but legal challenges to a strict interpretation of the Delaney Amendment have not succeeded.

There are circumstances in which the question of the applicability of the Delaney Amendment is without an unambiguous answer. Consider the case of indirect food additives. Such substances are known to be present in food, and thereby become food additives, only if they can be detected by means of chemical analysis. They are not directly added to food, but are present in packaging and other food-contact materials and may migrate into food from those materials. Whether and to what extent such migration occurs is critical to a determination of whether a component of a food-contacting substance is a food additive and, if such a component is carcinogenic, whether it is subject to the Delaney Amendment requirements. The prototypical case concerns residual monomers, such as vinyl chloride or acrylonitrile, that may be found at low levels in polymers used in contact with foods.

The FDA introduced risk assessment as a regulatory tool to deal with such agents. Migrants from food-contacting substances become food additives only if they can be detected in food. The FDA does not specify the detection limits or the analytical methods to be used for each compound. Instead, the Agency is satisfied if the petitioner uses methods capable of detecting residues at concentrations sufficient to create daily intakes corresponding to lifetime risks no greater than $10^{-6}$. Therefore, if a method of sufficient detection power shows that a carcinogen has not migrated from packaging or other food-contact materials, there is no need to apply the Delaney restriction because no additive is introduced (Rodricks and Taylor, 1989). The FDA has applied this same approach to deal with carcinogenic manufacturing by-products that are present as impurities in food additives. If the additive is not carcinogenic when tested, trace amounts of carcinogenic impurities are permitted if their lifetime cancer risks do not exceed the same one-in-one-million criterion.

## The food additive petition

Once safety data have been gathered for a potential new food additive, a food additive petition is prepared according to guidelines set out in Section 409 (b)(2) of the FDCA.

Five general areas of information must be presented in a petition.

1 The identity of the additive
2 The proposed use of the additive
3 The intended technical effect of the additive
4 A method of analysis for the additive in food
5 Full reports of all safety investigations that have been performed to support its use

In addition, a petitioner may be asked to submit a full description of methods, facilities, and controls used in the production of the additive, along with samples of the additive and of food in which the additive will be used. The following is a brief description of each section that is included in a food additive petition for a direct food additive. In the case of indirect additives, additional information on extraction and migration of the substance into foods will be required, as discussed earlier.

### *Identity of the additive*

Information that identifies the food additive needs to be highly specific and complete. There are a number of items that should be included in this section of the petition.

1 The common or usual name of the additive.
2 The formal names, such as those used by the Chemical Abstracts Services (CAS), together with the CAS number.
3 The chemical formula(s), structures, and molecular weights for each compound or simple mixture.
4 Information on the composition of mixtures.
5 Information on purity. Food grade specification should be proposed in the format of Food Chemicals Codex. The specification should include a description of the substance, and assay test, and limits of impurities.
6 Results of analyses for several typical lots or batches (usually five or more) of the additive. The purpose is to demonstrate specification conformance.
7 Methods of the analysis.
8 A description of the manufacturing process. This description should be detailed, and include lists of all substances used, a description of the reaction conditions, and the controls employed to ensure purity and uniformity of batches.
9 Chemical and physical properties of the additive. Data demonstrating degree of stability are also required.
10 Any other information which identifies or describes the additive or its preparation.

### *Use of the additive*

Information included in this section of the food additive petition should describe the amount of the substance added to food, the type of food, and the amount of the substance or its by-products consumed as food. This information should then be used to furnish the Agency with an estimate of human intake on a chronic daily basis (the EDI). The calculation of the EDI must be justified in a discussion and sources of information referenced or provided with the petition. If the petition is for a new use of a food additive covered under existing regulations, the anticipated increase in consumption due to the new use should be estimated and compared with the existing intake value. The EDI calculation should be performed for both the typical or average user and the high-end consumer. The high-end consumer is usually reflected by a 90th percentile consumer in an "eaters only" population. Also to be considered in the EDI calculation is any naturally occurring source of the food additive.

### Intended technical effect of the additive

In this section of the petition the technical effect intended by addition of the substance to food must be identified (e.g., flavoring, preservation). The FDA has a list of accepted effects; other effects can be defined in the petition. It is important to note that the food additive can be added only in an amount that is necessary to accomplish the intended effect. If there in no intended effect demonstrated, the substance is not allowed to be added to food.

Information should be included that demonstrates the minimum level required to accomplish the intended effect. This information is usually in the form of experimental data. If the petition covers an additive with a new technical effect, this information is extremely important.

### Analytical methodology

This section of the petition describes the methods that can be used for determining the quantity of an additive in food, as well as any substance formed in food because of its use. This information is required for all food additives.

Food additives are generally subject to a regulatory tolerance, the limitation on the amount of a substance that a food can contain. To determine the amount of a substance present in food, some practical analytical method must be presented in the petition. The method must be specific for the additive, precise, accurate, and reliable. Since it may need to be reproduced under a variety of laboratory conditions, the method must not be unduly complicated.

The suggested format for this section of the petition includes a brief summary description of the method followed by detailed procedures. The procedures should describe all aspects of the protocol, from selection and collection of samples through sample preparation and analytical determination. Examples of analytical results should be presented from samples of food containing known amounts of the additive to validate the recovery, repeatability, and specificity of the method. The results should be followed by a discussion of the method and any conclusion that can be reached. As estimate of the variability of the method should be included.

The analytical method may be extremely important depending on the nature of the substance being considered in the petition; if carcinogenic monomers migrate into food from polymers, the analytical results become critical to the success of the petition. Familiarity with and acceptance of an analytical method by organizations such as the Association of Official Analytical Chemists can be helpful.

### *Safety of the additive*

Full laboratory reports of all safety studies need to be submitted in this section of the petition. Conformance with Good Laboratory Practice regulations is essential.

The section described above is essential for any food additive petition. In addition to these requirements, a petitioner must submit information on the environmental impact of the substance. Depending on the specific intended effect of the food additive, other information such as nutritional attributes or special labeling needs will be requested by Agency.

## Gras reviews

When the food additive amendments to the FDCA were enacted in 1958, certain food ingredients with a long history of use were given exemptions from the premarket evaluation and approval that is required for food additives. Such compounds were labeled as GRAS under the conditions of their intended use. Any food ingredient can be classified as GRAS if it is generally recognized among scientific experts, qualified by scientific training and experience, to be safe under the conditions of its intended use.

On April 17, 1997, the FDA released its proposed rule to replace the current GRAS petition process for use of substances in food and animal feed with a notification procedure. The GRAS self-determination process remains unchanged.

Included in the proposed rule is a clarification of the requirements for determining GRAS status for a food ingredient. The key elements of a GRAS review are: (1) technical evidence of safety and (2) a basis to conclude that this evidence is generally known and accepted. Technical evidence utilizes one or two approaches: (1) scientific procedures or (2) common use in food prior to January 1, 1958.

The new GRAS notification process specifies both the format and scientific content of the submission to the FDA. Notification is not mandatory but available if the sponsor wishes to inform the FDA of its GRAS determination. The FDA's goal is, in part, to facilitate review and increase efficiency of resource allocation in the food area from GRAS reviews to more complex safety issues, such as those posed by novel food ingredients. Implied in this goal is the need for highly organized well-written and scientifically defensible notices that provide compelling arguments for the basis of the GRAS determination. The FDA indicated in the notice that although this notice was a proposal the FDA was no longer going to accept GRAS petitions. Effectively this means that all future GRAS reviews are going to rely on "self determinations" of GRAS status and the FDA may-or-may-not be informed of such a determination.

### *How many GRAS compounds are there?*

The FDA officially lists the compounds they consider GRAS in 21 CFR Parts 182, 184, and 186. A total of over 680 substances are listed broken down as shown in Table 5.5. Added to the 680 + compounds actually approved by the FDA are at least another 1,000 compounds that have been independently affirmed as GRAS by the Flavor and Extract Manufacturers Association (FEMA). The FDA has accepted the FEMA GRAS review process as consistent with the criteria set forth in the Food, Drug, and Cosmetic Act (Oser and Hall, 1977) and has essentially adopted the FEMA list as a de facto official FDA approved list of GRAS flavoring agents.

In contrast (Table 5.6), a relatively small number of compounds are approved for direct addition to food. There are about 280 direct food additives and 55 color additives that are approved by the FDA in 21 CFR Parts 73–75, 172 and 173. An additional 880 + synthetic and natural flavorings are approved under 21 CFR Parts 172.510 and 172.515. The largest numbers of compounds approved by the FDA are indirect additives that are used to make paper and plastic packaging. Exposure to these compounds occurs through migration out of the packaging and is therefore of an indirect nature. These number in the thousands and are listed in 21 CFR Parts 174–178.

The FDA has published a list of GRAS substances that include more that 600 ingredients. This list does not include every substance that could be recognized as GRAS. The FDA realized that is was impractical to list all substances that could be considered GRAS (Roberts, 1981).

In 1969, a presidential directive was issued requiring the FDA to review GRAS substances. The FDA used two major groups to compile and evaluate the large amount of relevant data. Information on usage rates and daily intakes were collected by the National Research Council. To evaluate the scientific literature pertaining to the safety of GRAS substances, the Federation of American Societies of Experimental Biology assembled an expert committee know as the Select Committee on GRAS Substances (SCOGS). Most of the originally listed substances were reaffirmed as GRAS by SCOGS. However, 6 per cent of those substances were conditionally approved, with additional research recommended (examples were carrageenan, oil of nutmeg, glutamates, caffeine), while 8 per cent were judged to have insufficient data available to affirm safety. For the substances not reaffirmed as GRAS, the FDA requested additional data from industry and, if it did not appear, rescinded the GRAS status of the listed substance (SCOGS, 1981).

When reviewing the GRAS list today, it is evident that most chemicals directly added to food are GRAS. A selected group of these substances is listed in Table 5.7. GRAS substances represent a highly diverse group of chemical structures with diverse toxicological characteristics. Most are present because they have a long history of use without significant reports of adverse health effects. The FDA cannot remove an agent from the GRAS list unless evidence appears showing that the substance is no longer safe for its intended use. The levels of addition of these substances to food are specified for only some listed substances. For most, usage levels are defined according to the amounts consistent with "good manufacturing practice". If there is a limit on use specified, however, new uses of the substance that result in an increased intake have to be justified based on scientific procedures recognized by experts.

It is difficult to generalize about the criteria used to affirm or judge GRAS status. Expert judgement is an important part of the process. However, the quantity and quality of the available toxicology data for each of the substances on the list vary greatly, and the decisions made by the experts cannot readily be shown to rest on a particular set of criteria. Therefore, it is not possible to set out a list of toxicological results that must be available before a substance can be listed as GRAS. It is clear, however, that whenever new uses of a GRAS substance are contemplated, the scientific experts conducting the review will usually consider whether new toxicology studies are to be conducted to ensure continued GRAS status. Again, manufacturers may self-affirm GRAS status or seek affirmation, through a formal petition process, from the FDA.

*Table 5.5* Number of GRAS compounds listed by categories in 21 CFR Parts 182, 184, and 186

| *Category* | *CFR citation* | *Approximate no. of substances* |
|---|---|---|
| Spices and other natural seasoning and flavoring | Part 182.10 | 83 |
| Essential oils, oleoresins, and natural extractives | Part 182.20 | 160 |
| Natural extractives (solvent free) in conjunction with spices | Part 182.40 | 5 |
| Certain other spices, seasoning, essential oils, spicesoleoresins, and natural extractives | Part 182.50 | 5 |
| Synthetic flavoring substances and adjuvants | Part 182.60 | 21 |
| Substances migrating from cotton and color fabrics used in dry food packaging | Part 182.70 | 23 |
| Substances migrating to food from paper and paperboard | Part 182.90 | 31 |
| Multiple purpose GRAS food substances | Part 182 Subpart B | 28 |
| Anti-caking agents | Part 182 Subpart C | 6 |
| Chemical preservatives | Part 182 Subpart D | 20 |
| Emulsifying agents | Part 182 Subpart E | Reserved |
| Dietary supplements | Part 182 Subpart F | 56 |
| Sequestrants | Part 182 Subpart G | 20 |
| Stabilizers | Part 182 Subpart H | 1 |
| Nutrients | Part 182 Subpart I | 16 |
| Direct food substances affirmed as generally recognized as safe | Part 184 | 188 |
| Indirect food substances affirmed as generally recognized as safe | Part 186 | 16 |

## Emerging issues

Recent advances in science and technology as well as the globalization of the marketplace have brought new challenges and problems to both the food industry and the FDA. A discussion of the regulations and issues associated with food additives would not be complete without exploring some of the emerging issues in food science. A recent FDA document entitled *Emerging Issues in Food Safety and Quality for the Next Decade* was published in February, 1991, and contains an examination of a number of areas that will be addressed by the Agency in the near future. In addition, the FDA has proposed changes to the Redbook. Issues of particular interest to scientists and regulators include: (a) products of biotechnology; (b) food ingredients with nutritional and "health-enhancing" properties; (c) new packaging and processing technologies; and (d) new toxicology requirements. Significant international developments are also surveyed.

### *Products of biotechnology*

Biotechnology is defined as the process of obtaining useful products from biological sources. It is a large field encompassing some traditional technologies such as fermenta-

*Table 5.6* Number of GRAS and food additives listed in 21 CFR

| *Category* | *CFR Citation* | *Approximate no. of substances* |
|---|---|---|
| Direct additives – excluding flavoring | 21 CFR Parts 172 and 173 | 288 |
| Color additives and diluents | 21 CFR Parts 73, 74, and 81 | 55 |
| Synthetic flavoring substances | 21 CFR Section 172.515 | 756 |
| Natural flavoring substances | 21 CFR Sections 172.510 | 131 |
| GRAS | 21 CFR Parts 182, 184 and 186 | 680 |
| Indirect food additives | 21 CFR Parts 174–178 | Thousands |

tion and antibody production. It also includes the newer technologies such as genetic engineering, protein engineering, and DNA-based diagnostics. It is these newer technologies that have created challenges for the FDA in the consideration of product approvals, and it is the safety of the products of genetic engineering that has provided most of the challenging questions. Interest in application of this new technology to foods is great because it may provide the world with more nutritious and abundant food that is less expensive than foods derived from traditional sources.

There are a number of different ways the techniques of biotechnology can be or have been applied in the food industry. One of the first was gene transfer into microorganisms for production of enzymes useful in food processing. Some of these products have already been approved and are currently in use (USFDA, 1991). Probably the most exciting advance in food science has been the production of food crops from transgenic plants, plants that have had gene transfer to transmit a desirable trait such as toleration of herbicides, resistance to viruses, or resistance to insect damage (by producing natural insecticidal substances). Such traits have already been engineered into food crops such as tomatoes and potatoes, and some such crops have recently been allowed into commerce (USFDA, 1994).

In evaluating the safety of such food products, there are a number of issues that arise:

1 The need for strict product specifications
2 Assurance of microbiological purity
3 Exposure of consumers to foreign DNA
4 Exposure to the population as consumption of these new products increases over time (Munro, 1989 as cited in USFDA, 1991)

*Table 5.7* GRAS substances listed by FDA

| *Spices and natural flavors* | *Multipurpose substances* | *Affirmed as GRAS by FDA regulations* |
|---|---|---|
| Anise | Acetic acid | Benzoic acid |
| Basil | Caffeine | Clove/its derivatives |
| Capsicum | Calcium carbonate | Ethyl alcohol |
| Elder flowers | Carmel | Garlic/its derivatives |
| Geranium | Carbon dioxide | Guar gum |
| Ginger | Glycerin | Potassium iodide |
| Licorice | Hydrogen peroxide | Sorbitol |
| Parsley | Lecithin | Dextrans (indirect additive) |
| Spearmint | Methylcellulose | |
| Vanilla | Sodium aluminum phosphate | |

The regulatory community has not yet determined exactly how it will deal with these new food products. In 1986, the US Office of Science and Technology, working through the Biotechnology Science Coordinating Committee (BSCC), published a comprehensive statement outlining the formation of the Coordinated Framework for Regulation of Biotechnology (FR 51:23302–23350). This was an announcement of policy and notice for public comment. According to this notice, the use of new manufacturing processes should not affect the legal status of previously approved foods or food additives unless the use of biotechnology changes their identities or creates adulterating impurities. The Coordinated Framework also states that the existing FDA statutes are appropriate for regulating products of biotechnology, while recognizing that additional regulation or different types of review may be necessary because these products may present unique risks. Since the publication of this statement in 1986, DDA has maintained that existing regulations and statutes are adequate to evaluate the safety of food additives produced by biotechnology and that new products can be reviewed on a case-by-case basis (USFDA, 1991).

In 1990, a second document emerged as a guide to regulatory policy. It was produced by a consortium of industries and other concerned citizens, and was entitled *Biotechnologies and Food: Assuring the Safety of Foods Produced by Genetic Modification* (IFBC, 1990). This comprehensive document was based on existing food laws and regulatory practice. It suggests a flexible, tiered-approach system that is guided by decision trees. Three different decision trees are proposed corresponding to three product categories: (a) microorganisms and their food products; (b) single chemical substances and simple mixtures; and (c) complex mixtures and whole foods. The decision tree structure is based on a series of questions concerning the genetic origin of the parent organism, the history of use of the relevant plasmids, and the composition and safety of the product. The answers to the questions for each branch of the decision tree result in three possible pathways: (a) readiness to submit for regulatory approval; (b) rejection of the material for regulatory submission; or (c) an indication to remand the material for further investigation.

Based on these published documents, both the legislative and executive branches of the US government have decided that new legislation for the products of biotechnology is unnecessary. The FDA seems to be moving toward regulation of the new products under existing statutes with special consideration to the unique nature of these products.

The Calgene FLAVR SAVR TM tomato was given regulatory sanction on May 17, 1994. An opinion letter was issued by the FDA to Calgene concerning the FLAVR SAVR TM tomato under 21 CFR 10.85:

> This is in response to your request, dated August 12, 1991, for consultation with the FDA concerning FLAVR SAVR TM tomatoes. You requested that the FDA issue an advisory opinion under 21 CFR 10.85 concerning whether FLAVR SAVR TM tomatoes are food and therefore, subject to the same regulation as other tomato varieties. This request is separate from your request for the evaluation of safety of the aminoglycoside-3 phosphotransferase II [APH(3′)II] protein used for selection of plant cells that incorporated the new genetic trait. Because the question concerning APH(3′)II is addressed in a separate rulemaking, this letter addresses only aspects of FLAVR SAVR TM tomatoes other then APH(3′)II.In the Federal Register of May 29, 1992 (57 FR 22984), the FDA issued a "Statement of Policy:

Foods Derived from New Plant Varieties", in which the FDA advised that requests for consultation with the Agency should be made consistent with the principles outlined in the policy. Therefore, we are treating your request as a consultation in accordance with the May 1992 policy statement.As noted in that statement (57 FR 22984–22990), the FDA has rarely had the occasion to review the regulatory status of foods derived from new plant varieties because these foods have been widely accepted as safe. The FDA regulations in 21 CFR 170.30(f) do, however, provide for review of the regulatory status of certain substances of natural biological origin in certain circumstances. Specifically, 21 CFR 170.30(f)(2) provides for the review of the regulatory status of any substance of natural biological origin with a history of safe use that has had "significant alteration of composition by breeding or selection." Based on the information that Calgene has submitted concerning the FLAVR SAVR TM tomato, we believe that this new variety has not been significantly altered with in the meaning of 21 CFR 170.30(f)(2), when compared to varieties of tomatoes with a history of safe use.

## *Food ingredients with nutritional and "health-enhancing" properties*

As discussed earlier in conjunction with human study requirements for establishing the safety of a food additive, a new category of food additive is emerging, the macronutrient substitutes. These are typically low-fat, low-calorie, low-cholesterol, and high-fiber substitutes for normal fats and carbohydrates, and are likely to be consumed in increasingly larger quantities by more health-conscious consumers. The potential magnitude of human intake of these substances is large; a fact that will present a major challenge to toxicologists who will be required to develop relevant study designs.

Most common is the development of substitutes for conventional macronutrients such as fat and sugars. An example is fluffy cellulose, a no-calorie fiber made from sources such as straw, citrus pulp, or sugar beets that can be used to dilute wheat flour without affecting taste of baked goods (USFDA, 1991).

Interesting new fat substitutes are also emerging. A recently approved product, Simplesse, is prepared by microparticulation of egg and milk proteins and is in use in frozen desserts. Another product currently under evaluation by the FDA is a nonabsorbable sucrose polyester fat substitute that contributes no fat, cholesterol, or calories to the consumer but behaves like conventional fats in food processing. Whether, to what extent, and how such substances are to be investigated by toxicologists remains an open and highly challenging question.

## *New packaging and processing technologies*

There has been an explosion of new packaging and processing techniques in the last two decades, aimed at retarding microbial growth and extending shelf life of food products. Another issue in the 1990s will most likely be changes made in response to environmental concerns.

One of the currently used techniques in food packaging is controlled or modified atmosphere packaging. This process extends the shelf life of many products such as fresh meats, poultry, fish, and fresh fruits and vegetables. Nitrogen, oxygen, and carbon dioxide are used to modify the packaging atmosphere. An important issue with this technology is whether the process retards growth of spoilage organisms while allowing the unrecognized growth of pathogenic anaerobic organisms.

An earlier packaging issue, migration of substances into foods, is again in the forefront with the development of new microwave packaging. For example, heat susceptors are increasingly used in packaging to promote the browning of foods in the microwave oven. These materials contain aluminum and high local temperatures can be generated upon exposure of the aluminum to microwave energy. These high temperatures can lead to disruption of the plastic food contact layer of the packages and the release of pyrolysis products to the food (USFDA, 1991).

In addition to issues surrounding food packaging, new interest is growing in irradiation as a food protection treatment. Radiation is currently used in a limited way in food processing to control insects and microbes in dried spices, herbs, and other dehydrated foods from plants; to control *Trichinella spiralis* contamination in fresh pork; retard growth and maturation of fresh fruits and vegetables; and for preventing sprouts in potatoes. There has been extensive research into the safety of food irradiation, and the process has been found to be safe at does up to 1 Mrad of gamma or high-energy X-irradiation (USFDA, 1991). However, some consumers have become concerned about the use of irradiation in the food industry, and consumer acceptance of the process may be problematical.

### New toxicology requirements

The FDA is in the process or rewriting its safety assessment guidelines for direct food additives (the Redbook). This process was initiated because of changes in the field of toxicology that identified new areas of concern, as well as a recognized need to harmonize testing guidelines among regulatory agencies and nations.

As discussed in this chapter, the Agency has proposed major changes in several areas. Most notable is the emphasis being placed on endpoints of neurotoxicity and immunotoxicity, as well as the utility of pharmacokinetic data for assessing toxicity. Preclinical study designs have also been expanded in the proposed guidelines to include additional biochemical, pathological, and histopathological analyses. All of these proposed changes will significantly affect both the time and cost of product development.

### International regulations and global harmonization

International harmonization of food additive regulations is a somewhat elusive goal at the present time. In many cases, European and Japanese requirements for food additives are different from those imposed in the US and defy easy classification. The first major difference between US and international regulations is that the only country with a GRAS list is the US. This means that compounds considered GRAS in the US may still need formal approvals in other countries. Japan has an informal GRAS approach in that they consider products that occur naturally, either in plants or through fermentation, as inherently safe. Thus, a natural compound that has undergone little testing in Japan could require investigation if it were to be exported to the US or to European countries.

Indirect food additive regulations also have little to no world-wide harmonization. The food packaging regulation requirements of the European Community (EC) appear to be less extensive than US guidelines. Nevertheless, there are individual European country requirements that must be met. The EC is in the process of trying to integrate the varied national enforcement approaches into a single standard. Such an EC indirect additive regulation is not in place currently, and it is unknown whether it will match the FDA's extensive indirect food additive regulations.

In the area of direct food additive regulation, there are some similarities between the European and Japanese regulations. There is universal acceptance that, for a major new food additive, adequate animal studies are necessary to address potential mutagenicity, chronic and subchronic toxicity, reproductive and developmental toxicity, and carcinogenicity. Although there is general consensus on the list of the desired toxicological studies, there is yet to develop a harmonious picture with respect to the study designs considered acceptable in individual countries.

One of the major FDA requirements is that all studies conducted for a food additive comply with GLP regulations. Japan does not have such a requirement, which may make it difficult for sponsors to use Japanese studies for an US application. Furthermore, some older studies conducted in Europe, which meet European regulations, may have been conducted without GLP compliance and may not be acceptable for US registration. GLP guidelines issued by the Organization for Economic Cooperation and Development (OECD) are the basis upon which most of the national requirements of the EC countries have been developed, and there is general agreement that adherence to them will be acceptable to the FDA. The FDA is likely to accept studies from European countries that have a reliable GLP monitoring program.

Although impressive gains are being made in the international harmonization of toxicity testing for pharmaceutical agents (through the International Conference of Harmonization (ICH)), food additive test guidelines vary between countries. For example, when the FDA released its Redbook II testing guidelines for public comment, significant problems were identified by EC members because of guideline incompatibility with current OECD and EC initiatives to harmonize food additive testing.

The clear differences in testing requirements internationally make it imperative that anyone hoping to enter the worldwide market with a new food additive should proceed cautiously.

## References

Frankos VH. FDA perspectives on the use of teratology data for human risk assessment. *Fund Appl Toxicol* 1985;5:615–625.

Harrass MC, Eirkson CE, Nowell LH. Role of plant bioassays in FDA review: scenarios for terrestrial exposures. In: Wang W, Gorsuch JW, Lower WR, editors. *Plants for Toxicity Assessment*. Philadelphia, PA: American Society for Testing Material, 1990. pp. 12–28.

International Food Biotechnology Council (IFBC). Biotechnologies and food: assuring the safety of foods produced by genetic modification. *Regulat Toxicol Pharmacol* 1990;12:S1–S196.

Kokoski CJ, Henry SH, Lin CS, Ekelman KB. Method used in safety evaluation. In: Branen AL, Davidson PM, Salminen S, editors. *Food Additives*. New York: Marcel Dekker, 1990. pp. 579–616.

Munro IC. Issues to be considered in the safety evaluation of fat substitutes. *Food Chem Toxicol* 1990;28:751–753.

Oser B, Hall RL. Criteria employed by the expert panel of FEMA for the GRAS evaluation of flavoring substances. *Food Cosmet Toxicol* 1977;15:457–466.

Pao EM, Burk MC. *Portion Sizes and Day's Intake of Selected Foods*. Washington, DC: US Department of Agriculture ARS-NE67, 1975.

Pao EM., Fleming KH, Guenther PM, Mickle SJ. *Foods Commonly Eaten by Individuals: Amount Per Day and Per Eating Occasion*. Washington, DC: US Department of Agriculture, 1982.
Pennington JAT. Revision of the total diet study food list and diets. *J Am Diet Assoc.* 1983;82:166–173.
Roberts HR. Food safety in perspective. In: Roberts HR, editor. *Food Safety*. New York: Wiley, 1981.
Rodricks JV, Taylor M. Comparison of risks management in US regulatory agencies. *J Hazard Mater* 1989;21:239–253.
Rodricks JV. Risk assessment and food safety. In: Layne S, Beugelsdijk B, Patel G, editors. *Firepower in the Lab: Advanced Automation in the Flight Against Biological Threat.* Washington, DC: National Academy of Sciences Press, 2001 (in press).
Select Committee on GRAS Substances (SCOGS). Evaluation of health aspects of GRAS food ingredients; lessons learned and questions unanswered. *Federation Proc* 1981;36:2519–2562.
US Department of Agriculture (USDA). *Continuing Survey of Good Intakes by Individuals (CSFII) and the Diet and Health Knowledge Survey (DHKS), 1994–1996.* Beltsville, MD: Beltsville Human Nutrition Center, 1997.
US Food and Drug Administration (USFDA). *Toxicological Principles for the Safety Assessment of Direct Food Additives and Color Additives Used in Food.* Washington, DC: Center for Food Safety and Applied Nutrition, 1982.
US Food and Drug Administration (USFDA). *Recommendations for Chemistry Data for Indirect Food Additive Petitions.* Washington, DC: Center for Food Safety and Applied Nutrition, 1988.
US Food and Drug Administration (USFDA). *Emerging Issues in Food Safety and Quality for the Next Decade*. Washington, DC: Center for Food Safety and Applied Nutrition, 1991.
US Food and Drug Administration (USFDA). *Toxicological Principles for the Safety Assessment of Direct Food Additives and Color Additives used in Food ''Redbook II''*. Washington, DC: Center for Food Safety and Applied Nutrition, 1993.
US Food and Drug Administration (USFDA). *Biotechnology of Food.* FDA Backgrounder BG94-4. Washington, DC: Center for Food Safety and Applied Nutrition, May 18, 1994.
US Food and Drug Administration (USFDA). *Preparation of Premarket Notifications for Food Contact Substances: Chemistry Recommendations.* Washington, DC: Center for Food Safety and Applied Nutrition, 1999a.
US Food and Drug Administration (USFDA). *Preparation of Premarket Notifications for Food Contact Substances: Toxicology Recommendations.* Washington, DC: Center for Food Safety and Applied Nutrition, 1999b.

## Further Reading

International Food Biotechnology Council. Biotechnologies and food: assuring the safety of foods produced bye genetic modification. *Regul Appl Toxicol* 1991;12(Supplement):S1–S196.
Kokoski CJ, Flamm WG. *Establishment of acceptable limits of intake.* Proceedings of the Second National Conference for Food Protection. Washington, DC: DHHS, 1984.
Rodricks J. Food constituents additives and contaminants. In: Lippmann M, editor. *Environmental Toxicants: Human Exposure and their Health Effects*. New York: Wiley-Intersciences, 2000. pp. 377–405.
Rulis A. US Food and Drug Administration. Scientific advisory opinion letter, DHHS to Donald L. Emlay, Calgene, Inc. under 21 CFR 10.85. Washington, DC: Center for Food Safety and Applied Nutrition, 1994.

Chapter 6

# Regulations affecting cosmetic and over-the-counter drug products

*Joseph C. DiNardo*

There have been many inaccurate statements about the regulation of cosmetic products and Over-The-Counter (OTC) drugs in the United States. The cosmetic industry has never been directly regulated (requiring pre-market approval) by any government agency and in 1972 the Food and Drug Administration (FDA) initiated a monograph process for OTC drugs which eliminated the need to have pre-marketing approval for products prior to being sold to consumers. The process of regulation for cosmetics and OTC drugs would appear to be better described as "self-regulated" and therefore, impacted by various guidelines, legislation, and regulatory bodies as opposed to governed by these ententes. A review of the laws that are currently in place, as well as a historical perspective, which allows for the current state of regulatory affairs is outlined.

## General regulations

There are a few laws that have been put in place mainly by the FDA and the Federal Trade Commission (FTC) that have had a significant impact on both the cosmetic and OTC industries. The primary laws are the Food, Drug, and Cosmetic Act of 1938 and the 1960 Amendment, the Fair Packaging and Trade Act of 1966 and 1973, the Over-The-Counter Drug Monograph Process introduced in 1972, and the 1916 Federal Trade Commission Act. With the exception of the latter, which is administered by the FTC, all are the responsibility of the FDA. In addition to these Acts, there have been laws brought about by individual members of the government (Delaney Amendment in 1958) and in more recent years by individual States, primarily due to the deregulation mode of the 1980s (California, New York, New Jersey, and Massachusetts to name a few). The major regulations in this category are the Volatile Organic Compounds (VOCs) and the Safe Drinking Water and Toxic Enforcement Act (Proposition 65). Lastly, there are several agencies that publish guidelines and/or safety data that also play a major role in the regulation of ingredients that are used in cosmetics and OTC drugs. The more common agencies are listed later in this chapter along with their web address (Table 6.1).

The Food, Drug, and Cosmetic Act of 1938 was considered the first step in the FDA's involvement in the cosmetic arena. It extended the FDA's control into cosmetics and shifted the responsibility for product safety to the manufacturer. In practical terms, it defined the difference between drugs and cosmetics whereby a drug was defined as "Any substance that can alter the structure or function of the body..." and a cosmetic as

*Table 6.1* Regulatory agency web sites

| *Agency* | *Acronym* | *Web site* |
|---|---|---|
| Agency for Toxic Substances and Disease Registry | ASTDR | http://www.atsdr.cdc.gov/ |
| Bureau of Alcohol, Tobacco and Firearms | ATF | http://www.atf.treas.gov/ |
| California Air Resource Board | CARB | http://www.arb.ca.gov/homepage.htm |
| Code of Federal Regulation | CFR | http://www.access.gpo.gov/nara/cfr/cfrretrieve.html#page1 |
| Consumer Product Safety Commission | CPSC | http://www.cpsc.gov/ |
| Cosmetic Toiletry and Fragrance Association | CTFA | http://www.ctfa.org/ |
| Department of Transportation | DOT | http://www.dot.gov/ |
| Environmental Protection Agency | EPA | http://www.epa.gov/ |
| Food and Drug Administration | FDA | http://www.the FDA.gov/ |
| Federal Trade Commission | FTC | http://www.ftc.gov/ |
| National Institute for Occupational Safety and Health | NIOSH | http://www.cdc.gov/niosh/homepage.html |
| National Technical Information Service | NTIS | http://www.ntis.gov/index.html |
| Organization for Economic Co-operations and Development | OECD | http://www.oecd.org/ehs/index.htm |
| Occupational Safety and Health Administration | OSHA | http://www.osha.gov/ |
| National Toxicology Program | NTP | http://ntpserver.niehs.nih.gov/ |

"A substance that is made for the purpose of beautifying and/or cleansing... ". These definitions may seem trivial and basic, but, the FDA has been able to control numerous claims made on cosmetic products claiming violation of the 1938 Act. The violations included "instant face lift" claims on products containing estrogen in the 1970s, "cell renewal" and "increased micro-circulation and oxygen" claims in the 1980s, and "anti-wrinkle and photoaging" claims in the 1990s. All regulatory letters sent by the FDA to cosmetic companies making these claims referred to the above definition of a drug and noted that the products in question are considered to be misbranded drugs, since they are altering structure and function, and recommended the submission of a New Drug Application (NDA). This sparked a number of companies to take legal action against the FDA since the average cosmetic company can usually take a product from conception to market place within 6–12 months compared to the 10–12 year NDA route recommended by the FDA. Needless to say the cosmetic industry in all cases modified product claims to reflect more cosmetic wording and continued to push the envelope of drug claims via new terminology call "cosmeceuticals" (not an approved term by the FDA).

The second point noted about the 1938 Act which impacts cosmetic companies is that, the FDA shifted the responsibility of product safety to the manufacturer by stating that products not tested prior to marketing must be labeled as follows: "*WARNING: The safety of this product has not been determined*". This has created great diversity among cosmetic companies in how products should be safety tested since no specific recommendations were outlined in the Act. In addition, this point has even become more elusive since the cosmetic industry has primarily eliminated animal testing in recent years due to anti-vivisectionist pressures and has adopted a variety of *in vitro* procedures to replace the basic acute toxicity studies previously conducted. This has prompted much discussion among industry groups as well as regulatory agencies to develop "guidelines" and with the exception of a few groups, which offer some information (the Cosmetic Toiletry and Fragrance Association CTFA) and the Organization for Economic Co-operations and Development (OECD), no standard guidelines for the safety testing of cosmetic products prior to marketing exists in the industry.

In 1960, the FD&C Act of 1938 was modified to broaden the FDA's scope and allow the agency to set limits on how much color could be safely added to products. This was primarily due to an incident in the 1950s, in which scores of children contracted diarrhea from Halloween candy and popcorn colored with large amounts of FD&C Orange No. 1. Additionally, the Act allowed the FDA to develop a list of compounds that were acceptable for use, devised categories that classified the colorants (food, drug and cosmetic = FD&C; drug and cosmetic = D&C; in addition to external, internal, and eye area classifications), and defined the types and levels of impurities allowed in order to achieve certification. Currently the FDA certifies well over 10 million tons of colors a year which is paid for via user fees by the manufacturer requesting certification for the lots of colors submitted. A list of accepted colors are provided in Tables 6.2 and 6.3. As an aside, one would be prudent to look at a few of the 21 CFR references noted in the color tables to become familiar with the types and levels of impurities allowed by the FDA, since many an EPA fine has been given to companies with unacceptable levels of cyanide, lead, arsenic, and other heavy metals due to colorants being dumped down the drain of the manufacturing plant.

The Fair Packaging and Labeling Act of 1966 and 1973 established a legal requirement to designate the individual chemicals used to make a product, the amount of product in the item being sold, and enforced the maker selling the product to designate their name, address, and where the product was made. The fact that individual chemicals had to be identified on the package required that some form of standardized nomenclature be developed so that the consumer would be able to recognize what was in the product. The CTFA put together such a book and updates it annually to be inclusive of the Domestic (CTFA) and International Nomenclature of Cosmetic Ingredients (INCI). The requirement regarding the amount of product in the package set up what is commonly called a net weight declaration which is usually expressed as: net weight for the amount of ounces and grams and weight for the amount of fluid ounces and milliliters. Lastly, the identity of seller was established to allow people to find the maker of the product in case of questions and/or problems about the product. Additionally, it was required for the maker of the product to identify where the product was made.

The OTC Drug Monograph process introduced in 1972, was done so with the hopes of setting standards for certain categories of products that would provide temporary relief of minor problems (acne, diaper rash, cough and cough medications, etc.) which did not require immediate care from a physician, unless the situation worsened within a specified time period after product use. Table 6.4 lists the categories currently available as well as various dates in which information was published in the Federal Register. Each monograph outlines the labeling requirements accepted for the specific categories identified; for example: *Statement of Identity* (BPO 10 Acne Medication), *Indications* (for the treatment of acne lesions), *Directions for Use* (apply twice a day to cleansed skin), *Warnings* (do not use this medication in conjunction with other acne medications), *Active Ingredients* (benzoyl peroxide).

*Table 6.2* Organic color additives approved for use in cosmetics, Part 74, Subpart C

| *21 CFR Section* | *Straight color* | *Approved* | *Use and restrictions* |
|---|---|---|---|
| 74.2101 | FD&C Blue No. 1 | 1982, 1993, 1994 | Cosmetics generally; allows $MnO_2$ in manufacture; eye area use (includes lake) |
| 74.2104 | D&C Blue No. 4 | 1977 | Externally applied cosmetics |
| 74.2151 | D&C Brown No. 1 | 1976 | Externally applied cosmetics |
| 74.2203 | FD&C Green No. 3 | 1982 | Cosmetics generally |
| 74.2205 | D&C Green No. 5 | 1982, 1994 | Cosmetics generally, eye area use |
| 74.2206 | D&C Green No. 6 | 1982 | Externally applied cosmetics |
| 74.2208 | D&C Green No. 8 | 1976 | Externally applied cosmetics (NTE 0.01% (by wt.) of finished cosmetic product) |
| 74.2254 | D&C Orange No. 4 | 1977 | Externally applied cosmetics |
| 74.2255 | D&C Orange No. 5 | 1984, 1982 | Externally applied cosmetics; restricted ingested uses |
| 74.2260 | D&C Orange No. 10 | 1981 | Externally applied cosmetics |
| 74.2261 | D&C Orange No. 11 | 1981 | Externally applied cosmetics |
| 74.2304 | FD&C Red No. 4 | 1976 | Externally applied cosmetics |
| 74.2306 | D&C Red No. 6 | 1983 | Cosmetics generally |
| 74.2307 | D&C Red No. 7 | 1983 | Cosmetics generally |
| 74.2317 | D&C Red No. 17 | 1976 | Externally applied cosmetics |
| 74.2321 | D&C Red No. 21 | 1982 | Cosmetics generally |
| 74.2322 | D&C Red No. 22 | 1982 | Cosmetics generally |
| 74.2327 | D&C Red No. 27 | 1982 | Cosmetics generally |
| 74.2328 | D&C Red No. 28 | 1982 | Cosmetics generally |
| 74.2330 | D&C Red No. 30 | 1982 | Cosmetics generally |
| 74.2331 | D&C Red No. 31 | 1976 | Externally applied cosmetics |
| 74.2333 | D&C Red No. 33 | 1988 | Externally applied cosmetics, mouthwashes, dentifrices, cosmetic lip products (NTE 3% (by wt.) of finished cosmetic product) |
| 74.2334 | D&C Red No. 34 | 1976 | Externally applied cosmetics |
| 74.2336 | D&C Red No. 36 | 1988 | Externally applied cosmetics; cosmetic lip products (NTE 3% (by wt.) of finished cosmetic product) |
| 74.2340 | FD&C Red No. 40 | 1975, 1994 | Cosmetics generally, eye area use (includes Al lake). No oxidizing or reducing agents that may affect integrity |
| 74.2602 | D&C Violet No. 2 | 1976 | Externally applied cosmetics |
| 74.2602a | Ext. D&C Violet No. 2 | 1976 | Externally applied cosmetics |
| 74.2705 | FD&C Yellow No. 5 | 1985, 1994 | Cosmetics generally, eye area use (includes lake) |
| 74.2706 | FD&C Yellow No. 6 | 1986 | Cosmetics generally |
| 74.2707 | D&C Yellow No. 7 | 1976 | Externally applied cosmetics |
| 74.2707a | Ext. D&C Yellow No. 7 | 1976 | Externally applied cosmetics |
| 74.2708 | D&C Yellow No. 8 | 1976 | Externally applied cosmetics |
| 74.2710 | D&C Yellow No. 10 | 1983, 1984 | Cosmetics generally, modification of uses and restrictions |
| 74.2711 | D&C Yellow No. 11 | 1976 | Externally applied cosmetics |

*Table 6.3* Color additives approved for use in cosmetics, Part 73, Subpart C

| *21 CFR Section* | *Straight color* | *Year approved* | *Uses and restrictions* |
|---|---|---|---|
| 73.2030 | Annatto | 1977 | Cosmetics generally including eye area use |
| 73.2085 | Caramel | 1981 | Cosmetics generally including eye area use |
| 73.2087 | Carmine | 1977 | Cosmetics generally including eye area use |
| 73.2095 | Carotene | 1977 | Cosmetics generally including eye area use |
| 73.2110 | Bismuth citrate | 1978 | Cosmetics intended for coloring hair on the scalp (with restrictions) |
| 73.2120 | Disodium EDTA-copper | 1974 | Coloring of shampoos that are cosmetics |
| 73.2125 | Potassium sodium copper chlorophyllin (chlorophyllin copper complex) | 1969 | Coloring dentifrices that are cosmetics (with restrictions) |
| 73.2150 | Dihydroxyacetone | 1973 | Externally applied cosmetics intended solely or in part to impart color to the human body |
| 73.2162 | Bismuth oxychloride | 1977 | Cosmetics generally including eye area use |
| 73.2180 | Guaiazulene | 1977 | Externally applied cosmetics |
| 73.2190 | Henna | 1965 | Coloring hair (with restrictions) |
| 73.2250 | Iron oxides | 1977 | Cosmetics generally including eye area use |
| 73.2298 | Ferric ammonium ferrocyanide | 1977 | Externally applied cosmetics including eye area use |
| 73.2299 | Ferric ferrocyanide | 1978 | Externally applied cosmetics including eye area use |
| 73.2326 | Chromium hydroxide green | 1977 | Externally applied cosmetics including eye area use |
| 73.2327 | Chromium oxide greens | 1977 | Externally applied cosmetics including eye area use |
| 73.2329 | Guanine | 1977 | Cosmetics generally including eye area use |
| 73.2396 | Lead acetate | 1981 | Cosmetics intended for coloring hair on the scalp (with restrictions) |
| 73.2400 | Pyrophyllite | 1973 | Externally applied cosmetics |
| 73.2496 | Mica | 1977 | Cosmetics generally including eye area use |
| 73.2500 | Silver | 1979 | Coloring fingernail polish (NTE 1% of final product) |
| 73.2575 | Titanium dioxide | 1973 | Cosmetics generally including eye area use |
| 73.2645 | Aluminum powder | 1977 | Externally applied cosmetics including eye area use |
| 73.2646 | Bronze powder | 1977 | Cosmetics generally including eye area use |
| 73.2647 | Copper powder | 1977 | Cosmetics generally including eye area use |
| 73.2725 | Ultramarines | 1976 | Externally applied cosmetics including eye area use |
| 73.2775 | Manganese violet | 1976 | Cosmetics generally including eye area use |
| 73.2991 | Zinc oxide | 1977 | Cosmetics generally including eye area use |

The guidelines are written in such a way as to enable the consumer to determine what the appropriate product is for the specific condition they are trying to relieve. This issue has recently been re-examined by the FDA and on March 17, 1999 the FDA published its final rule for the labeling of all OTC drug products. Cosmetic-drug products are included in the final rule as well as the general information that must accompany any OTC product being marketed. Lastly, certain monographs also

*Table 6.4* OTC monograph milestones

| *Drug category* | *Monograph date* |
|---|---|
| Acne | 8-16-91 |
| Alcohol | 3-13-95/11-18-96 |
| Anorectal | 9-2-93/6-3-94 |
| Antacid | 2-8-96/10-9-97 |
| Anthelmintic | 8-1-86 |
| Anticaries | 10-7-96/12-16-96 |
| Antidarrheal | 4-30-86 |
| Antiemetic | 4-30-87/4-11-94 |
| Antiflatulent | 6-4-74/3-5-96 |
| Antifungal | 9-2-93/9-23-93 |
| Antifungal – diaper rash | 12-18-92 |
| Antimicrobials – alcohol (topical) | 7-22-91/11-19-97 |
| Antimicrobials – antibiotics (first aid) | 2-14-96/11-15-96 |
| Antimicrobials – antiseptics (first aid) | 1-6-78/7-22-91 |
| Antimicrobials – antiseptics (health care) | 1-6-78/6-17-94 |
| Antimicrobials – diaper rash | 6-20-90 |
| Antimicrobials – mercurials | 7-22-91 |
| Antiperspirant | 8-20-82 |
| Aphrodisiac | 7-7-89 |
| Benign prostatic hypertrophy | 2-27-90 |
| Boil treatment | 11-15-93 |
| Camphorated oil | 9-21-82 |
| Category II/III ingredients (Phase I) | 11-7-90/1-30-92 |
| Category II/III ingredients (Phase II) | 5-10-93 |
| Cholecystokinetic | 6-10-83/2-28-89 |
| Conditions (panel report) | 10-3-96 |
| Corn/callus remover | 8-14-90 |
| Cough/cold – anticholinergic | 11-8-85 |
| Cough/cold – antihistamine | 1-28-94/4-9-96 |
| Cough/cold – antitussive | 6-3-94/4-9-96 |
| Cough/cold – bronchodilator | 10-20-93/5-20-96 |
| Cough/cold – combinations | 7-27-95 |
| Cough/cold – expectorant | 6-30-92/9-14-92 |
| Cough/cold – nasal decongestant | 8-23-94/3-8-96 |
| Dandruff/seborrhea/psoriasis | 12-4-91/1-28-94 |
| Daytime sedative | 6-22-79 |
| Deodorants (internal) | 5-11-90 |
| Digestive aids | 10-21-93 |
| Exocrine pancreatic insufficiency | 4-24-95 |
| External analgesic | 8-29-97/11-19-97 |
| External analgesic – diaper rash | 12-18-92 |
| External analgesic – fever blister/cold sore (ext) | 1-30-90 |
| External analgesic – insect bites and stings | 10-3-89 |
| External analgesic – male genital desensitizers | 6-12-92 |
| External analgesic – poison ivy/oak/sumac | 10-3-89 |
| Fever blister/cold sore (internal) | 6-30-92 |
| Hair grower/loss | 7-7-89 |
| Hexachlorophene | 9-27-72 |
| Hormone (topical) | 9-9-93 |
| Hypo/hyperphosphatemia | 5-11-90 |
| Ingrown toenail relief | 9-9-93 |
| Insect repellents | 6-17-85 |

Table 6.4 (*continued*)

| *Drug category* | *Monograph date* |
|---|---|
| Int. analgesic | 6-13-96/11-14-97 |
| Int. analgesic – leg muscle cramps | 8-22-94 |
| Labeling | 2-27-97/6-19-97 |
| Laxative | 9-2-97/9-9-97 |
| Menstrual products | 11-16-88 |
| Nailbitting/thumbsucking deter. | 9-2-93 |
| Nighttime sleep aid | 2-14-89/4-11-94 |
| Ophthalmic | 3-4-88/12-18-92 |
| Oral health care – antimicrobial | 2-9-94 |
| Oral health care – nonantimicrobial | 1-27-88 |
| Oral discomfort (relief) | 9-24-91/5-13-92 |
| Oral mucosal injury | 7-26-83 |
| Oral wound healing | 7-18-86 |
| Otic (earwax) | 8-8-86 |
| Otic (swimmer's ear) | 2-15-95/8-16-95 |
| Overindulgence remedies | 12-24-91/5-5-93 |
| Overindulgence remedies – prevention inebriation | 7-19-83 |
| Pediculicides | 12-14-83 |
| Poison treatment | 9-5-78/1-15-85 |
| Salicylanilides (TBS) | 10-30-75 |
| Silver | 10-15-96 |
| Skin bleaching | 9-3-82 |
| Skin protectant | 2-15-83/11-19-97 |
| Skin protectant – astringents | 10-21-93/6-3-94 |
| Skin protectant – diaper rash | 6-20-90 |
| Skin protectant – fever blister/cold sores | 1-31-90 |
| Skin protectant – insect bites/stings | 10-3-89 |
| Skin protectant – poison ivy/oak/sumac | 10-3-89 |
| Smoking deterrent | 6-1-93 |
| Stimulant | 2-29-88 |
| Stomach acidifier | 8-17-88 |
| Sunscreen | 10-16-96/2-26-97 |
| Sweet spirits of nitre | 6-27-80 |
| Vaginal contraceptive | 2-3-95/12-19-96 |
| Vaginal drug products | 2-33-94 |
| Weight control | 8-8-91/8-26-93 |
| Zirconium (aerosol) | 8-16-77 |

include specific guidelines for testing a final formulation for efficacy prior to marketing. Of these the most common are procedures for sunscreen products for the determination of Sun Protective Factor (SPF) values and antiperspirant efficacy studies for determining the amount of sweat reduction achieved.

The Federal Trade Commission Act was established to prohibit unfair and deceptive practices in advertising. What defines these practices is legal precedent based on previous FTC rulings and various legal conclusions made by judges. The National Advertising Division (NAD), a branch of the Better Business Bureau, is the overseer of these regulations and periodically reviews various claims made on TV and in magazines to assure compliance with the regulation. More commonly, companies that feel as if their competitors are using deceptive or unfair advertising contact NAD directly and

*Table 6.5* Volatile organic compounds (VOCs) regulations

| *Product category* <br> *Date adopted* | *California*[a] <br> *1/27/91 and 10/21/91* | *Massachusetts* <br> *11/18/94* | *New Jersey* <br> *11/6/95* | *New York* <br> *5/28/99* | *Oregon* <br> *5/18/95* | *Rhode Island* <br> *Pending* | *Texas* <br> *5/4/94* |
|---|---|---|---|---|---|---|---|
| Antiperspirants (aerosol) | 1-6-93 = 60% HVOC/20% MVOC <br> 1-1-97 = 40% HVOC/10% MVOC <br> 1-1-99 = 0% HVOC/10% MVOC | 10-1-95 = 60% HVOC | 5-1-96 = 60% HVOC | 1-1-94 = 60% HVOC/ 20% MVOC <br> 1-1-02 = 0% HVOC/ 10% MVOC | 1-1-96 = 60% HVOC | 3/98 = 60% HVOC/ 20% MVOC | 1-1-95 = 60% HVOC |
| Antiperspirants (nonaerosol) | 1-6-93 = 0% HVOC and MVOC | No limit | 0% HVOC | 1-1-94 = 0% HVOC and MVOC | 0% HVOC | 0% HVOC and MVOC | 0% HVOC |
| Deodorants (aerosol) | 1-6-93 = 20% HVOC/20% MVOC <br> 1-1-97 = 14% HVOC/ 10% MVOC <br> 1-1-99 = 0% HVOC/ 10%MVOC | 10-1-95 = 20% HVOC | 5-1-96 = 20% HVOC | 1-1-94 = 20% HVOC/ 20% MVOC <br> 1-1-02 = 0% HVOC/ 10% MVOC | 1-1-96 = 20% HVOC | 3/98 = 20% HVOC/ 20% MVOC | 1-1-95 = 20% HVOC |
| Deodorants (nonaerosol) | 1-6-93 = 10% HVOC and MVOC | No limit | 0% HVOC | 1-1-94 = 0% HVOC and MVOC | 0% HVOC | 0% HVOC and MVOC | 0% HVOC |
| Hair shines | 1-1-05 = 55% VOC (regulation passed 7/24/97) | | | Not regulated | Not regulated | Not regulated | Not regulated |
| Hair sprays | 1-6-93 = 80% VOC <br> 6-1-99 = 55% VOC | 10-1-95 = 80% VOC | 5-1-96 = 80% VOC | 1-1-94 = 80% VOC <br> 1-1-02 = 55% VOC | 1-1-96 = 80% VOC | 3/98 = 80% VOC | 1-1-95 = 80% VOC |

Table 6.5 *(continued)*

| *Product category Date adopted* | *California[a] 1/27/91 and 10/21/91* | *Massachusetts 11/18/94* | *New Jersey 11/6/95* | *New York 5/28/99* | *Oregon 5/18/95* | *Rhode Island Pending* | *Texas 5/4/94* |
|---|---|---|---|---|---|---|---|
| Hair mousse | 1-1-94 = 16% VOC | Not regulated | 5-1-96 = 16% VOC | Not regulated | 1-1-96 = 16% VOC | 3/98 = 16% VOC | 1-1-95 = 16% VOC |
| Hair styling gels | 1-1-94 = 6% VOC | Not regulated | 5-1-96 = 6% VOC | Not regulated | 1-1-96 = 6% VOC | 3/98 = 6% VOC | 1-1-95 = 6% VOC |
| Nail polish removers | 1-1-94 = 85% VOC<br>1-1-96 = 75% VOC | Not regulated | 5-1-96 = 85% VOC | Not regulated | 1-1-96 = 75% VOC | 3/98 = 85% VOC | 1-1-96 = 75% VOC |
| Shaving creams | 1-1-94 = 5% VOC | Not regulated | 5-1-96 = 5% VOC | Not regulated | 1-1-96 = 5% VOC | Not regulated | 1-1-95 = 5% VOC |
| Personal fragrance products | 1-1-95 = 20% or less fragrance 80% VOC<br>1-1-99 = 20% or less fragrance 75% VOC<br>1-1-95 = more than 20% fragrance 70% VOC<br>1-1-99 = more than 20% fragrance 65% VOC | Not regulated | Not regulated | Not regulated | Not regulated | Not regulated | Not regulated |

[a] Note: other categories noted only for California air fresheners 1-1-96 aerosol 30% VOC, 1-1-93 liquids/pump sprays 18% VOC, 1-1-93 solids/gels 3% VOC, dual purpose air freshener/disinfectant aerosols 1-1-94 60% VOC, heavy duty hand cleaner or soap 1-1-2005 = 8% VOC (regulation passed 7/24/97); insect repellents 1-1-94 = aerosols 5% VOC.

challenge the claims. This evokes a series of events to take place which requires the company being challenged, by NAD, to substantiate the claims made via submission of scientific documentation and/or consumer testing showing agreement with the claims in question. The NAD will evaluate both internally (via staff) and externally (through independent consultants) the information sent and determine if the claims are substantiated. Additionally, the challenge process can also be initiated by consumers who may feel that the product purchased did not live up to the claims made.

The most comprehensive, to date, of these local/state legislative/regulatory plans has been administered by the Air Resources Board of California (CARB). The California state implementation plan includes transportation, solvent handling and other controls. CARB has also set standards for the percentage of VOCs, included in or emitted from a number of consumer products including personal care products, such as, hair sprays, hair gels, hair mousses, antiperspirants, deodorants, nail polish removers, shaving creams, and personal fragrance products. Under the most recent plan, the state must accomplish further reductions of up to 60 tons per day of VOCs under a Midterm Measures program. The mid-term measures regulations were approved by CARB in June 1997 and regulate additional product categories, including hair shine products and waterless hand cleaners. In addition to California, five other states have adopted consumer product VOC rules (Table 6.5). New York and Massachusetts regulate hair sprays, antiperspirants and deodorants. New Jersey, Oregon and Texas regulate hair sprays, hair gels, hair mousses, antiperspirants, deodorants, nail polish removers, and shaving creams. Arizona, Connecticut, Georgia, and Washington have chosen not to regulate personal care products. Regulations in Rhode Island have passed, but implementation has been put on hold as a contingency measure.

The California Safe Drinking Water and Toxic Enforcement Act (Proposition 65) was put in place in 1986 by the California Environment Protection Agency's Office of Environment Health Hazard Assessment and identifies a number of chemicals that are known to the state of California to cause cancer or reproductive toxicity (Tables 6.6–6.8). If an ingredient listed in the tables is used in a product it must bare a label stating that the product contains ingredient(s) that are known to the State of California to cause cancer or reproductive toxicity. Additionally, several states have tried to introduce similar bills into legislature (Massachusetts and Texas more recently) some with inclusions to cover chemicals that have a potential to induce neurotoxicity, mutagenicity, and other irreversible chronic human health effects. To date, all have been unsuccessful. There are also several regulations that deal with deceptive packaging that impact cosmetic and OTC products. The most common is the California Slack Fill Law which addresses the issue of nonfunctional space which is explained as a package that is filled to substantially less than capacity. I mention this regulation because it is one that is forgotten by most companies, that is until they are fined several thousand dollars per violation by California. There are a number of examples given in the guidelines defining what is and is not deceptive.

## Overview of the FDA authority

Since the FDA is the main agency that impacts the cosmetic and OTC industries, it is worth looking deeper into the activity of this agency. First, it is important to know that the FDA is only able to regulate cosmetics after products are released to the market-

*Table 6.6* Proposition 65 compounds known to be carcinogenic (list updated as of June 15, 1999)

| *CTFA/INCI name* | *CAS number* | *Listing date* | *No significant risk level* |
|---|---|---|---|
| Acetaldehyde | 75-07-0 | April 1, 1988 | 90 mcg/day (inhalation) |
| BHA | 25013-16-5 | January 1, 1990 | 4,000 mcg/day |
| Caffeic acid | 331-39-5 | October 1, 1994 | None listed |
| Captan | 133-06-2 | January 1, 1990 | 300 mcg/day, USEPA value = 200 mcg/day |
| D and C Orange No. 17/ Pigment Orange 5 | 3468-63-1 | July 1, 1990 | None listed |
| D and C Red No. 8/ Pigment Red 53 | 2092-56-0 | October 1, 1990 | None listed |
| D and C Red No. 9/ Pigment Red 53:1 | 5160-02-1 | July 1, 1990 | 100 mcg/day |
| D and C Red No. 19/ Basic Violet 10 | 81-88-9 | July 1, 1990 | None listed |
| Dioctyl phthalate | 117-81-7 | January 1, 1988 | 80 mcg/day |
| Disperse Blue 1 | 2475-45-8 | October 1, 1990 | 200 mcg/day |
| Ethylene dichloride | 107-06-2 | October 1, 1987 | 10 mcg/day |
| Formaldehyde (gas) | 50-00-0 | January 1, 1988 | 40 mcg/day |
| Lead acetate | 301-04-2 | January 1, 1988 | 3 mcg/day |
| Nitromethane | 75-52-5 | May 1, 1997 | None listed |
| Phenacetin | 62-44-2 | October 1, 1989 | 300 mcg/day |
| Phenolphthalein | 77-09-8 | May 15, 1998 | None listed |
| Sodium *o*-phenylphenate | 132-27-4 | January 1, 1990 | 200 mcg/day |
| Potassium bromate | 7758-01-2 | January 1, 1990 | 1 mcg/day |
| Saccharin | 81-07-2 | October 1, 1989 | None adopted by state (proposed ranges 2,800—840,000 mcg/day) |
| Sodium saccharin | 128-44-9 | January 1, 1988 | None adopted by state (proposed ranges 2,800–840,000 mcg/day) |
| Selenium sulfide | 7446-34-6 | October 1, 1989 | None listed |

place. Neither cosmetic products nor cosmetic ingredients are reviewed or approved by the FDA before they are sold to the public. The FDA cannot require companies carry out safety testing of their cosmetic products before marketing. If, however, according to the FD&C Act a cosmetic product whose safety has not been substantiated, must bare the product label "WARNING: The safety of this product has not been determined". The FDA does not have the authority to require manufacturers to register their cosmetic establishments, file data on ingredients, or report cosmetic-related injuries. To keep abreast of such information, the FDA has reinstated the voluntary data collection program, as of January 1999, which allows cosmetic companies to participate in the program by forwarding this data to the FDA.

Recalls are voluntary actions taken by the cosmetic industry to call back products that present a hazard or that are somehow defective. The FDA is not permitted to require recalls of cosmetics but does monitor companies that conduct a product recall. If the FDA wishes to remove a cosmetic product from the market, it must first prove in a court of law that the product may be injurious to users, improperly labeled, or otherwise violates the law.

*Table 6.7* Proposition 65 chemicals known to have developmental toxicity (list updated as of June 15, 1999)

| *CTFA/INCI name* | *CAS number* | *Listing date* | *No significant risk level* |
|---|---|---|---|
| Dichlorophene | 97-23-4 | April 27, 1999 | None listed |
| Ethoxyethanol | 110-80-5 | January 1, 1989 | None listed |
| Ethoxyethanol acetate | 111-15-9 | January 1, 1993 | None listed |
| Methoxyethanol | 109-86-4 | January 1, 1989 | None listed |
| Toluene | 108-88-3 | January 1, 1991 | (Oral) 7000 mcg/day, OEHHA draft = (inhalation) 13,000 mcg/day |

The FDA inspects cosmetics manufacturing facilities, collects samples for examination, and can take action through the Department of Justice to remove adulterated and misbranded cosmetics from the market if the product is in conflict with the 1938 FD&C Act. Foreign products also need to follow these regulations and in the event they appear to be adulterated or misbranded the FDA can refuse entry of the products into the United States.

## Prohibited ingredients and other hazardous substances

With the exception a few prohibited ingredients, a cosmetic manufacturer may use almost any raw material as a cosmetic ingredient and market the product without an approval from the FDA. In addition, although not required by law or regulation cosmetic and fragrance manufacturers have voluntarily agreed to eliminate or to limit maximum use levels of certain ingredients that have been found to cause skin discoloration, redness and irritation, or other allergic reactions (CTFA, 2000, Research Institute for Fragrance Materials (RIFM)).

The following is a list of ingredients that have been prohibited or restricted for use in cosmetics, or that have otherwise been judged to be hazardous by the FDA. The Code of Federal Regulations (CFR) is cited where appropriate.

*Hexachlorophene (21 CFR 250.250).* Because of its neurotoxic effect and ability to penetrate human skin, hexachlorophene (HCP) may be used only when an alternative preservative has not been shown to be as effective. The HCP concentration of the cosmetic may not exceed 0.1 per cent, and it may not be used in cosmetics which in normal use may be applied to mucous membranes.

*Mercury compounds (21 CFR 700.13).* The use of mercury compounds as cosmetic ingredients is limited to eye area cosmetics at concentrations not exceeding 65 parts per million (ppm) (0.0065 per cent) of mercury calculated as the metal (about 100 ppm or

*Table 6.8* Proposition 65 chemicals know to have male reproductive toxicity (list updated as of June 15, 1999)

| *CTFA/INCI name* | *CAS number* | *Listing date* | *No significant risk level* |
|---|---|---|---|
| Ethoxyethanol | 110-80-5 | January 1, 1989 | None listed |
| Ethoxyethanol acetate | 111-15-9 | January 1, 1993 | None listed |
| Methoxyethanol | 109-86-4 | January 1, 1989 | None listed |

0.01 per cent phenylmercuric acetate or nitrate) and provided no other effective and safe preservative is available for use.

Mercury compounds are readily absorbed through the skin on topical application and have the tendency to accumulate in the body. They may cause allergic reactions, skin irritation or neurotoxic manifestations.

*Chlorofluorocarbon propellants (21 CFR 700.23 and 2.125).* The use of chlorofluorocarbon propellants (fully halogenated chlorofluoroalkanes) in cosmetic aerosol products intended for domestic consumption is prohibited.

Chlorofluorocarbon-containing cosmetic aerosol products may continue to be manufactured for export provided they are not in conflict with the laws of the country to which they are to be exported.

*Other prohibited ingredients.* The following additional substances are prohibited as cosmetic ingredients, (but not as unintentional contaminants of cosmetics manufactured in accordance with current good manufacturing practices):

1 Bithionol: photo-contact sensitization (21 CFR 700.11).
2 Halogenated salicylanilides (di-, tri-, metabromsalan and tetrachlorosalicylanilide): photocontact sensitization (21 CFR 700.15).
3 Chloroform: carcinogenic potential (21 CFR 700.18).
4 Vinyl chloride (aerosol products): carcinogenic potential (21 CFR 700.14).
5 Zirconium (aerosol products): pulmonary toxicity (21 CFR 700.16).
6 Methylene chloride: carcinogenic potential (21 CFR 700.19).

*Acetylethyltetramethyltetralin (AETT).* AETT should not be used in fragrance formulations and finished cosmetic products. In a subchronic toxicity study in rats conducted AETT was found to cause serious neurotoxic disorders and discoloration of internal organs. It was also determined to penetrate human skin. The fragrance industry voluntarily discontinued use of AETT in 1978.

*Musk ambrette.* Various tests and clinical experience have demonstrated that musk ambrette may cause photocontact sensitization. Other studies have indicated that musk ambrette may also cause neurotoxic effects. The International Fragrance Association has recommended that musk ambrette should not be used in products applied to the skin, particularly those products used on skin that is customarily also exposed to sunlight.

*6-Methylcoumarin (6-MC).* The fragrance ingredient 6-MC is a photo-contact sensitizer which may cause serious skin and systemic disorders in some consumers on contact in the presence of sunlight. In 1978, the FDA asked the manufacturers of suntan and sunscreen products to discontinue the use of 6-MC. Two firms voluntarily recalled products containing 6-MC from the market.

*Nitrosamines.* Cosmetics containing as ingredients amines or amino derivatives, particularly di- or triethanolamine, may form nitrosamines if they also contain an ingredient which acts as a nitrosating agent such as, for example, 2-bromo-2-nitropropane-1,3-diol (Bronopol, Onyxide 500), 5-bromo-5-nitro-1,3-dioxane (Bronidox C) or tris(hydroxymethyl)nitromethane (Tris Nitro), or if they are contaminated with a nitrosating agent, e.g., sodium nitrite. Amines and their derivatives are mostly present in creams, cream lotions, hair shampoos and cream hair conditioners. The nitrosation may occur during manufacture or during product storage.

Many nitrosamines have been determined to cause cancer in laboratory animals. They have also been shown to penetrate the skin. Nitrosamine contamination of cosmetics became an issue in early 1977. In a study of 29 cosmetic creams and lotions, *N*-nitrosodiethanolamine (NDELA) was determined in 27. The levels of NDELA contamination ranged from less than 10 parts per billion (ppb) to 50 ppm. Of the more than 300 cosmetic samples analyzed in 1978, 1979 and early 1980 in the FDA's laboratories, 7 per cent contained less than 30 ppb NDELA, 26 per cent contained 30 ppb to 2 ppm and 7 per cent contained between 2 and 150 ppm.

The FDA expressed its concern about the contamination of cosmetics with nitrosamines in a notice published in the Federal Register of April 10, 1979 (44 FR 21365). It stated that cosmetics containing nitrosamines may be considered adulterated and subject to enforcement action. In surveys of cosmetic products conducted in 1991-92, NDELA was found in 65 per cent of the samples at levels up to 3 ppm.

*Dioxane*. Cosmetics containing as ingredients ethoxylated surface active agents, i.e., detergents, foaming agents, emulsifiers and certain solvents identifiable by the prefix, word or syllable "PEG", "Polyethylene", "Polyethylene glycol", "Polyoxyethylene", "-eth-", or "-oxynol-", may be contaminated with 1,4-dioxane. It may be removed from ethoxylated compounds by means of vacuum stripping at the end of the polymerization process without an unreasonable increase in raw material cost.

In rodent feeding studies conducted for the National Cancer Institute, 1,4-dioxane was found to produce cancer of the liver and the nasal turbinates. It also caused systemic cancer in a skin painting study. Skin absorption studies demonstrated that dioxane readily penetrates animal and human skin from various types of vehicles. However, it was also determined that most of the dioxane applied to the skin in a vehicle evaporates into the environment and may not be available for skin absorption.

The contamination of ethoxylated surface-active agents with dioxane was first reported in 1978. Many of the raw materials analyzed since then have been found to contain dioxane; some contained as much as, or more than, 100 ppm. In finished cosmetic products containing ethoxylated surface-active agents, the incidence and level of dioxane contamination was significantly lower.

The following sections have additional information on special product types and/or requirements from the FDA's *Guide to Inspections of Cosmetic Product Manufacturers.*

## Tamper-resistant packaging

Cosmetic liquid oral hygiene products, e.g., mouthwashes and breath fresheners, and any kind of vaginal product introduced into interstate commerce after February 6, 1984, must be packaged in tamper-resistant packages if intended to be accessible to the public while held for retail sale (21 CFR 700.25).

A tamper-resistant package may be an immediate container and closure system or an outer (secondary) container system which has an indicator or barrier to entry and which provides visible evidence to consumers that tampering has occurred when its indicator or barrier to entry has been breached or is missing. To prevent substitution of the tamper-resistant feature after tampering, the indicator or barrier to entry must be distinctive by design (e.g., an aerosol container or breakable cap) or by use of an identifying characteristic (e.g., name, pattern or logo on cap or carton seal).

In addition, a tamper-resistant package other than an aerosol package must bear a prominently placed label statement which alerts the consumer to the tamper-resistant feature of the package and which is not affected when the tamper-resistant feature is breached or missing. Example: "For your protection, this bottle has an imprinted seal around the neck."

If cosmetics are manufactured requiring tamper-resistant packaging, check whether the indicator or barrier to entry is distinctive by design or bears an identifying characteristic which cannot readily be duplicated by the public and whether the label statement properly alerts the consumer to the tamper-resistant feature and is appropriately placed.

### *Adequacy of preservation*

Cosmetics need not be sterile, however, they must not be contaminated with microorganisms which may be pathogenic, and the density of nonpathogenic microorganisms should be low. In addition, cosmetics should remain in this condition when used by consumers. Some cosmetics, i.e., those containing more than about 10 per cent ethanol, propylene glycol, glycerol, etc., and cosmetics in self-pressurized containers, are self-preserving and are not likely to become contaminated with microorganisms.

The hazard of inadequately preserved cosmetics to human health has been amply demonstrated by reports of staphylococcal infections in hospitals from use of contaminated hand creams and hand lotions and the studies conducted on eye area cosmetics. Regardless of whether a cosmetic becomes contaminated during manufacture or during consumer use, the hazard is two-fold, namely, (1) the direct effect of microorganisms on human health and (2) the circuitous effect on human health due to product contamination and spoilage, product separation, or formation of harmful microbial metabolites.

Microbial contamination of cosmetics during manufacture was a major issue during the 1960s and early 1970s. Since then, significant progress has been made by the cosmetic industry towards implementation of sanitary manufacturing practices, more rigorous microbiological control, and the development of better-preserved cosmetic products. However, the problem of adequacy of preservation of cosmetics to prevent contamination during consumer use continues to be of concern to the Agency, particularly with respect to cosmetics coming into contact with the eye.

The studies conducted to determine the hazard associated with inadequately preserved eye area cosmetics revealed that microbial contamination of new mascaras was rare but that many became readily contaminated with the microorganisms found on the eyelids and fingers of consumers. If an inadequately preserved mascara becomes contaminated with *Pseudomonas aeruginosa* and the delicate cornea of the eye is scratched with the applicator, the eye may become infected. *P. aeruginosa* is an ubiquitous microorganism which may also occasionally be present on the skin. Corneal ulceration may lead to partial or total blindness in the injured eye. Several cases of corneal ulceration and blindness associated with *Pseudomonas* contaminated mascaras have been identified. Eye area cosmetics contaminated with *Staphylococcus epidermidis* or other cocci may cause conjunctivitis or blepharitis.

The issue of adequacy of preservation of eye area cosmetics was addressed in the Federal Register notice of October 11, 1977. The Agency announced its intention to propose regulations and invited interested persons to submit information on microbial testing methods and standards of performance suitable to ensure that such cosmetics do not become contaminated with microorganisms during manufacture and use by consumers. Since no useful information was received about such methods, standards for determining adequacy of preservation are now being developed for the Agency under contract. The notice also stated that "the FDA does not intend to await the completion of the rulemaking pronounced in this notice of intent before taking needed regulatory action."

In addition to the inspection of an establishment for sanitary storage and handling of raw materials and for sanitary manufacture of finished products, it should be determined, whether:

1 Each batch of cosmetic which is not self-preserving is tested for microbial contamination before a batch is released for interstate shipment, and
2 Each cosmetic, particularly each eye area cosmetic, has been tested during product development for adequacy of preservation against microbial contamination which may occur under reasonably foreseeable conditions of consumer use.

Review the qualitative and quantitative composition of the preservative system of each eye area cosmetic. Report findings in the EIR. When collecting surveillance samples, select eye area cosmetics over creams or cream lotions.

### Aerosol products

Chlorofluorocarbon propellants, vinyl chloride, and zirconium compounds are prohibited as ingredients of cosmetic aerosol products – 21 CFR 700.23, 700.14 and 700.16.

Cosmetic aerosol products must bear the following label statement (21 CFR 740.11(a)(1)): – Avoid spraying in eyes. Contents under pressure. Do not puncture or incinerate. Do not store at temperature above 120°F. Keep out of reach of children.

Hydrocarbon propellant-containing products also must bear the statement (21 CFR 740.11(b)(1)):

Warning – Use only as directed. Intentional misuse by deliberately concentrating and inhaling the contents can be harmful or fatal.

Exempt from the second warning requirement are:

1 Foam or cream products containing less than 10 per cent propellant.
2 Products in a container with a physical barrier that prevents escape of the propellant at the time of use.
3 Products of a net quantity of contents less than 2 ounces and equipped with a metering valve.
4 Products with a net quantity of contents less than 1/2 ounce.

Review labels of aerosol products for compliance with these requirements and report any deviation.

### Feminine deodorant sprays

Products whose labeling states or suggests that the product is for use in the female genital area or for use all over the body must bear the following label statement – 21 CFR 740.12:

Caution – For external use only. Spray at least 8 in. from skin. Do not apply to broken, irritated, or itching skin. Persistent, unusual odor or discharge may indicate conditions for which a physician should be consulted. Discontinue use immediately if rash, irritation, or discomfort develops.

Feminine deodorant sprays which are not packaged in self-pressurized containers, need not bear the sentence "Spray at least 8 inches from skin". If they are aerosol products, they must also bear the warning required at 21 CFR 740.11(a)(1). Additionally, if the propellant is a hydrocarbon, the label must bear the warning required at 740.11(b)(1).

Review labels for compliance with these requirements and report any deviation.

### *Children's foaming detergent baths (bubble bath products)*

The risk associated with certain conditions of use of foaming detergent bath products, i.e., bubble bath products, particularly excessive or prolonged exposure, has been known for some time. Over the years, the agency has received numerous complaints from consumers and physicians about itching, rashes and urinary tract disorders. Reports in the medical literature have mentioned that the adverse reactions have either subsided or disappeared when the use of bubble bath products was discontinued. Most adverse reactions appeared to have been caused by inadvertent product misuse which may not have occurred if consumers had been given proper directions for safe use of these products and had been cautioned about the possible adverse effects by means of mandatory label warning.

21 CFR 740.17 requires that children's foaming detergent bath products, i.e., children's bubble bath products, and all foaming detergent bath products not labeled as intended for use exclusively by adults, distributed after June 5, 1987, bear adequate directions for safe use and the following caution:

Caution – Use only as directed. Excessive use or prolonged exposure may cause irritation to skin and urinary tract. Discontinue if rash, redness, or itching occur. Consult your physician if irritation persists. Keep out of reach of children.

For the purpose of this regulation, a foaming detergent bath product (bubble bath product) is defined as any product intended to be added to the bath for the purpose of producing foam and containing a surface-active agent serving as a detergent or foaming agent.

Examples of label statements properly identifying a product as being intended for use exclusively by adults are: "Keep out of reach of children". or "For adult use only".

Determine whether bubble bath products bear adequate directions for safe use and the caution statement required by regulation. Review consumer complaints for adverse reactions associated with bubble bath products. Report findings in the EIR.

### *Hair dye products*

Hair dye products may be divided into three categories, i.e., permanent, semi-permanent and temporary hair colors. Permanent hair colors are the most popular hair dye products. They may be further divided into oxidation hair dyes and progressive hair dyes. Oxidation hair dye products consist of a solution of dye intermediates, e.g., *p*-phenyl- enediamine (which form hair dyes on chemical reaction), and preformed dyes, e.g., 2-nitro-*p*-phenylenediamine (which already are dyes and are added to achieve the intended shades), in an aqueous, ammoniacal vehicle containing soap, detergents and conditioning agents, and a solution of hydrogen peroxide, usually 6 per cent, in water or cream lotion. The ammoniacal dye solution and the hydrogen peroxide solution, often called the developer, are mixed shortly before application to the hair. The applied mixture causes the hair to swell, and the dye intermediates (and preformed dyes) penetrate the hair shaft to some extent before the chemical reaction forming the hair dye is complete.

Progressive hair dye products contain lead acetate as the active ingredient. Lead acetate is approved as a color additive for coloring hair on the scalp at concentrations not exceeding 0.6 per cent w/v, calculated as metallic lead (21 CFR 73.2396). Bismuth citrate, the other approved color additive (21 CFR 73.2110), is used to a much lesser extent. Progressive hair dyes change the color of hair gradually from light straw color to almost black by reacting with the sulfur of hair keratin as well as oxidation on the hair surface.

Semi-permanent and temporary hair coloring products are solutions (on rare occasions dry powders) of various coal-tar dyes, i.e., synthetic organic dyes, which deposit and adhere to the hair shaft to a greater or lesser extent. Temporary hair colors must be re-applied after each shampooing. The vehicle may consist of water, organic solvents, gums, surfactants and conditioning agents. The coal-tar dyes are either approved, listed and certified color additives or dyes for which approval and listing has not been sought. The dyes may not be nonpermitted metallic salts or vegetable substances.

If a hair dye product contains a nonapproved coal-tar color (but not a nonapproved metallic or vegetable dye), and even if this coal-tar color is known to cause adverse reactions under conditions of use, the product may not be considered adulterated if the label bears the caution statement provided in Section 601(a) of the FD&C Act and offers adequate directions for preliminary patch testing by consumers for skin sensitivity. The caution statement reads as follows:

Caution – This product contains ingredients which may cause skin irritation on certain individuals and a preliminary test according to accompanying directions should first be made. This product must not be used for dyeing the eyelashes or eyebrows; to do so may cause blindness.

If the label of a coal-tar color containing hair dye product does not bear the caution statement of Section 601(a) and the patch testing directions, it may be subject to regulatory action if it is determined to be harmful under customary conditions of use.

Several coal-tar hair dye ingredients have been found to cause cancer in laboratory animals, for example, 4-methoxy-*m*-phenylenediamine (4-MMPD, 2,4-diaminoanisole). Additionally, studies in humans and monkeys have demonstrated that 4-MMPD readily penetrates the skin. The Agency considered the risk associated with the use of hair dyes containing 4-MMPD a "material fact" which should be known to consumers and published in October 1979 a regulation requiring a label warning on hair dye products containing 4-MMPD which was to become effective on April 16, 1980. The regulation required that hair dyes containing 4-MMPD bear the following warning:

Warning – Contains an ingredient that can penetrate your skin and has been determined to cause cancer in laboratory animals.

Some hair dye manufacturers held that the potential risk was too small to be considered "material" and challenged the validity of the regulation in court. The Agency decided to reconsider its earlier position and entered into a consent agreement with the hair dye manufacturers. The effective date of the regulation has been stayed until completion of the assessment of the carcinogenic risk of 4-MMPD in accordance with scientifically accepted procedures.

In addition to 4-MMPD, the following other hair dye ingredients have been reported to cause cancer in at least one animal species in lifetime feeding studies: 4-chloro-*m*-phenylenediamine, 2,4-toluenediamine, 2-nitro-*p*-phenylene-diamine and 4-amino-2-nitrophenol. They also were found to penetrate human and animal skin.

Determine whether the manufactured hair dye products contain 4-MMPD, nonlisted metallic salts or vegetable substances as dye ingredients. Also identify and report hair dye products not bearing the caution statement of Section 601(a) and containing a nonlisted (nonapproved) coal-tar color.

### *Depilatories and hair straighteners*

Chemical depilatories are highly alkaline pastes, creams or cream lotions containing either alkali or alkali-earth sulfides (usually up to 35 per cent barium or strontium sulfide) or mixtures of alkali-earth hydroxides (usually 5–10 per cent calcium hydroxide) and salts of aliphatic mercapto acids (usually 2–5 per cent calcium thioglycolate). These ingredients cause degradation of hair keratin and deterioration of hair fibers to a jelly-like mass that can easily be removed by wiping or scraping. The pH of chemical depilatories usually falls between 10 and 12.5.

Hair straighteners are mostly creams or cream lotions containing either up to about 3 per cent sodium hydroxide or, as a kit, a lotion containing about 5 per cent calcium hydroxide and a solution of up to about 30 per cent guanidine carbonate. The pH is around 12. Some straighteners contain about 4 per cent ammonium thioglycolate as the active ingredient.

Because improperly formulated or incorrectly used depilatories or hair straighteners may cause serious skin irritation, they should be thoroughly tested for safety, be subjected to careful quality control during manufacture, and provide explicit warnings and directions for safe use.

Review labeling for appropriate warnings and directions for safe use, investigate quality control procedures, obtain information on consumer injury complaints, and report findings in EIR.

### *Hair shampoos, rinses, conditioners*

Hair shampoos contain anionic or ampholytic detergents serving as cleansing and foaming agents; rinses and conditioners may contain cationics (quaternary ammonium compounds) serving as antistatic agents. When inadvertently introduced into the eye, these surface active agents may cause stinging, mucosal irritation or even corneal damage, and products contaminated with microorganisms may cause infection. If the cornea has been scratched or otherwise damaged, pathogenic microorganisms, particularly *Pseudomonas aeruginosa*, may cause corneal ulceration and blindness. Cosmetic hair products may be adequately preserved with, for example, formaldehyde releasing preservatives to prevent microbial contamination. See section on adequacy of preservation above.

Investigate what product testing has been performed to determine the type and degree of irritation that may occur when coming into contact with the eye. Examine labels for appropriate warnings and directions for use. Also investigate what antimicrobial testing has been, and is being, carried out to assure that marketed products are not contaminated and will not become contaminated during normal use. Report findings in the EIR.

### *Permanent wave neutralizers*

Permanent wave neutralizers containing either sodium bromate or potassium bromate, can also be purchased in supermarkets, drugstores, and mass merchandise stores. Some beauty supply outlets also sell permanent wave kits labeled "For Professional Use Only" to the general public.

Toxic effects from ingestion of sodium bromate and potassium bromate include nausea and vomiting accompanied by abdominal pain and diarrhea, anemia, destruction of the red blood cells, decreased blood pressure, convulsions, coma, respiratory depression, and possibly death.

In response to documented reports of a number of cases of accidental ingestion by young children of bromate neutralizer solutions, the Consumer Product Safety Commission, under the Poison Prevention Packaging Act of 1970, published a final rule on December 18, 1990, requiring that home permanent wave neutralizers, in liquid form, containing in a single container more than 600 mg of sodium bromate or more than 50 mg of potassium bromate be packaged in child-resistant packaging.

Determine if the firm manufactures home permanent wave neutralizers containing sodium bromate or potassium bromate products and if child-resistant packaging is used for such products. Collect samples and report findings in the EIR if child-resistant packaging is not being used.

### *Products containing estrogenic hormones, placental extract or vitamins*

Products containing estrogen, estrone, estradiol, progesterone, placental extract or vitamins may be considered drugs, misbranded drugs, or misbranded cosmetics, particularly if the label declaration is supplemented with statements implying prevention or treatment of disease or effect on the structure or any function of the human body. See Federal Register Notice of the proposed rule of October 28, 1977 (42 FR 56757) and 21 CFR 201.300.

The estrogen content of an OTC product, be it a drug or a drug as well as cosmetic, may not exceed 10,000 IU per ounce, and users must be directed to limit the amount of product applied daily so that no more than 20,000 IU of estrogen or equivalent be used per month. Some estrogen-containing products have been claiming to prevent or reduce wrinkles, treat seborrhea, or stimulate hair growth. The Advisory Review Panel on OTC Miscellaneous External Drug Products has concluded that there are inadequate data to establish the safety of these products and that they are ineffective and may therefore be misbranded, even if marketed as cosmetics without making medicinal claims (advance notice of proposed rulemaking, Federal Register of January 5, 1982, 47

FR 430). In a Final Rule, published in the Federal Register of September 9, 1993, 58 FR 47608, the FDA accepted this panel's recommendation and determined that all topically-applied hormone containing drug products for OTC human use are not Generally Recognized As Safe (GRAS) and effective and are misbranded. In a companion Notice of Proposed Rulemaking published concurrently (58 FR 47611), the Agency proposed to allow as safe the use in cosmetic products of pregnenolone acetate up to a level of 0.5 per cent or progesterone up to a level of 5 mg/ounce (product label to specify upper level of consumer usage of finished product containing one of the foregoing substances not to exceed 2 ounces/month). The agency also concluded, however, that any use of natural estrogens in cosmetic products makes the product an unapproved new drug. This proposal also designated any cosmetic using the term "hormone" in the text of its labeling or in its ingredient statement as making an implied drug claim, subjecting such a product to regulatory action under Sections 502 and 505 of the Act.

In addition to being considered misbranded drugs, products claiming to contain placental extract may also be deemed to be misbranded cosmetics if the extract has been prepared from placentas from which the hormones and other biologically active substances have been removed and the extracted substance consists principally of protein. The FDA recommends that this substance be identified by a name other than "placental extract" and describing its composition more accurately because consumers associate the name "placental extract" with a therapeutic use of some biological activity.

Cosmetics declaring ingredients as vitamins, for example, tocopherol as vitamin E, convey the misleading impression that these ingredients and products offer a nutrient or health benefit and may therefore be deemed misbranded. The second edition of the CTFA Cosmetic Ingredient Dictionary, the recognized source of cosmetic ingredient names, lists vitamin ingredients by their respective chemical names.

Review labeling for declaration of estrogenic hormones, placental extract, vitamins and the label statements associated with these ingredients. Report findings in the EIR.

### *Nail builders, hardeners, enamels*

Nail builders (elongators, extenders) have been involved in numerous reports of irritation, inflammation and infection of the nail bed and nail fold as well as in complaints of discoloration, splitting and loss of fingernails. The products are marketed as kits consisting of a powder (a mixture of methyl methacrylate polymer and peroxide catalyst) and a liquid (a mixture of methacrylate ester monomer and promoter). Ultraviolet light-curing products consist of a single unit containing methacrylate ester monomers, polyurethane and a curing agent (e.g., hydroxycyclohexyl phenyl ketone). The methacrylate monomers currently used in nail builders are mostly ethyl, hydroxy-ethyl, butyl, isobutyl, hydroxypropyl or other esters of methacrylic acid. Methyl methacrylate is now rarely used because a court ruling in an injunction proceeding against a former manufacturer of nail builders and numerous seizures and recalls of methyl methyacrylate containing products. The currently used esters of methacrylic acid may be as harmful as methyl methacrylate.

When a nail builder is manufactured, determine which ester of methacrylic acid is used in the liquid component, review the firm's consumer complaint files, and report findings in the EIR.

Nail hardeners often contain formaldehyde as the active ingredient. Formaldehyde has been reported to be irritating to the skin or cause allergic reactions. In the past, the FDA has not objected to its use as an ingredient of nail hardeners provided the product:

1 Contained no more than 5 per cent formaldehyde.
2 Provided the user with nail shields which restrict application to the nail tip (and not the nail bed or fold).
3 Furnished adequate directions for safe use; and
4 Warned consumers about the consequences of misuse and potential for causing allergic reactions in sensitized users.

The safety of formaldehyde as a cosmetic ingredient was reviewed in 1984 by a panel of scientific experts appointed by the Cosmetic, Toiletry and Fragrance Association, a trade association representing a major portion of the cosmetic industry. The panel reported that available toxicological data and other information were insufficient to conclude that cosmetics containing formaldehyde in excess of 0.2 per cent are safe. (J Am Coll Toxicol 1984;3(3):157–184).

Ascertain the concentration of formaldehyde, inspect the nail shields for proper design and construction. Review labeling for appropriate warnings and directions for use, and review consumer complaint files for the kinds and numbers of adverse reactions associated with this product.

Nail enamels usually consist of nitrocellulose and aryl-sulfonamide-formaldehyde resin as film formers, toluene or ethyl or butyl acetate as solvents, and phthalate, citrate or phosphate esters as plasticizers. Adverse reactions associated with nail enamels are not uncommon. The formaldehyde resin or residual formaldehyde may elicit allergic reactions in already sensitized consumers, the solvents or plasticizers may be irritating, and the deposited film may cause irritation and inflammation because of its occlusiveness or lack of flexibility. Nail enamels marketed as hardeners have had a particularly high rate of adverse reactions. Their high resin content or low concentration of plasticizer causes them to be particularly occlusive and inflexible. Another frequent complaint is flammability during and shortly after application.

### *Artificial or sculptured fingernail glue removers*

Household glue removers containing acetonitrile used in removing or debonding glues for artificial or sculptured fingernails may be purchased in supermarkets, drugstores, and mass merchandise stores. Even products labeled "For Professional Use Only" are available for purchase by the general public in retail and "wholesale" beauty supply establishments.

Acetonitrile is toxic by ingestion, inhalation and skin absorption. In response to documented reports of a number of cases of accidental ingestion by young children of sculptured nail removers containing acetonitrile, the Consumer Product Safety Commission, under the Poison Prevention Packaging Act of 1970, published a final rule on December 18, 1990, requiring that household glue removers, in liquid form, containing more than 500 mg of acetonitrile in a single container be packaged in child-resistant packaging.

Determine if the firm manufactures acetonitrile-containing glue remover products and if child-resistant packaging is used for such products. Collect samples and report findings in the EIR if acetonitrile-containing glue remover products without child-resistant packaging are found.

### Soap

Products that are "soap" are exempt from the provisions of the FD&C Act because soap is excluded from the definition of the term "cosmetic" in Section 201(i) of the Act. The FDA interprets the term "soap" to apply to products:

1. Intended for cleansing the human body;
2. Labeled, sold and represented solely as soap; and
3. Consisting primarily (i.e., the bulk of its nonviolative matter serving as the detergent) of an alkali salt of fatty acids – 21 CFR 701.20.

Products consisting primarily of alkali salts of fatty acids and intended not only for cleansing but for other cosmetic uses, i.e., products intended also for beautifying, promoting attractiveness or altering the appearance, must comply with the regulatory requirements applicable to cosmetics and must, for example, bear ingredient declarations as required at 21 CFR 701.3. They may also be regulated as drugs if intended to cure, treat or prevent disease or to affect the structure or any function of the human body.

Also cosmetics are products not consisting predominantly of alkali salts of fatty acids (i.e., products consisting predominantly of synthetic detergents or combinations of significant proportions of both alkali salts of fatty acids and synthetic detergents) and/or significant levels of other functional additives whose intended purpose in the composition of matter is other than cleansing (i.e., moisturizers, emollients, humectants, anti-irritants, etc.).

Products consisting predominantly of synthetic detergents may be identified in labeling as "soap" if they are intended for cleansing the human body and have the characteristics consumers generally associate with soap.

Products intended solely for cleansing the human body and having characteristics consumers generally associate with soap may be identified in labeling as soap even though they do not consist of detergent ingredients which are predominantly alkali salts of fatty acids. These products are also regulated as cosmetics.

Determine whether the manufactured soap products are cosmetics or are not subject to the provisions of the FD&C Act. Review labeling accordingly for compliance with regulatory requirements.

### Suntan products

Suntan products generally are sunscreening preparations which, when applied to the skin, permit penetration of sufficient erythemal ultraviolet radiation to produce a perceptible erythema for best tanning results. When used as directed, consumers may remain in the sun for a predetermined time period without risking a sunburn. The sunburn protection is provided by sunscreen active ingredients, e.g., cinoxate, homosalate, padimate o (octyldimethyl PABA).

Suntan products claiming to prevent sunburn are drugs, in addition to being cosmetics. They are also considered drugs if they contain a sunscreen and are represented exclusively for the production of a tan and do not refer in labeling to sunburn protection. See Federal Register notice of proposed rulemaking of August 25, 1978 (43 FR 38206). In the tentative Final Monograph for Sunscreen Drug Products for OTC Human Use, published in May 12, 1993 Federal Register (58 FR 28194), the FDA proposed to regulate all topical products for which tanning claims are made in conjunction with the use of a sunscreen active ingredient inherently as drugs. Exceptions to this, however, were also proposed if the sunscreen ingredient is used in the topical product to achieve a legitimate "qualified cosmetic benefit", such as to protect the integrity of the product formulation's composition of matter (i.e., color, fragrance, or lipids). Such products may be regarded as cosmetics if the term "sunscreen" is not used in label copy, if no SPF value is declared, and if the sunscreen ingredient is disclosed only in the product labeling by its accepted cosmetic name in the cosmetic product ingredient statement (c.f. 21 CFR 701.3).

For enforcement purposes, the following particulars may serve as a guide for determining whether a product may be treated as a cosmetic or drug or a cosmetic that is also a drug. The product should be regulated as a drug if:

1 The labeling bears any direct or implied statement that the product screens out ultraviolet sunlight, prevents or treats sunburn, helps prevent wrinkles, or prevents premature aging of the skin;
2 The label bears a number representing the sun protection factor (SPF) value; or
3 The sunscreen ingredient is declared as an active drug ingredient and is listed before the listing of the cosmetic ingredients – Section 502(e)(1) of the FD&C Act and 21 CFR 701.3(d).

Products regulated as cosmetics should, and those regulated as drugs must, bear adequate directions for safe use. Section 502(f)(1) of the Act and 21 CFR 201.5. As an example, the labels should state the maximum safe sun exposure period under conditions of prescribed use. Suntan and sunscreen products also must bear warning statements as necessary or appropriate to prevent health hazards – 21 CFR 740.1 and Section 502(f)(2) of the Act. The need for an appropriate warning applies particularly to suntan products not containing a sunscreen ingredient or providing only marginal sunburn protection, such as those with SPF values of less than 4. In the May 12, 1993, Federal Register Notice (see above), the FDA proposed to amend the cosmetic regulations to include a new cosmetic warning statement under 21 CFR 740.19, as follows:

Suntanning preparations. The labeling of suntanning preparations that do not contain a sunscreen ingredient must display the following warning:

Warning – This product does not contain a sunscreen and does not protect against sunburn.

Note that this proposed warning is not yet enacted in final form.

Other "suntan" products of interest are capsules intended for ingestion and containing mostly beta carotene and canthaxanthin. These color additives enter the blood stream and are partially deposited in skin tissue, giving the skin a tan-like color. Neither color additive is approved for this particular use, and products containing them are considered adulterated. Some reports of adverse reactions associated with "tanning pills" have mentioned stomach cramps, hepatitis, nausea, diarrhea, and deposition of the color in the retina of the eye.

In recent years, "suntan accelerators" have appeared on the market. They claim to enhance tanning by stimulating and increasing melanin formation. Because their intended purpose is to affect a function of the human body, they may be considered drugs. One type suntan accelerator is based on bergapten (5-methoxypsoralen) which is found in bergamot oil and is a well-known phototoxic substance, responsible for Berloque Dermatitis. Bergapten increases the skin's sensitivity to ultraviolet light, intensifies erythema formation, and stimulates melanocytes to produce melanin. It has also been reported to be photocarcinogenic in animals. The other kind of suntan accelerator contains tyrosine, alone or in combination with other amino acids, as the "active" principle. Tyrosine is the starting compound of the melanin synthesis in the skin. Its use is based on the assumption that it penetrates the skin, increases tyrosine content of the melanocytes, and thus enhances melanin formation. The effect has not been documented in the scientific literature. In fact, an animal study reported a few years ago demonstrated that ingestion or topical application of tyrosine has no effect on melanogensis. The Agency has recently concluded that "suntan accelerators" are unapproved new drugs within the meaning of Section 201(p) of the FD&C Act, and has issued warning letters to several major manufacturers of these products whether a manufactured suntan-sunscreen product has been tested for safety as well as for sunscreening effectiveness in accordance with the procedure proposed in the 1978 Federal Register Notice (43 FR 38206) and to obtain the respective SPF value.

Review labels accordingly for appropriate directions for safe use and warnings and for ingredient labeling in compliance with 21 CFR 701.3. Particularly check whether products not containing sunscreen ingredients warn consumers about the risk of sunburn. Inquire about consumer complaints of adverse reactions and lack of sunscreening effectiveness. Report findings in the EIR.

Determine if the firm manufactures and/or distributes tanning products intended for ingestion and containing canthaxanthin, beta carotene or other color additives. Collect samples and report findings to CFSAN with a recommendation for regulatory action.

## References

Cosmetic Toiletry and Fragrance Association (CTFA). *International Cosmetic Ingredient Dictionary and Handbook*, 8th ed. Washington, DC: CTFA, 2000.

## Further reading

Federal Food, Drug, and Cosmetic Act; Public Law No. 75-717, 1938.
Color Additive Amendments to the Food, Drug, and Cosmetic Act; Public Law No. 86-618, 1960.
Fair Packaging and Labeling Act; Public Law No. 89-755, 1966.
Federal Trade Commission Act; Public Law No. 63-203, 1914.
Food and Drug Administration Web Site. http://www.the FDA.gov/
Federal Trade Commission Web Site. http://www.ftc.gov/
California Air Resource Board Web Site. http://www.arb.ca.gov/homepage.htm
Cosmetic Toiletry and Fragrance Association WebSite. http://www.ctfa.org/

Chapter 7

# OTC drugs and nutraceuticals

*Charles B. Spainhour*

## OTC drugs

### *United States*

Over the last half-century and especially over the last 10 years, the "right" of people to diagnose and treat themselves for maladies has become prominently recognized. The demand for medicines which can be self-chosen and self-administered has become not only accepted but embraced by pharmaceutical companies. This has led to a booming market in Over-The-Counter (OTC) products and the interchangeability between prescription drugs and OTC drugs. The whole concept of self-treatment is still growing and developing and assuredly many changes and innovations lie ahead for the health care system, pharmaceutical companies and regulators.

To fully understand the regulation of OTC products, a historical review is probably the best approach to take. If we go back a long time ago, a mechanism was provided for the review of new drugs with the passage of the Federal Food, Drug and Cosmetic Act in 1938. In this piece of legislation, prescription or ethical pharmaceuticals and OTC drugs were not differentiated from each other until the passage of the Durham–Humphrey Amendment in 1951. Product safety was the only concern and there was little to no regulatory control of OTC drugs, unless the path of litigation was pursued over issues of misrepresentation (mislabeling) or concerns for public safety (Federal Register, March 25, 1960).

With the passage of the Kefauver–Harris Amendment in 1962, all drugs were not only required to be safe, but also effective. Accordingly, the Food and Drug Administration (FDA) was forced to go back and re-evaluate all earlier submitted new drug applications. This is because all of these drugs had been first approved with safety as the primary concern. Efficacy now needed to be confirmed. As can be easily imagined, this was a task of monumental proportions. To complete such a review process within any sort of reasonable time frame would have been impossible. Fortunately, many of the OTC products existing at the time were redundant in nature, had long histories of use in the population, were composed of similar if not identical ingredients and were associated with few adverse events when used properly and responsibly. With all of this in mind, the FDA opted to attack the review process from the perspective of reviewing and evaluating active ingredients as opposed to reviewing and evaluating individual products.

The FDA announced in January 1972, its plan of attack to re-evaluate drug products

(Stringer, 1999). In this proposal, the agency stated its intent to establish a group of expert advisory review panels and requested the submission of data pertinent to OTC products that were already being sold on the market. Typically the requests for information included but were not limited to: pharmacology data, medical data, indication data, efficacy data, toxicology data, human safety data, labeling information and quantitative formulation data. In accordance with the stated position of an active ingredient review, a classification scheme was implemented for use by the panels. This taxonomic approach categorizes ingredients into three different groups, categories I, II and III. The first category of active ingredients are those that are generally accepted as being safe, effective and not misrepresented. The second category of active ingredients are those that are generally accepted as being either not safe or effective or would result in misrepresentation. The third and final category of active ingredients is for those ingredients that cannot be classified in either of the first two categories, because there is insufficient data to do so.

Individual panels were autonomous, but functioned under the leadership and guidance of the FDA. They were given complete authority to review scientific data, schedule and convene open sessions and seek consultation with other relevant scientific authorities. The mandate given to the panels was to address the efficacy and safety of each drug product using uniform and consistent standards and sound scientific principles, while adhering to the FDA's stated position. The roles and contributions of individual ingredients in combination products were to be ascertained. Benefits and risks of products or components of products were to be defined and clarified. Finally, truthfulness of labeling was to be evaluated.

The 1972 plan also originally identified or proposed 26 different OTC drug categories, which were to be matched with a similar number of corresponding expert advisory panels. However, after additional consideration, the FDA reduced the number of panels to 17. At least part of the logic behind such a reduction was that there were a number of ingredients, which were used for multiple indications. In order to keep the number of expert advisory panels to a minimum and keep proper focus of discussion, the number of different use categories was defined.

Each expert advisory group conducts its review process work independently and not to any specific predetermined deadlines. However, at the conclusion of the review of each use category, the pertinent expert advisory review panel issues a report to the FDA. This report is quite extensive and complete and contains the conclusions agreed upon by the panel as well as any recommendations that the panel feels are important. The key feature of each report is what is referred to as a recommended monograph. By definition, a monograph is a book, article, etc., written about a particular subject. The subject in this case is the use category. The monograph states the conditions under which each active ingredient is efficacious, safe and most accurately and credibly represented with regard to its use. There is a wealth of other scientific information about the use category that is included in the complete report, but which is excluded from the monograph. There are two types of such information. The first type is the identification of all active ingredients, marketing and labeling claims which could lead to a lack of safety, efficacy or truth in marketing and labeling. The second is the specific identification of those active and inactive ingredients and processes and wordings of advertising and labeling claims, which were excluded from discussion in the monograph, because of a lack of availability of relevant and useful data. Such a paucity of data would preclude an adequate evaluation of any considerations of efficacy or safety as well as association with the claimed indication.

After a thorough and complete review of the report submitted by the advisory review panel, the FDA then publishes a proposed monograph. The proposed monograph is then for a 60-day period open to evaluation and comment by any interested individual or group. Following this, the comments themselves are then open for review over a subsequent additional 30-day period. Finally, when all comments collected over the combined 90-day period have been evaluated and the proposed monograph modified, a draft final monograph is published. This draft final monograph is also open to a period of scrutiny and comment by any interested individual or group. At the conclusion of this 30-day period, all questions, comments, objections or points-raised are reviewed by the FDA for relevance and validity. Finally, an oral hearing on the draft final monograph is scheduled and convened by the commissioner. At the conclusion of this hearing, after all arguments for and against the final draft have been heard and reviewed, a final monograph is prepared and published.

Even though the monograph becomes finalized, it can still be modified via several different approaches. An individual or group may file a formal petition requesting change or amendment to the final monograph. An extreme, but not unique alternative is litigation. Finally, the commissioner may alter, at his or her own discretion, the final monograph.

Not uncommonly, drugs available by prescription only are converted to OTC availability. There are two basic ways in which this can be accomplished. In the first, the commissioner can at his or her discretion or in response to a petition make such a change in status (21 CFR 310 Subpart C). Although it is true that some drugs have had their statuses changed via this mechanism, it is no longer a commonly used method. In the second approach, new drugs can be converted from availability by prescription only to OTC availability at the request of the applicant. Such a request for a change in status is effected by either filing a New Drug Application (NDA) or a supplement to the NDA that had been previously filed. The supplemental document or new NDA must provide a compelling argument for the safety of the drug or product under consideration, especially when used without the supervision of a physician.

It is important to understand the power and significance of the final monograph. Any and all products that differ in any way from the guidelines and standards set forth in the relevant final monograph are subject to confiscation. Furthermore, individuals associated with the actual or potential sales of such a non-compliant product are subject to legal action, up to and including federal prosecution. However, any product which does differ from the standards and guidelines set forth in the final monograph may still be marketed and sold as an OTC preparation by seeking approval through the filing of an NDA relevant to the difference(s) (21 CFR 330.11). With regard to this NDA "product switching" approach, it is very important to recognize that for OTC drugs, the FDA considers the relevant NDA to be a petition to amend the final monograph. Accordingly, if for whatever reason, an NDA filed in support of a switch from prescription to OTC availability is not approved, "the petition" may still be granted and marketing of the product allowed via modification of the final monograph.

NDAs filed in support of a switch from prescription to OTC availability or seeking approval of a difference from the relevant final monograph must contain certain critical information and adhere to some specific guidelines (21 CFR 330.10(a)(12)(ii)). First, it must be shown that the product, for which approval is being sought, meets the standards of the final monograph, except as pertains to the specifically identified difference. Second, all clinical testing referenced in the NDA has to have been performed specifically in pursuit of an NDA. Third, the proposed product cannot have been marketed for the indication for which approval is being sought.

Inactive ingredients in OTC products must be "only suitable inactive ingredients, which are safe in the amounts administered" (Federal Register, March 29, 1974, final order 330.1(e)). Furthermore, these inactive ingredients cannot interfere with a product's efficacy or with the procedures used and required to determine that a given product meets the claimed standards of biological potency, chemical purity, chemical concentration and identity. The inactive ingredients must not only be safe, but also serve a useful purpose (Federal Register, April 12, 1977, p. 19156).

As the market for OTC products continues to grow and develop, the FDA has responded in a supportive fashion. Historically, the agency has not given consideration to any marketing data generated from foreign countries. A similar lack of recognition has also been given to historical information relative to new concentrations of products already sold in the domestic market and for information concerning components that have been used for condition(s) other than those specified in an OTC monograph. However, the FDA has announced that it would consider the expansion of the list of criteria useful in performing evaluations of active ingredients in new OTC products (Federal Register, October 3, 1996, p. 51625). Under this proposal, such new criteria would potentially focus on the following: combinations of active ingredients, proposed conditions for treatment, dosage concentrations, routes of administration and dosage forms. Furthermore, evaluations for potential new OTC products and their ingredients would include the time interval and history of use, the extent of use and the basis of use.

### *Europe*

The "right" of people to diagnose and treat themselves for maladies has developed in Europe in parallel fashion to what has been observed in the US. European regulators depend on political support. A good example of this was in 1996, when the European Parliament and the Council of Ministers adopted resolutions that supported and recognized the importance of proper management of self-medication products and facilitated access to all EU markets (Official Journal of the European Communities, 1996). A similar view with regard to the importance of OTC drugs has also been promulgated by the World Health Organization (WHO, 1998). What all this means is a significant growth in the OTC market in the European Community (EC). Whereas at one time, OTC products were chiefly the interest of relatively small companies, now large international companies are investing heavily in and dissecting out their OTC-related activities.

In order to facilitate economic intercourse, the EC formulated and developed a detailed program for the pharmaceutical sector (Council directive 92/26/EEC, 1992, concerning the classification for the supply of medicinal products for human use, April 30, 1992). On March 31, 1992, the EC established standards for prescription medicines and common requirements for members of the European Union (EU). In this directive, it was stated that a medicinal product is subject to medical prescription when: (1) the material presents a danger, even when used correctly, if not used under medical supervision; (2) it is frequently used incorrectly and thereby presents a danger to humans; (3) the material contains substances that have actions or side effects that require further evaluation; (4) its normal route of administration is parenteral. Furthermore, the directive goes on to state that medicinal products that are not classified as prescription drugs need to be classified as nonprescription drugs. One possible interpretation of this definition and position is that the EC would prefer to see products available OTC rather than available by prescription by a medical professional. Finally, it was enunciated that all medicinal products must be examined every 5 years or earlier if compelling data requiring such is presented to the regulatory authority. Where the directive fell short is in not attempting to harmonize differences between member states, but still leaving the responsibility for status inter-conversion to each member state.

A common problem in the EU concerning the conversion of a product's status from prescription availability to OTC availability is whether or not the change refers to the product or the substance. For most, but not all (UK and Germany) member sates, the prescription to OTC availability change is made on a product basis. Typically, conversion from prescription to OTC status is made on the basis of a variety of considerations. These would include: safety data available from current or prior prescription use, status conversion data from other countries and efficacy data relevant to the indication if the indication differs from that originally approved for the prescription medication. Complicating matters can be differences within the EU relative to the legal classification of various pharmaceutical ingredients. Additionally, there are often other complicating differences such as dosage, dosage form and specified indications.

Probably the most significant development in the arena of OTC products in the EU has been the release of the European guidelines on the conversion of drugs from prescription availability status to OTC availability (European Commission, 1998). Of not insignificant importance are the opening statements of this directive, which recognize that there are considerable differences between drug availability and drug classification between various member states in the EC and that it is very important to reduce or eliminate such differences and to harmonize positions. This guideline was intended for use by those entities seeking to make application to change the classification of a medicinal product and also to facilitate harmonization within the EU. The EC chose to opt for clarification in this directive, without forcing a strict listing of products and categories on member states. Accordingly, this directive does not address differences in the rules and regulations for drug products available by OTC means. Despite its deficiencies, the directive does however, outline the criteria necessary for converting the status of a drug from prescription availability to OTC availability. Essentially, the switch of the status of a drug from prescription to OTC requires profiles of a product's safety and its potential for use and misuse. Note that no proof of efficacy is required. However, history, history of use, extent of use and the usage pattern are other facts considered in any request for switch of status. This directive represents an important consensus between all involved authorities and is a good beginning. What remains to be determined now is how individual member states will each interpret and apply the directive. Only the future can reveal that.

In conclusion, suitable guidelines are presently in place, which address the classification of prescription and nonprescription ingredients (Dechamp, 1999). However, significant differences still remain within the EU as a result of previous individual national or member state evaluations and because of political differences. Therefore, complete harmonization has not occurred and more than likely will not occur in the near future. However, the process of harmonization and current directives in place should provide a solid platform for future work and development.

### *Japan*

The demand for increasing availability of OTC products is prominent in Japan also. Regulations and guidelines pertinent to marketing OTC drug products in Japan are covered under the registration of proprietary drugs as described in the Japanese Technical Requirements For New Drug Registration (Japanese Ministry of Health and Welfare, 1997).

Reviews for the approval of proprietary drugs are based on the nature and characteristics of proprietary products. The ingredients used must be well within a well-defined range of efficacy and safety. A similar position is taken with regard to the concentrations or amounts of ingredients. Generally, drugs or products with strong action are not considered appropriate proprietary drugs. The dosage and administration of a product must be able to be competently determined and applied by the general public. Finally, allowed indications are restricted to prophylaxis, the treatment of mild disease states and the promotion of good health. Admittedly, the promotion of good health is a very vague concept most appropriately interpreted as meaning that a substance is possibly helpful and definitely not harmful. It should be obvious from this position, that any conditions which are typically diagnosed and treated by physicians are not candidates for indications for proprietary drug products.

Approval standards concerning indications, ingredients, quantities, dosages, administration and effects for each therapeutic classification are prepared according to the opinions of the Central Pharmaceutical Affairs Council (CPAC). Currently, approval standards have been prepared for cold remedies, antipyretics, analgesics, antitussives, expectorants, purgatives, gastrointestinal agents, antivertigo remedies, ophthalmologics, vitamins, anthelmintics, rhinitis (nasal and oral administration), enemas and external hemorrhoids. The authority to approve these classes of proprietary products has been delegated to the prefectural governors.

The approval review process for proprietary drugs is tied very closely with a classification scheme discriminating between new and "other" proprietary drugs.

New proprietary drugs are those drugs, which have not been approved previously as proprietary drugs. These drugs are reviewed on an individual basis by the Committee on Non-Prescription Drugs, the Subcommittee on Non-Prescription Drugs, the Subcommittee on Chinese Medicines and Animal and Plant Origin Products and possibly others of CPAC to determine if they should be approved. Since new proprietary drugs should have indications, effects, routes of administration, etc., identical to those approved for ethical drugs with the same ingredients, applications as ethical drugs are considered. New proprietary drugs are broken down into three different categories: (1) drugs with new active ingredients, these are referred to as "direct OTC" drugs; (2) drugs with approved ingredients used for the first time as active ingredients in proprietary drugs, these are referred to as "switch OTC" drugs; (3) drugs whose active ingredients are already approved, but now present with different combinations of active ingredients, indications, effects, etc.

"Other" drugs are those which are not new proprietary drugs. A further sub-classification is made for "other" proprietary drugs into those where approval authority has been transferred to prefectural governors and those for which the approval authority remains with the Minister of Health and Welfare (MHW). In the latter case, "other" drugs are categorized as: (4) drugs the same as new proprietary drugs, whose periods for observation for adverse reaction surveillance are over; (5) drugs with special dosage forms not in accordance with the normal approval standards; (6) drugs which meet the approval standards for proprietary drugs, but whose approvals are issued by the MHW. Other drugs, which do not fit into categories (1–5) above are included in this category. Proprietary drugs for which approval authority has been transferred to prefectural governors include: drugs for *tinea app.*, anthelmintics, cold treatments, drugs for rhinitis (oral and nasal administration), external hemorrhoid preparations, antipyretics, antitussives, ophthalmics, expectorants, analgesics, purgatives, vitamins and vitamin preparations, antivertigo drugs and enemas.

The review process for proprietary drugs is, according to category: (1) a review is performed by CPAC after hearings with the MHW, (2–3) CPAC performs a review only after the Drug Organization has reviewed issues of indication, effects, dose equivalence and dose administration with those of an approved drug, (4–6) The Drug Organization conducts an equivalence review, hearings at the MHW and consultation with CPAC as necessary.

As one can easily see, the approval process for OTC products in Japan is the most complex of the three geographic and regulatory entities presented (Japanese Ministry of Health and Welfare, 1995, 1997).

### *Closing statement*

Not unsurprisingly, the key to putting OTC products on the market is the ability to establish efficacy and safety. Whereas the inter-conversion of drug or product status from prescription availability to OTC availability, might appear to be simpler and even more common, the same volume of data demonstrating product safety and efficacy is ultimately required. When putting together a developmental package for a new OTC product for any regulatory agency in the world, the most responsible and successful approach is to treat a potential OTC drug or product just as one would an ethical pharmaceutical. To this end, applications should include:

1 History, origin and background information.
2 Chemical structure, name, enumeration of physico-chemical properties.
3 Analytical and/or bioanalytical testing methods.
4 Formulation information.
5 Stability test results.
6 General pharmacology profile, including efficacy testing.
7 ADME information: absorption, distribution, metabolism, excretion, elimination and biological equivalence.
8 Safety pharmacology profile: CNS, CV, renal, GI and pulmonary systems.
9 Toxicity profile: acute, subchronic, chronic, reproductive, carcinogenicity, mutagenicity, hypersensitivity and antigenicity testing.
10 Other specific specialty tests as appropriate.
11 Clinical trial results.

An excellent discussion on navigating the drug development maze for human pharmaceuticals is contained in Chapter 2. The reader is urged to consult this reference to fully appreciate the requirements for developing OTC products. Lest we forget, never underestimate the importance of opening a dialogue early with the regulatory agencies of concern.

## Nutraceuticals

What are nutraceuticals? Generally speaking, nutraceuticals are nutritionally or medicinally enhanced foods. The term nutraceuticals was originally used and defined in 1979 by Stephen DeFelice, the chairman and founder of the Foundation for Innovation in Medicine (FIM, Cranford, NJ). According to DeFelice, nutraceuticals are defined as "food, or parts of food, that provide medical or health benefits, including the prevention and treatment of disease" (Brower, 1998). Not unexpectedly, other terms were soon to follow this proposed definition. These included medical or functional food and dietary supplements. Still today, there is disagreement as to what each of these terms actually means. Medical foods, functional foods and dietary supplements are all specific types of nutraceuticals. Functional foods are foods, which are specifically created or supplemented to impart improved nutritional value to a food. They are freely available as OTC purchases. An example of one might be a genetically altered peanut with diminished allergenicity. Medical foods are those foods designed for consumption or enteral administration under the supervision of a physician. These foods are intended for the specific dietary management of a disease or condition and are available via prescription only. An example might be a bottle of soda pop, which contains an antibiotic. Finally, dietary supplements are materials produced as the result of synthesis, partial synthesis, purification, isolation, or culture, which provide health benefits. These products are available as OTC products and examples might be tyrosine, carnitine or choline (Glinsmann, 1996).

There are many reasons for the burgeoning popularity and prevalence of nutraceuticals, especially dietary supplements. People want to control their own destiny, and this feeling applies to health also. In some cases the use of products of this type is deeply rooted in culture. Take for instance the Japanese, they have a long history and tradition of using food for health purposes (The Japanese Standards for Herbal Medicines, 1993). Baby boomers have become disillusioned with the health care system and want more proactive control over their own health rather than merely reaction (Meyer, 1997; Brower, 1998). Place all of these feelings and philosophies against a backdrop of rising health care costs, and it is easy to appreciate at least in part why people now want to eat healthier.

The hope is that the use of nutraceuticals will provide a sufficient degree of prophylaxis to significantly decrease the costs associated with expensive visits to a physician and trips to the local pharmacy for expensive prescription drugs. A quarter of a century ago, eating healthy just meant eating a balanced diet with proper representation from each of the major food groups. Today, in all kinds of stores there are myriad new foods, which claim healthful effects. Many of these types of products are even considered to be legitimate by insurance companies and the expense of their purchase covered in health care plans.

The availability of these products is not serendipitous. Instead, their appearance is the result of careful market research by many companies, large and small, old and new. Products associated with 'healthy' eating represent a supreme opportunity, because of the attractive economic returns and an extremely lax global regulatory environment. For now, nutraceuticals, do not *require* an expensive and time-consuming process to gain marketing approval. However, with a plethora of products starting to flood the marketplace, concerns are starting to surface as to whether or not nutraceuticals, at least in some cases, are crossing the lines of demarcation between foods and drugs. This is probably best exemplified in a well-known case involving Merck, Pharmanex, the FDA and Pharmanex's product, Cholestin (Brower, 1998).

The term food generally means those commodities used for food or drink. The term drug generally refers to commodities that are intended for the diagnosis, cure, palliation, treatment or prevention of disease. A product can be both, but applicable food and drug laws regulate food and drug products. Foods are typically considered to be safe for the average person. Drugs, alternatively, are not necessarily safe and are approved on the basis of an acceptable benefit-to-risk ratio. This approach to regulation has become increasingly more difficult to apply, since foods are becoming progressively more visible as a result of their claimed health effects. Indeed, some foods now even contain therapeutic components (Glinsmann, 1996; Brower, 1998). Such foods might be most appropriately regulated through a combination of both food and drug guidelines and laws. Such a lack of regulatory clarity and direction is consistently found not only in the US, but also Europe and Japan.

A major question to ask and answer before proceeding with the development of any potential nutraceuticals is: "Is it a food, is it a drug or is it a supplement?" (Love, 1998). Keep in mind that it is the intended use of a proposed nutraceutical product rather than the type of ingredient in the nutraceutical product that determines the applicable review process. Drugs must be proven to be safe and effective for a particular indication before marketing. Food additives incur a FDA premarket review of ingredient safety. All ingredients must be Generally Recognized As Safe (GRAS) or specifically authorized as a food additive. Dietary supplements are an extremely diverse group of substances, and include, but are not limited to: amino acids, organ preparations, minerals, various extracts and concentrates, metabolites, enzymes, herbs, vitamins, botanicals or combinations of any of these ingredients. Dietary supplements have no FDA premarket review of ingredients or finished products. They also incur no regulations concerning good manufacturing practice, identification, characterization or standardization of ingredients, efficacy or safety.

If a nutraceutical is truly a drug, then the appropriate guidelines as set forth in Chapter 2 on human pharmaceuticals should be consulted. Generally speaking, global submission type developmental packages for this specific type of nutraceutical should minimally include:

1. History, origin and background information.
2. Chemical structure, name, enumeration of physico-chemical properties.
3. Analytical and/or bioanalytical testing methods.
4. Formulation information.
5. Stability test results.
6. General pharmacology profile, including efficacy testing.

7 ADME information: absorption, distribution, metabolism, excretion, elimination and biological equivalence.
8 Safety pharmacology profile: CNS, CV, renal, GI and pulmonary systems.
9 Toxicity profile: acute, subchronic, chronic, reproductive, carcinogenicity, mutagenicity, hypersensitivity and antigenicity testing.
10 Other specific specialty tests as appropriate.
11 Clinical trial results.

If the nutraceutical is a food or food additive, then appropriate regulations on foods need to be consulted (Love, 1998) or Chapter 5 on food additives in this book. In general, a petition to be submitted for the US, Europe or Japan for this specific type of nutraceutical should minimally include the following information:

1 A complete description and characterization of the chemical and or compositional identity of the additive. This section should also contain impurity and stability profiles and methods of analysis.
2 A complete discussion of the background and theory behind the proposed use of the additive. It is important to include an "estimated daily intake" calculation for the additive.
3 A complete description of the intended technical effect of the additive. Suitable documentation must also be provided substantiating the minimal amount of additive required to provide the intended effect.
4 Documentation of a sensitive, accurate, specific, precise and reliable method of analysis of the additive in the food. The method should be simple and facile to perform.
5 A safety profile in support of the additive's use. This safety profile should minimally include:

   a. Safety pharmacology profile: CNS, CV, renal, GI and pulmonary systems.
   b. Toxicology profile consistent with "concern level" as determined from "structure category" and level of dietary exposure (mg/kg per day).
   c. "Structure category" (A, B or C) and "concern level" are determined from the Redbook II (USFDA, 1993) once the level of exposure has been calculated.

As stated previously, dietary supplements have a very broad definition and this allows for a multitude of substances with various functional effects spanning the entire spectrum of health use. Accordingly, this only makes more complex the attempts to harmonize the regulation of development of these substances. From a global perspective the situation is even more complicated. Dietary supplements have been routinely used as part of the diet for centuries in Europe and particularly Japan and are deeply rooted in folklore and tradition. The use of these substances by the respective indigenous populations is perceived as being a right of the people and any attempt at governmental regulation would be construed as infringement on that right. Hence, there are currently no published guidelines for the development of dietary supplements in either Europe or Japan.

The culture entwining the use of dietary supplements is different in the US, so the US government attempted to address the growing use of and problems associated with these types of nutraceuticals with the passage of legislation. In 1990, the Nutritional Labeling Enforcement Act exempted medical foods from the health claim and labeling requirements applied to foods sold to healthy people. Then the Dietary Supplement Health Education Act of 1994 permitted unprecedented claims to be made about a food's or dietary supplement's ability to affect "structure and/or function" of the body. Although this latter piece of legislation was intended to create guidance in the field of integrative or alternative medical treatment research, it ended up creating significant controversy as to the necessary requirements for the approval of dietary supplement nutraceuticals with regard to medical and health claims. This is because the act still did not provide any published guidelines for the development of dietary supplements in the US. The FDA should soon be providing more guidance and clarification with regard to the statement of claims for dietary supplements. Last year, the agency stated its intent to soon define the criteria for structure/function claims and describe the various means by which a dietary supplement could make or imply a disease claim prohibited under the Dietary Supplement Health Education Act. Hopefully when they appear, these new rules will provide the strongly needed clarification in the US to the currently vague differences between an unapproved and implied health claim and a legitimate structure/function claim. It is hard to say what exactly lies ahead in Europe and Japan as regards this issue.

From a regulatory viewpoint, how a dietary supplement product is treated is determined in a large part by how it is labeled and what claims are stated on the label or in the course of marketing the material. Therefore, it behooves one to be very scrupulous with regard to the specific wording in a statement of claims for a potential product. Such claims and verbage in the claims can define the entire course of performance of future safety evaluation studies. Such a course can range from the performance of no studies to a volume of work equivalent to an NDA.

With regard to the development of dietary supplements it is important to understand and appreciate the equilibrium between governmental regulation, corporate financial pressures and ethics and liability. Governmental regulation is typically reactive and not proactive. Therefore, not unexpectedly there is a lack of formal guidance from governmental agencies with respect to the marketability of dietary supplements unless they creep into the areas of drugs or food additives. This is not surprising since historically the use of dietary supplements has not been problematic. However the dietary supplement market has begun to explode with products. This is because businesses have finally recognized significant market opportunity in the sales of substances that have long been used by a variety of racial and ethnic groups for health purposes. The goal of business is to achieve profit margins on their products. Therefore the cost to put products on the market is not an insignificant issue. Companies must conform to governmental regulations and guidelines while concurrently keeping a watchful eye on the costs associated with the marketing and sales of a product. Yet the issue is still more complex in that a business must also be concerned with its image and any potential liability resulting from the use or misuse of a product. From a litigious perspective, a company should be able to demonstrate "good faith" in its performance of research, development and safety evaluation for a potential product before it is put on the market. In the development of dietary supplement nutraceuticals it is essential to focus on the interrelationships that exist between law, profit and liability and not just law, profit or liability.

For the global marketing of a potentially simple, pure dietary supplement nutraceutical with no health benefit claims, the only information that is required from a regulatory perspective is a description of the product. However, one should consider a benefit-to-risk ratio in having or not having available the following additional minimal information:

1 Basic general pharmacology profile.
2 Acute toxicology profile.
3 Antigenicity and hypersensitivity testing.
4 Other specific tests as appropriate.
5 Clinical trials? (see below).

For the global marketing of a potentially simple, pure dietary supplement nutraceutical with limited health benefit claims, the following minimal amount of information in support of a product is recommended:

1 Product description. This does not have to be detailed and does not need to include the chemical structure, composition, purity, stability analysis or formulation information. If the product is a standardized formulation, it should be mentioned.
2 Definition of the target condition.
3 Documentation of the prevalence of the target condition.
4 Documentation of the structural or functional benefit or effect. This can assume the format of testimonials.
5 A statement of an effective dose.
6 Data or literature describing a mechanism of action or possible mechanism of action. A basic or general pharmacology profile could be included.
7 Toxicity profile: acute, subchronic, hypersensitivity and antigenicity testing.
8 Other specific specialty tests as appropriate.
9 Clinical trials? (see below).

For the global marketing of a potential simple or complicated, pure dietary supplement nutraceutical with far reaching health benefit claims, the minimal amount of information that should be generated in the course of a development program would be similar if not identical to that previously suggested for an OTC drug or nutraceutical viewed as a drug.

For two of the paradigms described above for dietary supplements, clinical trials were mentioned as potential components of development packages and their potential inclusion represents a good example for the choices that need to be made in the developmental decision tree for nutraceuticals. Even though such studies are not required to corroborate claims of efficacy or safety with regard to a compound's stated structural or functional benefits, they can provide a useful function. Many companies are now considering and some even electing to perform such studies to provide stronger support to their claims and secure proprietary positions for their products.

Post-marketing surveillance takes on different forms, depending upon the type of nutraceutical. Adequate systems currently exist for foods, food additives and drugs (Love, 1998). However the monitoring of dietary supplements is a complex affair because there are a variety of factors that influence post-marketing safety. These factors include, but are not limited to a lack of adequate scientific data on: efficacy and safety, widespread use throughout the population, chronic use, abuse, biochemical, physiological or pathological synergy, allergy and sensitivity. This is in part a result of these products being sold via catalog sales, the internet, super markets, health food stores and other small establishments, where no surveillance is conducted in contrast to the local pharmacy. Remember, that relative to dietary supplements, the burden of proof of significant risk of a product used according to label directions lies with the FDA before any action can be taken.

Probably the most important and lasting message to close this section with is that although there are always faster and more inexpensive ways to put products on the market, there are still only few responsible ones (Hathcock, 1993). Following the latter path, ensures efficacy, safety and no loss of credibility (Borins, 1998). Adjustments to this approach, made on the basis of economy of cost should be implemented with a full understanding of the ramifications of such action.

## References


Borins M. The dangers of using herbs; what your patients need to know. *PostGrad Med* 1998;104(1):91–100.

Brower V. Nutraceuticals: poised for a healthy slice of the healthcare market? *Nat Biotechnol* 1998;16:728–731.

Council directive 92/26/EEC. Concerning the classification for the supply of medicinal products for human use, April 30. *Official J Eur Commun* 1992:No. L 113.

Dechamp, J-F. Rx to OTC switch plays a crucial role in Europe. *Regulat Affairs Focus* 1999;4(3):14–18.

European Commission. On changing the classification for the supply of a medicinal product for human use. *The Rules Governing Medicinal Products In The European Community, Volume III B: Guidelines*. European Commission, Directorate General III-E-3, September 29, 1998.

Glinsmann, WH. Functional foods in North America. *Nutr Rev* 1996;54(11):S33–S37.

Hathcock, JN. Safety limits for nutrient intakes: concepts and data requirements. *Nutr Rev* 1993;51(9):278–285.

*The Japanese Standards For Herbal Medicines*. Tokyo: Yakuji Nippo, Ltd., 1993.

Love L. Food and drug administration update; the MedWatch Program. *Clin Toxicol.* 1998;36(3):263–267.

Meyer, GS (editor). The evolving health care market-exploring alternatives. *Nutrition* 1997;13(6):589–590.

Official Journal of The European Communities. May 13, 1996. No. C 141. (http://europa.eu.int).

Japanese Ministry of Health and Welfare (editors). *Japanese Guidelines For Nonclinical Studies Of Drugs Manual 1995*. Tokyo: Yakuji Nippo, Ltd., 1995.

Japanese Ministry of Health and Welfare (editors). Registration of proprietary drugs. Chapter 4. *Japanese Technical Requirements for New Drug Registration 1997 with Relevant Japanese and ICH Guidelines Collected In Appendix*. Tokyo: Yakuji Nippo, Ltd., 1997. pp. 74–85.

"Redbook II" (*Toxicological Principles for the Safety Assessment of Direct Food Additives and Color Additives Used in Food*). US Food and Drug Administration, Center for Food Safety and Applied Nutrition, 1993.

Stringer, S. Over the counter drug regulation encapsulated. *Regulat Affairs Focus* 1999;4(3)8–11.

WHO. Contribution to updating the WHO guidelines for developing national drug policies, action programme on essential drugs. *WHO Expert Committee on National Drug Policies*. Geneva: WHO, June 19–23, 1998. pp. 49–50 (WHO/DAP/95.9).

Chapter 8

# Consumer products

## Nonpersonal care products

*Shayne C. Gad*

Citizens in western societies are increasingly aware of a need to reside within a safe environment. Access to safe medicines, foods, and cosmetics, and access to nonhazardous products available for our daily use are expected aspects of our lives. In this quest for safety, we have demanded action on the part of our governing officials, and the response has been a plethora of laws and regulations designed to increase safety of the products we buy. The Food and Drug Administration (FDA), for example, was created in response to concerns about the safety of our food and medicines. There are, however, a large number of consumer products with which people come into direct contact that are not regulated by the FDA. These regulations fall under the purview of the Consumer Product Safety Commission (CPSC).

The CPSC is an independent commission (established in 1972) which reports to both the executive and legislative branches of the federal government. Administrators (of which there have been three to five at any given time) are appointed by the White House and approved by Congress. Currently there are approximately 500 employees of the CPSC with half of those working in the Washington area while the remainder are in regional offices. It is estimated that there are approximately five to ten professional toxicologists with the Commission at this time. It has jurisdiction over $\sim$ 15,000 types of consumer products. The agency's web site is http://www.cpsc.gov.

The potential chemical hazards associated with consumer products include flammability, explosive potential, corrosivity, and other forms of reactivity. Assessments of human toxic hazards associated with chemical exposures are conducted by professionals with an understanding of the applicable toxicological and chemical information pertinent to the products under consideration. This chapter will focus on issues related to the assessment of toxic hazards associated with the use of consumer products.

By definition, consumer products are any manufactured goods (e.g., household detergents, electric appliances, clothing, cleaners, toys, cosmetics, personal care products) which are sold off the shelf or over the counter to consumers and which require no product-specific license or application to market. In an effort to regulate labeling and other activities designed to enhance the safe use of these products, the CPSC was created by the Consumer Product Safety Act (CPSA), enacted in 1972 (Federal Hazardous Substance Act Regulations, 1992). The intentions of the CPSA were to protect the public against unsafe consumer products, to develop uniform safety criteria for consumer products, to provide the consumer with information so that informed purchase choices can be made, and to pursue research with the stated goal of preventing unsafe products from entering the marketplace.

The CPSA requires adherence to these regulations, but unlike the FDA, CPSC is largely reactive and relies on criminal and civil prosecution to enforce regulations. With specific regard to chemical toxicity hazards, the Federal Hazardous Substance Act (Part 1500 of the regulations), enacted in 1958 and amended several times since, gives the CPSC authority to ban or otherwise regulate products containing potentially dangerous chemicals such as certain paint solvents, cleaning fluids, charcoal lighter fluids, and certain other products. Under these regulations, the CPSC can require special warning labels or an outright ban of certain toxic substances from consumer use.

What is the role of toxicology in these efforts, given the stated goals of consumer protection regulations? The CPSC is responsible for maintaining product safety information, setting forth safety criteria, and banning unsafe products. To the extent that a product contains chemicals which may be "bioavailable," toxicity information and assessments are required for the CPSC to meet these mandates. It is important to note that the CPSC does not require premarket clearance for a specific product. If a product does contain a chemical that is known to be potentially hazardous, however, the agency is empowered to establish compliance standards. These standards may require labeling or specific packaging designs when a product component is regarded as hazardous. There are regulations limiting the sale of products containing asbestos or formaldehyde. In other cases, the regulations require explicit warning labels on products such as paints and cleaning agents which contain certain agents with an elevated hazard potential.

There are specific testing guidelines for development of acute toxicity information within CPSC regulations for those products with inadequate safety information. It is noted that organizations selling consumer products and the CPSC rely primarily on existing toxicity data of product components obtained from published literature.

## Art materials

One category of consumer product for which "bioavailable" chemicals are major components is art products. Regulatory oversight of labeling of these materials is the responsibility of the CPSC. The required product labeling is written by suppliers of art materials following guidelines specified by CPSC, and is intended to provide purchasers and users of these products information that is specific regarding the potential hazards associated with their use. For art materials as well as with virtually all consumer products, appropriate labeling includes warnings of immediate or short-term hazards associated with contact with the product. Regulations addressing label specifications are part of the Federal Hazardous Substance Act Regulations (1992) Federal Hazardous Substance Act (1992).

Over the last 30 years, there has developed increased recognition that many chemicals have the potential to induce delayed adverse effects with long-term consequences. These toxicities include carcinogenicity, birth defects, and impaired reproductive capabilities. Clearly, a concern for these deleterious effects is justified, and scrutiny of all appropriate safety information on product constituents is needed. While there have been substantial amounts of animal test information developed concerning chemicals and their capacities to produce adverse effects, interpretation of this information and its application to the development of appropriate warnings on labeling requires judgment from qualified professional toxicologists, a fact recognized in the applicable guidelines for performing label development.

## Applicable laws and regulations

### *Acute effects*

Warning labels which address possible deleterious effects from short-term exposures (such as skin or eye irritation) are generally based upon either human experience with these chemicals or from animal testing. In the SPSC regulations addressing label development and acute hazards, there are very specific definitions of terms such as *irritancy, toxic,*and *highly toxic*. For the most part, chemical designations are derived from animal tests which have been conducted according to proscribed protocols and which allow quantitative determinations of toxicity classes. For instance, when a dose of 200 mg/kg of a substance applied to the skin of rabbits causes more than half of a group of ten animals to die within 14 days, this chemical is considered to be highly toxic. Once the toxicity class had been established, certain signal words are then mandated by regulation as warnings on the label. In the case of a highly toxic substance, the signal word is *DANGER*.

Other categories of defined hazards include designations of *hazardous substance, corrosive, strong sensitizer, flammable, extremely flammable,*and *poison.*The signal word for hazardous substances is *DANGER*and, for compounds with lesser hazard potentials, *WARNING*or *CAUTION.*These products are those that have drawn the greatest pressure to adopt *in vitro* test methods from animal rights groups. Industry groups such as the Soap and Detergent Association (SDA) have attempted to validate and promote such methods.

### *Chronic effects*

As a result of the absence of guidelines for providing label warnings to consumers which address long-term hazards from art products, a group of art material trade organizations (the Art Supplies Labeling Coalition) provided support for the development of a consensus standard which would address chronic hazards. The task was undertaken by a committee of the American Society for Testing and Materials (ASTM), and resulted in a guideline designated ASTM Standard D-4236 or just D-4236. The first edition was developed in 1983, and the most recent version in 1991 is designated D-4236-91 (ASTM, 1991).

In 1988, the US Congress passed Public Law 100-695, which amended the Federal Hazardous Substances Act to include the requirement that chronically hazardous art materials should be appropriately labeled. The law indicates that this labeling is to be specifically performed according to the D-4236 guidelines. For all existing art materials, labeling changes were needed so as to include warnings of chronic hazards; these labeling modifications were to be made by November 1990. Parenthetically, art materials represent the first class of consumer products for which chronic warnings are federally mandated.

The adverse effects considered "chronic" are traditionally those such as carcinogenicity and teratogenicity. However, D-4236 also includes findings such as permanent eye damage, specific organ damage, certain neurological and hematological effects, sensitization, and potential for excretion in human milk as chronic effects.

As with labeling for acute hazards, label development for chronic effects requires a thorough understanding of testing information pertaining to potential adverse effects. From this step, selections of appropriate chronic hazard and precautionary labeling statements are relatively straightforward since the guidelines are quite explicit, providing many specific statements for various toxicity categories. For chronically hazardous products, the signal word for labeling is *WARNING*. However, when the material possesses both acute and chronic hazard properties, the designated signal word is derived from the acute hazard. It should also be noted that many precautionary statements that are prescribed for labels for chronic hazards (ASTM, 1991) are well suited for acute effects as well.

## Health-based risk assessments

### *Professional qualifications*

These D-4236 guidelines for chronic hazards indicate that product labeling will appropriately reflect a hazard when "in an opinion of a toxicologist, [the product] has the potential to produce a chronic adverse health effect". It also indicates that the person (toxicologist or physician) should be appropriately trained and experienced in addition to be being certified by a nationally recognized professional board. A number of toxicological as well as occupational medicine certifications exist to provide the accreditation needed to qualify individuals for this task.

In the authors' experience, knowledge and familiarity with matters dealing with chemical and physical properties of art material components is of great benefit when considering the exposure potential of individual components (discussed further below). Issues such as chemical phases, viscosities, pH, volatilities, and dilutions are important factors in exposure assessments, and experience in chemistry contributes immeasurably to informed assessments of potential risks associated with working with these products.

### *Assessment of art material hazards*

On November 18, 1988, the President signed into law the Labeling of Hazardous Art Materials Act (Public Law 100-695). This law requires that all art materials be reviewed to determine the potential for causing a chronic hazard and that appropriate warning labels be put on those art materials found to pose a chronic hazard. The term "art material" includes "any substance marketed or represented by the producer or repackager as suitable for use in any phase of the creation of any work of visual or graphic art of any medium" (15 U.S.C. 1277(b)(1). The law applies to many children's toy products such as crayons, chalk, paint sets, modeling clay, coloring books, pencils, and any other products used by children to produce a work of visual or graphic art.

The Labeling of Hazardous Art Materials Act (LHAMA) amended the Federal Hazardous Substances Act (FHSA) by adding Section 23 and designating the ASTM Standard Practice for Labeling Art Materials for Chronic Health Hazards (ASTM D-4236-88) as a regulation under Section 3(b) of the FHSA. The requirements of the LHAMA became effective on November 18, 1990. These requirements apply to art materials that are intended for use in the household or by children, which are initially introduced into interstate commerce on and after November 18, 1990.

The Commission believes that under the broad statutory definition of "art material" three general categories can be seen:

1 Those products which actually become a component of the work of visual or graphic art, such as paint, canvas, inks, crayons, chalk, solder, brazing rods, flux, paper, clay, stone, thread, cloth, and photographic film.
2 Those products which are closely and intimately associated with the creation of the final work of art, such as brush cleaners, solvents, ceramic kilns, brushes, silk screens, molds or mold making material, and photo developing chemicals.
3 Those tools, implements, and furniture that are used in the process of the creation of a work of art, but do not become part of the work of art. Examples are drafting tables and chairs, easels, picture frames, canvas stretchers, potter's wheels, hammers, chisels, and air pumps for air brushes.

The Commission does not believe that Congress intended products in the third category to be considered "art materials." Therefore, as an enforcement policy, the Commission is not requiring that products failing in this third category comply with the standard for art materials. However, manufacturers still have the responsibility under the FHSA to assure that these products comply with any FHSA labeling or other requirements due to chronic toxicity or other hazards.

Parents and others buying art materials, school supplies and toys such as crayons, paint sets, or modeling clay should be alert and purchase only those products which are accompanied by the statement "Conforms to ASTM D-4236".

The LHAMA does not change the fact that products which are hazardous are banned for distribution to young children, whether the hazard is based on chronic toxicity, acute toxicity, flammability, or other hazard identified by the FHSA. There is an exception for art materials if they meet all three of the exemption criteria of Section 2(q) of the FHSA in that they: (1) require the inclusion of the hazardous substances for their functional purpose, (2) bear labeling giving adequate directions and warnings for safe use, and (3) are intended for use by children who have attained sufficient maturity, and may reasonably be expected, to read and heed such directions and warnings.

The assessments should proceed in a manner appropriate for all chemical products according to the following concept:

TOXICITY and EXPOSURE ⇒ CHEMICAL HAZARDS

Typically, art products are composed of mixtures of ten or more chemicals, and frequently exceed twenty. Rarely are these complex art materials tested for either acute or chronic toxic effects; therefore, the toxicologist must assess the hazard potential of complex mixtures based upon:

1 Toxic properties of components
2 Concentrations of each component in the product
3 Chemical/physical properties of the product which may alter components' toxic or exposure potentials
4 Exposure scenarios during use of product.

Obviously, the lack of empirical information on the intact product may raise questions of the accuracy of such an evaluation. However, this problem is best approached by carefully recording the bases and rationales used in arriving at conclusions on the hazards of these mixtures. If each factor is carefully considered and documented so that risk assessments can be reconstructed at later times, conflicts can be avoided. Of course, if new information indicates that a component(s) possesses previously unknown toxic properties, a reassessment of the product may be indicated and a change in labeling may be required. The following expands on the issues to be considered in the evaluation of art materials.

### *Toxic properties of components*

A valuable source of information on chemicals which constitute the art material are the component suppliers' Material Safety Data Sheets (MSDSs), which all art material producers should possess and make available to the toxicologist developing the labeling. Because MSDSs have been substantially upgraded since enactment of the Occupational Safety and Health Administration (OSHA) Hazard Communication Standard, these documents should include both acute and chronic toxicity information, physical and chemical properties as well as other information needed to provide an understanding of the hazard contribution that chemical components may make to the final art product. Important chemical properties to look for in an MSDS include pH of aqueous solutions, corrosiveness, solubility and reactivity (particularly in water), and volatility. In the event that there is uncertainty regarding the adequacy of the information in the MSDS, contact should be made with the supplier to determine what information sources were used to prepare the MSDS and how recently these sources were searched.

In addition to the MSDSs from suppliers of the component chemicals, art material manufacturers should also prepare MSDSs for their art products. It is important that reviewing toxicologists examine the art product MSDS to confirm that there is concordance between information included in that document and that which is to be incorporated into the product label. It is the obligation of the toxicologist to alert the art material producer when the product's MSDS is not consistent with findings leading to label development.

When there is inadequate information from MSDSs for the investigator to perform the hazard assessment, alternative sources of information are needed. One printed source for acute and chronic effects of chemicals is the Registry of Toxic Effects of Chemical Substances (RTECS) published by the National Institute of Occupational Safety and Health (NIOSH). For carcinogens, documents such as the International Agency for Research in Cancer monographs or National Toxicology Programs (NTPs) Annual Reports on Carcinogens provide lists of known or suspect human carcinogens. Other textual sources include Larson *et al.* (1994); Kapp (1999). There are also national data banks available via computer terminals (Medline®, Toxline®, and others that are specific for cancer, birth defects, and mutations). In the event little or no information exists for a chemical, knowledge of the toxic properties of related chemicals may suffice, depending on the exposure potential.

When high exposure potentials are projected for the use of certain art products, thorough literature searches are important. In the event adequate data do not exist, it may be appropriate for the art material supplier or producer to conduct specific toxicity testing on their product. If, on the other hand, a component is present at a low concentration (e.g., $< 1$ per cent), it becomes very unlikely that this substance contributes significantly to the overall hazard of the product. One important exception to a lack of toxicity contributions from minor components is for chemicals which are sensitizers. Such agents may cause a formulated art product to exhibit allergic properties, particularly in previously sensitized individuals. The toxicologist must try to balance all available information on these products to develop rational and defensible proposals for the appropriate labeling.

#### CONCENTRATIONS OF COMPONENTS IN THE PRODUCT

It is absolutely necessary that the toxicologist have access to the exact composition of the art material to be evaluated. Obviously, judgement on chronic risks would be substantially different for a product which contains greater than 10 per cent of a teratogen or a carcinogen than when the component concentrations is less than 1 per cent. Similarly, sodium hydroxide or an acid can be viewed as very hazardous at high concentrations, whereas dilution or neutralization may cause the component to be completely innocuous. Measurements of pH determinations of the art material would be important in such cases.

It should be remembered that while a "minor component" is most often defied in concentration terms (i.e., $< 1$–5 per cent), its toxic properties need to be considered before concluding that a component contributes in a "minor" way to potential toxic properties of the art material. As mentioned previously, a low concentration of a skin allergen may contribute in a "major" manner to the overall toxic potential of a product.

Parenthetically, composition information on art materials is usually considered proprietary by the manufacturer, and it is necessary that the toxicologist provide assurances to its client (usually in the form of a written agreement) that strict confidentiality will be maintained.

*Chemical/physical properties of the art product which may alter components' toxic properties* Individual ingredients of art materials possess a wide range of characteristics needed to supply the desired properties of art materials which are composed of water and oil solutions and suspensions of pigments, dyes, solvents, oils, clays, plastic monomers and resins, stabilizers, and other constituents. Toxicologists should factor into their assessment the components' potential toxic activities in the matrix of these mixtures. Illustrations of this concept are given below.

## Exposure scenarios during use of art products

One of the most important aspects in the understanding of chemical hazards and the subsequent development of appropriate labeling is a careful assessment of exposure potential to the art material under consideration. Exposure factors to be scrutinized include:

1. Intended use of art product
2. Use conditions (heat, spraying, brushing, etc.)
3. Likelihood of skin, eye, respiratory exposure
4. Age group of users (children versus adults)
5. Percent of toxic component(s) in product
6. Other components which may enhance or attenuate exposure to toxic components

7 Physical properties (liquid viscosity, volatility; solids/powders, particle size).
8 Art materials pose a variety of exposure potentials for individuals using these products.

For example, materials such as paints, coatings, and lacquers are applied in thin layers, spread (with a brush or spray) over flat areas which present large surface areas conducive to the release of volatile components into the breathing zone of the user. In such cases, precautionary warnings that encourage good ventilation to minimize exposures to these volatiles are clearly in order. In these use situations, however, concern about potential eye or respiratory irritation from solid or liquid particulates is less of an issue unless the product is applied in a manner (e.g., spraying) that generates aerosols or mists. Under such circumstances, proper respiratory protection may also be indicated.

Skin and eye exposure potential is high for many art materials (e.g., paints), particularly those available for use by children. Brush bristles can propel fine particles into the air during brush application of these products. Art materials should be carefully evaluated for irritating and sensitizing properties. In the event that a product contains chemicals which are irritants or sensitizers, appropriate label warnings and recommendations for skin (gloves, long-sleeved shirts) and/or eye (safety or chemical goggles) protection need to be considered. Again, sound judgements are needed to estimate the irritating/sensitizing potential of the intact art material; secondly, an informed judgment is required on the need for warnings, precautionary statements, or other information in the product's label.

Art materials intended for use by children present a special situation. For instance, very young children are prone to chew or eat nonfood items, lick fingers, etc., which results in a high probability or oral contact with these products. In addition, children take fewer precautions in avoiding dermal contact with art products. Therefore, the toxicologist should carefully consider the greater exposure potential inherent in products which are considered for use by children and must recognize these differences when considering labeling for products for this population versus the adult population. This fact would tend to lead to more conservative assessments for product labeling for the former group. Also, statements should be considered which indicate that certain products should not be used by children and/or should be kept out of reach of children.

## Representative components in art materials

The following is a short list of representative components found in art materials and issues to be considered in the evaluation of products containing these chemicals. The following assumes that the chemicals are major components in the product, and it is foreseeable that exposures to these agents are both likely and could occur at high levels.

### *Ethyl acrylate*

This material is volatile and is listed as having irritating effects upon contact with the skin, eyes, and following inhalation of vapors. It has also been listed as a carcinogen by the National Toxicology Program.

For acute hazards of art products containing ethyl acrylate, it is important to include hazard and precautionary statements in the labeling which convey the following:

1 Breathing of vapors/mists as well as skin and eye contact may be irritating and harmful.
2 May be harmful if swallowed.
3 Keep out of reach of children.
4 Use in an adequately ventilated area.
5 Use of impermeable gloves and protective goggles recommended.

Warnings for chronic hazards do not appear to be indicated. The findings related to cancer were stomach tumors found in rodents following long-term oral gavage dosing to this irritating chemical. The compound is not known to be a human carcinogen, nor is it known to have significant mutagenic activity. Given the low opportunity for significant oral exposure, its designated use only by adults, and warnings presented under Acute Hazards, warnings for chronic hazards are not warranted.

### *Titanium dioxide*

The material is a white powder commonly used in paints. Assuming it was contained in a liquid matrix (dry spray applications were not employed), the component would have low potential to become airborne to present a hazard from eye or respiratory exposures. In this case, no acute or chronic warnings would be needed. If, however, the material was contained within a dry formulation and aerosol generation was likely, precautions for the avoidance of eye and respiratory exposures would be indicated. Chronic hazards have not been associated with exposure to this chemical.

### *Toluene*

Toluene is a volatile solvent which is associated with skin and eye irritation as well as nervous system depression following short-term exposures. Following inhalation exposures to long-term high levels (abuse conditions), the chemical can produce chronic neurotoxic responses in humans.

For acute hazards, the five points presented above for ethyl acrylate are appropriate for this chemical. A reasonable chronic hazard statement would be: "Repeated exposures to intoxication levels of this chemical may cause permanent neurological (brain) damage".

### *Epoxy resins*

It is well known that these partially reacted products contain free epoxy groups which are both chemically and biologically reactive. Product safety information usually highlights both their acute irritating effects to the eyes and skin (and possibly the respiratory tract), as well as their potential for inducing allergic responses to exposed body surfaces. Acute warnings should be the same as those given for ethyl actylate (above):

A chronic warning would be: "May produce allergic reactions by ingestion, inhalation, or skin contact".

*Statement of conformance* As part of the labeling format, it is considered important to include a statement indicating that the label has been developed according to CPSC and ASTM guidelines. This statement of conformance informs the purchaser of the art product that the labeling is consistent with standards developed for this purpose. A statement to this effect signed by the examining toxicologist provides the art product supplier the documentation needed to support the claim made on the product's label. Certifying organizations (through art material trade associations) have also been established to provide development of labeling for art materials.

## References

Federal Hazardous Substance Act Regulations. Hazardous substance and articles: administration and enforcement of regulations. *Code of Federal Regulations* 1992;16(1500).

American Society for Testing and Materials (ASTM). Committee D-1 on Paint and Related Coating and Materials. Standard practice for labeling art materials for chronic health hazards, designation D-4236-91. *Annual Book of ASTM Standards.* Philadelphia, PA: ASTM, November 1991.

Larson DI, King TO, Nelson RP, Gad SC. Information sources: building and maintaining data files. In: Gad SC, editor. *Pharmaceutical Safety Evaluation.* New York: Van Nostrand Reinhold, 1994. pp. 28–46.

Kapp R. Available toxicology sources and their use. In: Gad SC, editor. *Product Safety Evaluation Handbook* 2nd ed., Marcel Dekker, New York 1999.

Chapter 9

# Agricultural chemicals

## Regulation, risk assessment, and risk management

*James T. Stevens and Charles B. Breckenridge*

### Introduction

More than 3,000 years ago, the Chinese used elemental sulfur as a fumigant (Ecobichon, 1993). Nearly 2,500 years later in the 1800s, it was discovered that arsenic could be used as an insecticide (Stetter, 1993). The botanicals nicotine and strychnine were established as possessing rodenticidal properties in the 1700s; by the mid1800s, rotenone and pyrethrum were considered insecticides (Hayes, 1991). Bordeaux mixture (copper sulfate, lime, calcium hydroxide, and water), Paris Green (copper arsenite), and calcium arsenite were used extensively in the late 1800s against the Colorado potato beetle (Stetter, 1993).

The first synthetic pesticide, dichlorodiphenyltrichloroethylene or DDT, was created by Zeidler in 1874 (Mellanby, 1992). However, its insecticidal properties were not fully recognized until 1939 (Spindler, 1990). DDT has been responsible for saving more human lives than any other pesticide in history (Simmons, 1959). In less than 20 years, DDT effectively eradicated malaria as a threat to 953 million people (WHO, 1967). DDT has also been used to control outbreaks of louseborne typhus, the plague, yellow fever, viral encephalitis, shigellosis, cholera, and tularemia (Simmons, 1959).

The agricultural chemistry industry grew rapidly following World War II, with a focus on new chemistry arising from the targeted application of principles of chemistry to the mechanism(s) of specific and selective pest control (Cremlyn, 1978). Sales in the US grew until the early 1980s, then they gradually declined, and they now hover at a little over $1 billion per year (Gianessi, 1986; Aspelin, 1995,1996).

The benefits derived from the use of pesticides during the twentieth century have been impressive. Agriculturists have increased crop yields by as much as 50 per cent while reducing the number of people involved in commercial agriculture by 69 per cent (Avery, 1993). Today, less than 2 per cent of the population produces more than sufficient food for the entire population. However, even with the extensive use of pesticides, an estimated one-third of the world's food crops every year are destroyed by pests, and the annual crop loss in the US alone is nearly $20 billion (Ballantine, 1992).

Despite the clear benefits derived from pesticide use, the potential hazard from inappropriate use has long been recognized, and pesticide uses are tightly regulated and monitored. The risks associated with pesticides are evaluated by assessing the toxicity of each chemical and estimating the magnitude of exposure from sources in the workplace, in the environment, in food, and in water. The need to balance the risk of damage to the environment and to man against the benefits associated with using pesticides is clear (BIBRA, 1988).

## History of pesticide legislation in the US

In the US, the regulation of pesticides is covered by several legislative acts and enforced by several federal and state agencies.

### *Federal Insecticide, Fungicide, and Rodenticide Act*

The Federal Insecticide, Fungicide, and Rodenticide Act (FIFRA) was passed in 1947 (US Congress, 1947). The legislation was administered by the US Department of Agriculture (USDA) and mandated proper labeling. FIFRA has been amended several times since that time, and its registration provisions have been strengthened. Conner *et al.* (1991) provide a detailed review of FIFRA, its history, and regulations.

### *Federal Food, Drug and Cosmetic Act*

Pesticides are also regulated under the Federal Food, Drug, and Cosmetic Act (FFDCA). FFDCA was amended with Section 408 (the Miller Amendment) establishing pesticide tolerances on foodstuffs in 1954 and Section 409, which established tolerances for food additives in processed foods, was added in 1958 (US Congress, 1958).

Section 409 of FFDCA contained the Delaney Clause, which forbids use of carcinogens as food additives. Although Section 409 directly addresses food additives only, interpretations of this section by the Environmental Protection Agency (EPA) have determined that pesticides, which concentrate during food processing, require food additive tolerances. Thus, the Delaney Clause applies only to pesticides, which concentrate during the processing of food. Otherwise, only Section 408 applies and food additives tolerances are not required.

### *Food Quality Protection Act of 1996*

The Food Quality Protection Act (FQPA) of 1996 (Public Law 104-170, 1996) amends the Food Drug and Cosmetic Act (US Congress, 1958) and FIFRA (US Congress, 1947). FQPA directs the EPA to consider a number of factors in making risk assessments as part of the tolerance setting procedure. This act provides for single, health-based standard, eliminating longstanding problems posed by multiple standards for pesticides in raw and processed foods and eliminating the Delaney Clause itself for chemicals that were carcinogenic in animal bioassays. FQPA requires the EPA to consider not only dietary, but also nonoccupational sources of exposure, drinking water, residential exposure (aggregate exposure), and exposure to other pesticides with a common mechanism of toxicity (cumulative risk) when setting tolerances (Breckenridge *et al.*, 1999). Most of these provisions reflect concerns that children may be especially susceptible to pesticide exposure and embody key recommendations of the National Academy of Sciences report, "Pesticides in the Diets of Infants and

Children" (NRC, 1993). The FQPA directs the EPA to set a tolerance for pesticide residues in food. The EPA is directed to use of an extra 10-fold safety factor to account for increased susceptibility of children, including effects of in utero exposure, if indicated by animal data. Finally, the EPA must define techniques for evaluating the potential for endocrine disrupting effects. A multi-stakeholder legislatively mandated group called the Endocrine Disruptors Screening and Testing Advisory Committee (EDSTAC) delineated new screens and revised tests to attempt to achieve this objective (EPA, 1998a,b).

FIFRA (US Congress, 1947) was written to contain an open-ended provision for the consideration of pesticide benefits when setting tolerances. FQPA places specific limits on benefits considerations. Benefits cannot be considered for pesticides that result in reproductive or other threshold effects. In addition, the FQPA of 1996 re-authorizes FIFRA and requires tolerances to be reassessed as part of the re-registration program. Additionally, FQPA expedites reviews of safer pesticides to help bring them to market sooner, replacing older and potentially more risky chemicals. Minor-use pesticides and anti-microbial pesticides are given greater attention under this new law.

The Delaney Clause of Section 409 of the FFDCA forbidding substances identified as having the potential to cause cancer in laboratory animals in food remained a block in the EPA risk characterization of pesticides shown to be carcinogenic in animals. The EPA's attempt at regulating animal carcinogens using mathematical modeling was the basis for the book entitled *Regulating Pesticides in Food: The Delaney Paradox* (Thornaton *et al.*, 1987). The legality of using the linearized multi-stage model's Q1* (hazard indicator) to characterize risk as identified in the EPA guidelines for risk assessment (McGaughy, 1986) was challenged by the Natural Resources Defense Council (NRDC). In fact, the EPA was sued and the federal courts agreed that as long as the Delaney Clause remained in place, the approach suggested by the EPA risk assessment guidelines was illegal (NRDC, 1987). This impasse set the scene for the passage of new legislation.

## Required toxicology studies

As public concern increased with regard to environmental and health issues as well the need for greater regulatory efficiency, the EPA was formed in 1970 (Conner *et al.*, 1991). Although the formation of the EPA did not significantly alter the process of pesticide registration, it did centralize it in a single agency (Stevens *et al.*, 1995; Stevens, 1997). In 1972, the EPA promulgated guidelines to evaluate mammalian and environmental hazards, environmental fate, environmental chemistry, and pesticide residues in food and water (US Congress, 1972) as given in Table 9.1.

The EPA guidelines have been repeatedly updated and harmonized to be consistent with the toxicology study requirements for registration of pesticides around the world (OECD, 1981[1]; EPA, 1982a,b; MAFF, 1985; EEC, 1994). This direction towards a consistent regulatory approach continued with the 870 series guidelines (EPA, 1998a) which harmonize the testing requirements for the EPA's Office of Pesticide Programs (OPP) and Office of Toxic Substances (OTS). OPPTS 870 Series Guidelines are delineated for mammalian toxicity in Table 9.2. The OECD is also in the process of harmonizing its guidelines to comply with the 870 Series Guidelines (OECD, 1998).

The 870 series of guidelines formalizes testing requirements in the areas of immu-

*Table 9.1* US Environmental Protection Agency pesticide assessment guidelines (US Congress, 1972)

| *Data required* | *Subdivision* |
|---|---|
| Product chemistry | D |
| Hazard evaluation for wildlife and aquatic organisms | E |
| Hazard evaluation for humans and domestic animals | F |
| Product performance | G |
| Experimental use permits | I |
| Hazard evaluation for nontarget plants | J |
| Reentry protection | K |
| Hazard evaluation for nontarget insects | L |
| Biorational pesticides | M |
| Environmental fate | N |
| Residue chemistry | O |
| Spray drift evaluation | R |
| Application exposure monitoring | U |

notoxicity and neurotoxicity, including a developmental neurotoxicity guideline.

The mammalian toxicology study requirements are presented in Table 9.3.

In general, the most extensive data needs have been established for food-use pesticides; fewer requirements are in place for limited-use products.

The EPA has the responsibility to ensure that pesticide products do not adversely effect the environment. This objective is accomplished through requirements for studies in a number of different species as well as through the generation of information on expected environmental concentrations. Test species have been selected based on their ability to be maintained in the laboratory and to be representative of an important species native to the US. Expected environmental concentrations are often calculated from mathematical models that have been developed and continuously upgraded to improve their precision. The environmental toxicology requirements are delineated in Table 9.4.

The need for additional studies will continue to expand as new test systems are developed. New test species currently being examined for future requirements include microorganisms, reptiles, amphibians, additional insect species, and wild mammals. The list of studies currently required by the EPA in ecological toxicology (EPA, 1996b) is presented in Table 9.5.

Population effects are evaluated in higher-tier analysis involving terrestrial organisms (birds and mammals) and aquatic organisms (fish, invertebrate animals, plants).

## Good laboratory practice standards

Good Laboratory Practices (GLPs) apply to all studies conducted to satisfy guideline requirements for product registration in the US (EPA, 1979), for OECD (OECD, 1989), and Japan (MAFF, 1984). These standards were put in place to assure that all the required laboratory and field studies have been established that all items in involving the design, conduct and reporting are clearly documented. GLPs encompass all aspects of these studies, including personnel training, facilities, equipment and methods. These provisions are intended to assure the quality and integrity of data for the

*Table 9.2* Series 870 – OPPTS health effects test guidelines (EPA, 1996a)

| *Number* | *Name* | *Model* | *OECD no.* |
|---|---|---|---|
| *Group A – acute toxicity test guideline* | | | |
| 870.1000 | Acute toxicity testing – background | | None |
| 870.1100 | Acute oral toxicity | Rats | 401 |
| 870.1200 | Acute dermal toxicity | Rats /rabbits | 402 |
| 870.1300 | Acute inhalation toxicity | Rats | 403 |
| 870.2400 | Acute eye irritation | Rabbits | 405 |
| 870.2500 | Acute dermal irritation | Rabbits | 404 |
| 870.2600 | Skin sensitization | Guinea pigs | 406 |
| *Group B– subchronic toxicity test guidelines* | | | |
| 870.3100 | 90-Day oral toxicity in rodents | Rats/mice | 408 |
| 870.3150 | 90-Day oral toxicity in nonrodents | Dogs | 409 |
| 870.3200 | 21/28-Day dermal toxicity | Rats/rabbits | 410 |
| 870.3250 | 90-Day dermal toxicity | Rats | 411 |
| 870.3465 | 90-Day inhalation toxicity | Rats | 413 |
| 870.3700 | Prenatal developmental toxicity study | Rats/rabbits | 414 |
| 870.3800 | Reproduction and fertility effects | Rats | 416 |
| *Group C – chronic toxicity test guidelines* | | | |
| 870.4100 | Chronic toxicity | Rats/ dogs | 452 |
| 870.4200 | Carcinogenicity | Rats/mice | 451 |
| 870.4300 | Combined chronic toxicity/carcinogenicity | Rats | 453 |
| *Group D – genetic toxicity test guidelines* | | | |
| 870.5100 | Bacterial reverse mutation test | *S. typhimurium* | 471, 472 |
| 870.5140 | Gene mutation in *Aspergillus nidulans* | | None |
| 870.5195 | Mouse biochemical specific locus test | Mice | None |
| 870.5200 | Mouse visible specific locus test | Mice | None |
| 870.5250 | Gene mutation in *Neurospora crassa* | | None |
| 870.5275 | Sex-linked recessive lethal test in *Drosophila* | | 477 |
| 870.5300 | In vitro mammalian cell gene mutation test | | 476 |
| 870.5375 | In vitro mammalian chromosome aberration test | | 473 |
| 870.5380 | Mammalian spermatogonial chromosomal aberration | Mice | 483 |
| 870.5385 | Mammalian bone marrow chromosomal aberration test | Mice | 475 |
| 870.5395 | Mammalian erythrocyte micronucleus test | Mice | 474 |
| 870.5450 | Rodent dominant lethal assay | Mice | 478 |
| 870.5460 | Rodent heritable translocation assays | Mice | None |
| 870.5500 | Bacterial DNA damage or repair tests | | None |
| 870.5550 | Unscheduled DNA synthesis in mammalian cells | Rats | 482 |
| 870.5575 | Mitotic gene conversion in *Saccharomyces cerevisiae* | | 481 |
| 870.5900 | In vitro sister chromatid exchange assay | | 479 |
| 870.5915 | In vivo sister chromatid exchange assay | None | |
| *Group E – neurotoxicity test guidelines* | | | |
| 870.6100 | Acute and 28-day delayed neurotoxicity of organophosphorus substances | Hens | 418, 419 |
| 870.6200 | Neurotoxicity screening battery | Rats | 424 |
| 870.6300 | Developmental neurotoxicity study | Rats | None |
| 870.6500 | Schedule-controlled operant behavior | Rats | None |
| 870.6850 | Peripheral nerve function | Rats | None |

Table 9.2 (*continued*)

| *Number* | *Name* | *Model* | *OECD no.* |
|---|---|---|---|
| 870.6855 | Neurophysiology: sensory evoked potentials | Rats | None |
| *Group F – special studies test guidelines* | | | |
| 870.7200 | Companion animal safety | | None |
| 870.7485 | Metabolism and pharmacokinetics | Rats | 417 |
| 870.7600 | Dermal penetration | Rats | None |
| 870.7800 | Immunotoxicity | Rats | None |

registration of a crop protection chemical. These regulations establish certain requirements for study conduct and management. Each study must be conducted under the direction of a Study Director who signs specific compliance statements for the studies.

## Registration process and procedures

### *Petition for registration*

A registrant must submit a formal petition of a given crop protection chemical to the EPA requesting that registration of each pesticide use is granted. This petition must contain data to address all required areas, including directions for use, product chemistry, toxicology, residue chemistry, environmental fate, re-entry intervals, spray drift, and wildlife and aquatic toxicology. The data package must fulfill the requirements as specified in the Code of Federal Register, 40 CFR, Part 158 (EPA, 1984). Study reports, indicating their conduct in compliance with GLPs, are submitted to the Registration Division of the Office of Pesticide Programs at the EPA.

### *Reregistration*

Consistent with the concept that science continues to advance, there are new approaches and enhanced methods of detection for potential hazards. Therefore, guidelines promulgated in the 1990s may be replaced 10 years later with guidelines based on more sensitive and superior techniques. To respond to these changes, EPA registration guidelines must also change. Hence, pesticides registered in the past may not have been tested as rigorously as newer pesticides. Congress addressed this issue in its 1972 amendment to FIFRA and specified that older pesticides would be reevaluated and re-registered and required replacement of old studies with new studies or new types of studies (US Congress, 1972). The need for re-registration was reinforced by the FQPA (Public Law 104-170, 1996). Re-registration entails the re-review of all existing data to compare it to current requirements, and if deficiencies are present, requires that relevant studies be conducted again. Older registered products would be reassessed, for example, to assure compliance with the 870 Series (EPA, 1996a). An additional byproduct of re-registration has been the publication of Acceptance Criteria by the EPA (EPA, 1989); these criteria identify the critical information that must be submitted to the Agency so an adequate review can be conducted.

*Table 9.3* Mammalian toxicology data requirements for pesticide registration[a]

| *Kind of data required* | *General use patterns* | | | | | | | | |
|---|---|---|---|---|---|---|---|---|---|
| | *Terrestrial* | | *Aquatic* | | *Greenhouse* | | *Forestry* | *Domestic outdoor* | *Indoor* |
| | *Food Crop* | *Nonfood* | *Food crop* | *Nonfood* | *Food crop* | *Nonfood* | | | |
| *Acute testing* | | | | | | | | | |
| Acute oral toxicity – rat | [R] | [R] | [R] | [R] | [R] | [R] | [R] | [R] | [R] |
| Acute dermal toxicity | [R] | [R] | [R] | [R] | [R] | [R] | [R] | [R] | [R] |
| Acute inhalation toxicity – rat | [R] | [R] | [R] | [R] | [R] | [R] | [R] | [R] | [R] |
| Primary eye irritation – rabbit | [R] | [R] | [R] | [R] | [R] | [R] | [R] | [R] | [R] |
| Primary dermal irritation | [R] | [R] | [R] | [R] | [R] | [R] | [R] | [R] | [R] |
| Dermal sensitization | [R] | [R] | [R] | [R] | [R] | [R] | [R] | [R] | [R] |
| Acute delayed neurotoxicity – hen | [R] | [R] | [R] | [R] | [R] | [R] | [R] | [R] | [R] |
| *Subchronic testing* | | | | | | | | | |
| 90-Day feeding studies – rodent and nonrodent | [R] | CR | [R] | CR | [R] | CR | CR | C | CR |
| 21-Day dermal | CR | CR | CR | CR | CR | CR | CR | CR | CR |
| 90-Day dermal | CR | CR | CR | CR | CR | CR | CR | CR | CR |
| 90-Day inhalation – rat | CR | CR | CR | CR | CR | CR | CR | CR | CR |
| 90-Day neurotoxicity – hen | CR | CR | CR | CR | CR | CR | CR | CR | CR |
| *Chronic testing* | | | | | | | | | |
| Chronic feeding in two species – rodent and nonrodent | [R] | CR | [R] | CR | [R] | CR | CR | CR | CR |
| *Oncogenicity*in two species – rat and mouse preferred | R | CR | R | CR | R | CR | CR | CR | CR |
| Teratogenicity in two species | [R] | CR | [R] | CR | [R] | CR | CR | CR | CR |
| Two generation reproduction | [R] | CR | [R] | CR | [R] | CR | CR | CR | CR |

Table 9.3 (*continued*)

| *Kind of data required* | *General use patterns* | | | | | | | | |
|---|---|---|---|---|---|---|---|---|---|
| | *Terrestrial* | | *Aquatic* | | *Greenhouse* | | *Forestry* | *Domestic outdoor* | *Indoor* |
| | *Food Crop* | *Nonfood* | *Food crop* | *Nonfood* | *Food crop* | *Nonfood* | | | |
| *Mutagenicity testing* | | | | | | | | | |
| Gene mutation | [R] | R | [R] | R | [R] | R | R | R | R |
| Structural chromosomal aberration | [R] | R | [R] | R | [R] | R | R | R | R |
| Other genotoxic effects | [R] | R | [R] | R | [R] | R | R | R | R |
| *Special testing* | | | | | | | | | |
| General metabolism | R | CR | R | CR | R | CR | CR | CR | CR |
| Dermal penetration | CR | CR | CR | CR | CR | CR | CR | CR | CR |
| Domestic animal safety | CR | CR | CR | CR | – | – | CR | CR | – |

[a] The table was adapted from the 40 CFR, Part 158 (EPA, 1984) without footnotes. R = required; CR = conditionally required; [ ] = needed for experimental use permits.

*Table 9.4* Environmental toxicity study data requirements for pesticide registration[a]

| *Kind of data required* | *General use patterns* | | | | | | | | |
|---|---|---|---|---|---|---|---|---|---|
| | *Terrestrial* | | *Aquatic* | | *Greenhouse* | | *Forestry* | *Domestic outdoor* | *Indoor* |
| | *Food crop* | *Nonfood* | *Food crop* | *Nonfood* | *Food crop* | *Nonfood* | | | |
| *Avian and mammalian testing* | | | | | | | | | |
| Avian oral $LD_{50}$ (preferably mallard or bobwhite) | [R] | [R] | [R] | [R] | CR | CR | [R] | [R] | CR |
| Avian dietary $LC_{50}$ (preferably mallard and bobwhite) | [R] | [R] | [R] | [R] | CR | CR | [R] | [R] | CR |
| Wild mammal toxicity | CR | CR | CR | CR | – | – | CR | CR | – |
| Avian reproduction (preferably mallard and bobwhite) | CR | CR | CR | CR | – | – | CR | CR | – |
| Simulated and actual field testing in mammals and birds | CR | CR | CR | CR | – | – | CR | CR | – |
| *Aquatic organism testing* | | | | | | | | | |
| Freshwater fish $LC_{50}$ (preferably rainbow and bluegill) | [R] | [R] | [R] | [R] | CR | CR | [R] | [R] | CR |
| Acute $LC_{50}$ freshwater invertebrates preferably *Daphnia*) | [R] | [R] | [R] | [R] | CR | CR | [R] | [R] | CR |
| Acute $LC_{50}$ estuarine And marine organisms | CR | CR | CR | CR | – | – | CR | CR | – |
| Fish early life stage and aquatic invertebrate life-cycle | CR | CR | CR | CR | – | – | CR | CR | – |
| Fish – life-cycle | CR | CR | CR | CR | – | – | CR | CR | – |
| Aquatic organism accumulation | CR | CR | CR | CR | – | – | CR | CR | – |
| Simulated or actual field testing – aquatic organisms | CR | CR | CR | CR | – | – | CR | CR | – |

[a] The table was adapted from the 40 CFR, Part 158 (EPA, 1984) without footnotes. R = required; CR = conditionally required; [ ] = needed for experimental use permits.

*Table 9.5* Series 850 – ecological effects test guidelines (EPA, 1996b)

| *OPPTS number* | *Name* | *Existing numbers* | | |
|---|---|---|---|---|
| | | *OTS* | *OPP* | *OECD* |
| 850.1000 | Special considerations for conducting aquatic laboratory studies | None | None | None |
| *Group A – aquatic fauna test guidelines* | | | | |
| 850.1010 | Aquatic invertebrate acute toxicity, test, freshwater daphnids | 797.1300 | 72-2 | None |
| 850.1020 | Gammarid acute toxicity test | 795.120 | None | None |
| 850.1025 | Oyster acute toxicity test (shell deposition) | 797.1800 | 72-3 | None |
| 850.1035 | Mysid acute toxicity test | 797.1930 | 72-3 | None |
| 850.1045 | Penaeid acute toxicity test | 797.1970 | 72-3 | None |
| 850.1055 | Bivalve acute toxicity test (embryo larval) | none | 72-3 | None |
| 850.1075 | Fish acute toxicity test, freshwater and marine | 797.1400 | 72-1, 3 | 203 |
| 850.1085 | Fish acute toxicity mitigated by humic acid | 797.1460 | None | None |
| 850.1300 | Daphnid chronic toxicity test | 797.1330 | 72-4 | 202 |
| 850.1350 | Mysid chronic toxicity test | 797.1950 | 72-4 | None |
| 850.1400 | Fish early-life stage toxicity test | 797.1000 | 72-4 | 210 |
| 850.1500 | Fish life cycle toxicity | none | 72-5 | None |
| 850.1710 | Oyster BCF | 797.1830 | 72-6 | None |
| 850.1730 | Fish BCF | 797.1520 | 72-6, 165-4 | 305 |
| 850.1735 | Whole sediment acute toxicity invertebrates, freshwater | None | None | None |
| 850.1740 | Whole sediment acute toxicity invertebrates, marine | None | None | None |
| 850.1790 | Chironomid sediment toxicity test | 795.135 | None | None |
| 850.1800 | Tadpole/sediment subchronic toxicity test | 797.1995 | None | None |
| 850.1850 | Aquatic food chain transfer | None | 72-6 | None |
| 850.1900 | Generic freshwater microcosm test, laboratory | 797.3050, 797.3100 | None | None |
| 850.1925 | Site-specific aquatic microcosm test, laboratory | 797.3100 | None | None |
| 850.1950 | Field testing for aquatic organisms | None | 72-7, 165-5 | None |
| *Group B – terrestrial wildlife test guidelines* | | | | |
| 850.2100 | Avian acute oral toxicity test | 797.2175 | 71-1 | None |
| 850.2200 | Avian dietary toxicity test | 797.2050 | 71-2 | 205 |
| 850.2300 | Avian reproduction test | 797.2130, 797.2150 | 71-4 | 206 |
| 850.2400 | Wild mammal acute toxicity | none | 71-3 | None |
| 850.2450 | Terrestrial (soil-core) microcosm test | 797.3775 | none | None |
| 850.2500 | Field testing for terrestrial wildlife | none | 71-5 | None |
| *Group C – beneficial insects and invertebrates test guidelines* | | | | |
| 850.3020 | Honey bee acute contact toxicity | None | 141-1 | None |
| 850.3030 | Honey bee toxicity of residues on foliage | None | 141-2 | None |
| 850.3040 | Field testing for pollinators | None | 141-5 | None |

Table 9.5 (*continued*)

| *OPPTS number* | *Name* | *Existing numbers* | | |
|---|---|---|---|---|
| | | *OTS* | *OPP* | *OECD* |
| *Group D – non-target plants test guidelines* | | | | |
| 850.4000 | Background – non-target plant testing | None | 120-1 | None |
| 850.4025 | Target area phytotoxicity | None | 121-1 | None |
| 850.4100 | Terrestrial plant toxicity, Tier I (seedling emergence) | None | 122-1 | None |
| 850.4150 | Terrestrial plant toxicity, Tier I (vegetative vigor) | None | 122-1 | None |
| 850.4200 | Seed germination/root elongation toxicity test | 797.2750 | 122-1 | None |
| 850.4225 | Seedling emergence, Tier II | 797.2750 | 123-1 | None |
| 850.4230 | Early seedling growth toxicity test | 797.2800 | 123-1 | None |
| 850.4250 | Vegetative vigor, Tier I I | 797.2750 | 123-1 | None |
| 850.4300 | Terrestrial plants field study, Tier III | None | 124-1 | None |
| 850.4400 | Aquatic plant toxicity test using *Lemna* spp.Tiers I and II | 797.1160 | 122-2, 123-2 | None |
| 850.4450 | Aquatic plants field study, Tier III | None | 124-2 | None |
| 850.4600 | Rhizodium-legume toxicity | 797.2900 | None | None |
| 850.4800 | Plant uptake and translocation test | 797.2850 | None | None |
| *Group E – toxicity to microorganisms test guidelines* | | | | |
| 850.5100 | Soil microbial community toxicity test | 797.3700 | None | None |
| 850.5400 | Algal toxicity, Tiers I and II | 797.1050 | 122-2, 123-2 | None |
| *Group F – chemical-specific test guidelines* | | | | |
| 850.6200 | Earthworm subchronic toxicity test | 795.150 | None | 207 |
| 850.6800 | Modified activated sludge, respiration inhibition test for sparingly soluble chemicals | 795.170 | None | 209 |
| *Group G– field test data reporting guidelines* | | | | |
| 850.7100 | Data reporting for environmental chemistry methods | None | None | None |

The EPA has separated the re-review process into phases based on potential hazards. Initial phases dealt with those compounds that by virtue of their toxicity or exposure were more likely to present a risk to the public. Pesticides that were considered less likely to present a risk to humans or the environment were to be reregistered later. An additional feature of FIFRA Section 3(c)(2)(B) authorizes the EPA to issue a "Data-Call-In", or DCI, to require registrants to submit additional data to the Agency. The DCI provision allows the Agency to require submission of additional data at any time.

## Standard evaluation procedures

The EPA developed standard evaluation procedures to aid reviewers in their evaluation of submitted data, as the agency forecasted the need to conduct these reviews in a timely manner and within a defined framework. These can be obtained from the National

Technical Information Service as FIFRA Accelerated Reregistration Phase 3 Technical Guidance (EPA, 1989).

## Special reporting requirements

### *Flagging criteria*

Any person submitting studies to support a FIFRA requirement must submit a statement concerning adverse findings. Adverse effects seen in reproduction studies must be highlighted to draw agency attention to these results. The specific studies and requirements are shown in Table 9.6.

Flagging statements are specified under FIFRA for the person submitting the study to the Agency. The Study Director usually signs the flagging statement. The statement reads, "I have applied the criteria of 40 CFR 158.34 for flagging studies for potential adverse effects to the results of the attached study. This study neither meets nor exceeds any of the applicable criteria." Of course, if the studies have identified reportable adverse effects the last sentence is changed to say, "...study meets or exceeds the criteria numbered [see codes in Table 9.6]".

### *FIFRA 6(a)(2) notification*

As an agricultural chemical is in the early development phases prior to submission of its registration petition, it is governed by the Toxic Substance Control Act (TSCA) (EPA, 1985). The TSCA has certain reporting requirements in its Section 8(e). TSCA 8(e) specifies that qualified personnel have a personal and professional responsibility to report adverse findings to the EPA within 15 working days of the discovery (EPA, 1978a,b). Individual and corporate penalties for violation are substantial. Toxicologists are specifically mentioned as "qualified". Thus, they must report information that reasonably supports the conclusion that substantial risk of injury to health or the environment may occur as a result of usage.

Once a petition for registration or the request for an experimental use permit, is submitted to the EPA, another statute, FIFRA Section 6(a)(2), becomes applicable (EPA, 1995). FIFRA Section 6(a)(2) has a section requiring reporting of adverse effects (EPA, 1997a).

## Principles of pesticide risk assessment

The principles of pesticide risk assessment are identical to those associated with other types of products, including pharmaceuticals and industrial chemicals. However, some differences may arise with regard to the manner in which studies are conducted, as well as the procedures used in evaluating and interpreting the results. The underlying principles remain the same, as defined in a paradigm described by the Office of Science and Technology (OSTP, 1985) as a four-step procedure consisting of hazard identification, dose response, exposure assessment, and risk characterization.

In order to assess hazard, which is a function of dose, animal studies are conducted with different exposure levels, or doses, as well as varying durations of exposure. The dose response element is characterized by increasing the dose until toxic effects are

*Table 9.6* Flagging criteria for adverse effect identification in toxicity studies[a]

| *Toxicity studies* | *FIFRA guideline number* | *Criteria* | *Report code* |
|---|---|---|---|
| Oncogenicity (or combined oncogenicity/chronic feeding study) or | 83-2 | Treated animals show any of the following: | |
| Subchronic feeding study | 82-1 | An incidence of neoplasms in male or female animals which increases with dose; or | 1 |
| | | A statistically significant ($P < 0.05$) incidence of any type of neoplasm in any test group (male or female animals at any dose level) compared to concurrent control animals of the same sex; or | 2 |
| | | An increase in any type of uncommon or rare neoplasms in any test group (male or female animals at any dose level) compared to concurrent control animals; or | 3 |
| | | A decrease in the time to development of any type of neoplasms in any test group (male or female animals at any dose level) compared to concurrent control animals | 4 |
| Teratogenicity | 83-3 | When compared with concurrent controls, treated animals show a dose-related increase in malformations (or deaths) on a litter basis in the absence of significant maternal toxicity at the same dose levels | 5 |
| Neurotoxicity | 81-7 | When compared with controls, treated animals show a response indicative of acute delayed neurotoxicity | 6 |
| Chronic feeding study or combined chronic feeding/ oncogenicity study | 83-1 | Cholinesterase inhibition noel less than 10 times the current existing ADI; or | 7 |
| | | General (systemic) toxicity noel less than 100 times the current existing ADI | 8 |
| Reproduction study | 83-4 | Reproductive effects noel less than 100 times the current ADI | 9 |
| Subchronic feeding study | 82-1 | Cholinesterase inhibition noel less than 100 times the current existing ADI; or | 10 |
| | | General (systemic) toxicity noel less than 1,000 times the current existing ADI | 11 |

[a] Table adapted from *Code of Federal Register* 40CFR 158 (EPA, 1984).

observed; the cumulative effects of a chemical are defined by increasing the duration. There are attempts to set dose levels for administration to the test animals for a given required study within the range of anticipated human exposure or its multiples (safety testing). EPA policy requires that the Maximum Tolerated Dose (MTD) be used in toxicity studies. The extent to which the toxic effect level is exceeded depends upon the type of study; this will be discussed later. With the establishment of the toxic response and the no-observable-effect level (NOEL), acceptable exposure levels can be determined (EPA, 1982a). The NOEL, using the most sensitive parameter in the most sensitive species, is divided by a suitable safety factor (now commonly referred to as an uncertainty factor) to provide a reference dose (RfD). This RfD serves as a hazard-based standard, and is essentially equivalent to the Acceptable Daily Intake (ADI) that is used by the WHO/FAO Joint Meeting on Pesticide Residues (JMPR), which is used by the Codex Alimentarius Commission and national governments to set international food standards and safe intake levels for protection of the consumer (JMPR, 1979).

The exposure assessment considers the duration of exposure; routes of exposure (ingestion, inhalation, external contact); rates of absorption through the skin, lungs, or intestinal tract; distribution of the chemical to various tissues; chemical modification by bodily processes (metabolism and excretion); and the possibility of accumulation in tissues. Risk characterization, in its simplest form, may be expressed as a comparison of the RfD to the actual exposure levels to establish Margins of Safety (MOS) or according to the most recent terminology, the Margins of Exposure (MOE).

### *Dose response and hazard identification*

The sixteenth century physician, Paracelsus (Klaassen and Doull, 1980), who is heralded as the Father of Toxicology, stated that "All substances are poisons; there is none that is not a poison. The right dose differentiates between a poison and a remedy." This maxim encompasses the concept of dose response. Therefore, hazard identification requires that the dose response be evaluated for a single dose or multiple levels. In testing scenarios, these regimens are generally classified relative to the duration of exposure: acute, short-term, intermediate-term (subchronic) and chronic exposure. The mammalian hazard test requirement can be grouped according to regimens as given in Table 9.7.

#### *Acute exposure*

Acute exposure usually refers to a single exposure. Dosing a group of laboratory animals with a single dose of a chemical and observing the animals for signs of toxicity or irritation assesses acute toxicity. Acute toxicity studies are conducted by administering the chemical orally, dermally, or by inhalation to determine the dose that causes mortality in 50 per cent of the animals tested. Acute studies are conducted to evaluate the irritation potential of the chemical after application to the skin and eyes. Finally, the potential of the chemical to cause an allergic reaction (i.e., sensitization) is determined. Because this reaction is an all or nothing response, chemicals are classified as either sensitizers or nonsensitizers. Acute oral and inhalation studies are usually conducted with rats, and dermal toxicity, eye, and skin irritation studies are conducted using rabbits or rats. Sensitization studies actually are repeated dose studies involving an

*Table 9.7* The Series 870 mammalian hazard tests grouped according to duration of exposure[a]

| *Exposure duration* | *Guideline no.* | *Study* |
|---|---|---|
| Acute | 870.1100 | Acute oral toxicity |
| | 870.1200 | Acute dermal toxicity |
| | 870.1300 | Acute inhalation toxicity |
| | 870.2400 | Acute eye irritation |
| | 870.2500 | Acute dermal irritation |
| | 870.6100 | Acute delayed neurotoxicity |
| | 870.6200 | Acute neurotoxicity |
| | 870.7600 | Dermal penetration |
| Short-term | 870.2600 | Skin sensitization |
| | 870.6100 | 28-Day delayed neurotoxicity |
| | 870.3200 | 21/28-Day dermal toxicity |
| | 870.6200 | 28-Day neurotoxicity |
| | 870.3700 | Prenatal developmental toxicity study |
| | 870.7200 | Companion animal safety |
| | 870.7485 | Metabolism and pharmacokinetics |
| Intermediate | 870.3100 | 90-Day oral toxicity in rodents |
| | 870.3150 | 90-Day oral toxicity in nonrodents |
| | 870.3250 | 90-Day dermal toxicity |
| | 870.3465 | 90-Day inhalation toxicity |
| | 870.6200 | 90-Day neurotoxicity |
| | 870.3800 | Reproduction and fertility effects |
| Chronic/carcinogenicity | 870.4100 | Chronic toxicity (dogs, rats) |
| | 870.4200 | Carcinogenicity |
| | 870.4300 | Combined chronic toxicity/carcinogenicity |

[a] Group D – genetic toxicity test guidelines could be classified as acute or short-term.

initiation and a challenge dose. These studies are carried out using guinea pigs. This combination of tests is used to establish the product labels for all crop protection chemicals (Stevens *et al.*, 1995). The criteria used are presented in Table 9.8.

This core group of tests must be performed on the technical grade of material, as well as on all formulated variants of a given pesticide. The guidelines for the conduct of these acute studies are specified for oral, dermal, inhalation, eye and skin irritation, and sensitization as 870.1100, 870.1200, 870.1300, 870.2400, 870.2500, and 870.2600, respectively (EPA, 1996a). The acute neurotoxicity test from the neurotoxicity screening battery (870.6200) should be included in the acute grouping.

### *Short-term and intermediate-term*

Short-term and intermediate tests formerly referred to as subchronic studies (EPA, 1996a) are conducted for periods of time that can range from 2 weeks to 90 days. "Subchronic" usually refers to study duration (fraction of a lifetime). In these studies, laboratory animals are exposed to continuous or to repeated doses of a product. The typical route of exposure is through ingestion (admixed with feed or administered by gavage). Animals are observed for signs of toxicity, effects on body weights, on body weight gain, and food consumption as well as clinical pathology. At study termination, organs and tissues are examined and histopathologic changes are evaluated. Results of all observations, measurements, and examinations performed during the study period

*Table 9.8* US EPA acute toxicology classification scheme (Stevens *et al.*, 1995)

| *Toxicology category* | *Signal word* | *Oral $LD_{50}$ (mg/kg)* | *Dermal $LD_{50}$ (mg/kg)* | *Inhalation $LC_{50}$ (mg/l)* | *Eye irritation* | *Skin irritation* |
|---|---|---|---|---|---|---|
| I | Danger[a] | Up to 50 | Up to 200 | Up to 0.2 | Corrosive. Corneal opacity not reversed in 7 days | Corrosive |
| II | Warning | From 50 to 500 | From 200 to 2,000 | From 0.2 to 2.0 | Corneal opacity reversed in 7 days; irritation persisting 7 days | Severe irritation at 72 h |
| III | Caution | From 500 to 5,000 | From 2,000 to 5,000 | From 2.0 to 20 | No corneal opacity; irritation reversed within 7 days | Moderate irritation at 72 h |
| IV | Caution | Greater than 5,000 | Greater than 5,000 | Greater than 20 | No irritation | Mild or slight irritation at 72 h |

[a] The word "Poison" is used on the label if the "Danger" category is based on oral, dermal or inhalation toxicity.

are compared between the animals dosed with the chemical and the control (untreated) animals. Data from subchronic studies are used to identify potential target organs (organs exhibiting adverse effects because of chemical exposure) and to select dose levels for lifetime chronic toxicity and/or oncogenicity studies.

Dermal toxicity is evaluated by applying the chemical to the skin for 6 h a day for 21–28 days in rats (21 days) or rabbits (28 days). The guideline for the conduct of this study is 870.3200. Twenty-eight-day range-finding studies in the rat, mouse, and dog are used to select doses for the intermediate or chronic studies. The 28-day neurotoxicity test from the neurotoxicity screening battery (870.6200) is included in the short-term grouping.

The oral feeding studies involve feeding rats, mice, or dogs diets containing the chemical for various lengths of time. Rat and mouse feeding studies are conducted for 90 days. The guidelines for the rat, mouse, and dog are 870.3100 for the rat and mouse, and 870.3150 for the dog. In all cases, animals are divided into test groups, 10–50 rats or mice and four to six dogs. At least four test groups are used in each study, one receiving no chemical (controls) and three groups receiving low, medium, or high concentrations of chemical in their diets. In these studies, urinalysis, hematology, and clinical chemistry parameters are evaluated, and gross and microscopic pathological examinations are performed on up to 50 tissues. Maximally tolerated doses are tested in order to demonstrate toxicity (up to 1,000 mg/kg per day in the diets). It is possible to determine whether a chemical damages or alters any organ or tissue, and to establish levels of the chemical which produce no observable effects (the NOEL), and the lowest level at which effects are noted (the LOEL). Also included with the intermediate duration tests are 90-day dermal toxicity (870.3250), 90-day inhalation toxicity (870.3465), and the 90-day neurotoxicity tests (870.6200).

### *Developmental and reproductive toxicity*

Developmental toxicology studies for FIFRA are conducted using protocols essentially identical to the Segment II studies used with pharmaceuticals. These teratology studies are performed in two species, usually rats and rabbits, to evaluate effects in the fetus following chemical exposure to the maternal system during major periods of organogenesis (EPA, 1982a). Following cesarean delivery, fetuses are examined for external, internal, and skeletal abnormalities, and results compared between treated and control animals. The harmonized prenatal developmental toxicity study guideline is 870.3700 (EPA, 1996a). In addition to the developmental toxicity studies, there is a guideline (870.6300) for a developmental neurotoxicity study. The protocol for this study is currently under discussion (Crofton, 2000).

Multigeneration reproduction studies are conducted to evaluate effects of the chemical on estrous cycles, mating behavior, pregnancy of parental animals, postpartum behavior, as well as the number, weight, survival, and growth of offspring over at least two generations of rats. This study involves feeding diets containing the chemical to young adult male and female rats for approximately 3 months prior to mating. The females are allowed to produce a litter of offspring that are then reared to adulthood. The animals are fed diets containing the chemical during this entire period. After reaching sexual maturity, the second-generation animals are allowed to mate. A multigeneration or two-generation reproduction study is conducted in rats following Guideline 870.3800 (EPA, 1996a).

### *Chronic toxicity/carcinogenicity*

Since individuals may be exposed to low levels of crop protection chemicals in their diet or water over a portion of their life span, studies to evaluate lifetime exposure are conducted in animal bioassays. The lifetime studies for the rat and mice are 24 and 18 months, respectively. An important aspect of these studies is an assessment of the potential of the chemical to cause cancer. An increase in the number of tumors and/or earlier onset for the occurrence of tumors due to treatment will identify a chemical as a potential carcinogen. For laboratory studies, mice and rats are divided into at least three treatment groups and a control group with a minimum of 50 animal/sex per group. These groups of mice and rats are fed selected concentrations of the test chemical in their diet for 18 months and 24 months, respectively.

Following lifetime feeding studies at the prescribed treatment levels, veterinary pathologists examine approximately 50 tissues from each animal for the presence of tumors or other evidence of tissue damage. A significant increase in the incidence of any tumor that is higher than that observed in control animals may be considered treatment-related.

The levels of the test chemical administered in the diet are generally selected from repeated-dose feeding studies of at least 90 days in duration, and are normally used to establish the NOEL, LOEL and the MTD (Farber, 1987). The MTD is defined as the highest concentration of a test chemical that can be administrated and tolerated without causing the death of the animal. A default assumption often used to define that an MTD has been achieved has been set at a reduction in body weight gain of 10 per cent or less (Foran *et al.*, 1999).

Chronic toxicity or carcinogenicity studies are conducted over a major portion of a lifetime. The clear exception to this approach is the chronic dog study that generally lasts only 1 year. Similar parameters are evaluated as in the subchronic studies in these dog studies. A NOEL is defined and the highest dose or feeding level should confirm the MTD seen in the subchronic study. At termination, reproductive and other organs are examined, and results of all observations, measurements, and examinations are compared between the treated groups and the control group.

The Group C Chronic Toxicity Test Guidelines (EPA, 1996a) include the chronic toxicity in the rat and dog (870.4100); carcinogenicity in the rat and mouse (870.4200) and the combined chronic toxicity/carcinogenicity in the rat (870.4300).

### *Genotoxicity*

Weisburger (1975) noted that certain chemical carcinogens are capable of interacting directly with genetic material such as DNA. Based upon this association, several short-term tests to identify the alteration of genetic material or mutation were introduced into hazard testing for crop protection chemicals. These include tests to examine the possible interaction with (1) genes (gene mutation tests), (2) chromosomes (clastogenic tests), and (3) directly with DNA (classified as other tests).

Mutagenicity studies are among the first to be conducted with an experimental product, and are often used by industry as a screening tool. These tests are performed in various *in vivo* and *in vitro* test systems (bacteria, liver cells, sperm cells, etc.) to evaluate the potential effects of the chemical on the genetic material. Mutagenic changes in reproductive cells (sperm or egg) may retard fetal development and result in congenital abnormalities, while mutagenic changes in somatic cells may lead to cancer. The Genetic Toxicity Test Guidelines have been ascribed to 870.5140, 870.5195, 870.5200, 870.5250, 870.5275, 870.5300, 870.5375, 870.5375, 870.5380, 870.5385, 870.5450, 870.5460, 870.5500, 870.5550, 870.5575, 870.5900 and 870.5915 (Dearfield, 1990; EPA, 1996a).

### *Metabolism*

The metabolism requirements for pesticide registration with the EPA include determinations of absorption, distribution, metabolism, and excretion. Kinetic models and studies on the mechanism of toxicity represent important aspects of pesticide safety evaluation. Metabolism and pharmacokinetics are given in Guideline 870.7485 (EPA, 1996a).

### *Neurotoxicity*

Neurotoxicity refers to any adverse effects on the structure or function of the nervous system related to exposure to a chemical substance. The EPA has revised the hazard evaluation requirements listed in Subdivision F of the Pesticide Assessment Guidelines to include neurotoxicity testing for new and existing chemicals (EPA, 1996a). The Group E Neurotoxicity Test Guidelines include: acute and 28-day delayed neurotoxicity of organophosphorus in the hen (870.6100), neurotoxicity screening battery that includes acute, 28-day, and 90-day neurotoxicity studies (870.6200), developmental neurotoxicity study (870.6300), schedule-control operant behavior (870.6500), peripheral nerves function (870.6850), and neurophysiology sensory evoked potentials (870.6855).

### *New requirements*

A mandate of the FQPA (Public Law 104-170, 1996) was that the EPA develop a valid screening and testing program to detect chemicals which may cause disruption of the endocrine system. These screens and tests should use the animal models currently used in the hazard identification. In order to address this requirement, the EPA formed the Endocrine Disruptors Screening and Testing Advisory Committee (EDSTAC). This committee was established under the provisions of the Federal Committee Advisory Act (FACA) to advise the EPA on a strategy for screening and testing chemicals and pesticides for their potential to disrupt endocrine functions in humans and wildlife (EPA, 1996d). This strategy is aimed at reducing or mitigating risk to human health and the environment. The EDSTAC (EPA, 1998a,b) recommended the Tier 1 (screens) *in vivo* and *in vitro* assays as follows. In vitro assays include an estrogen receptor binding or reporter gene assay, an androgen receptor binding or reporter gene assay, and a steroidogenesis assay with minced testis. Assays (*in vivo*) include a rodent 3-day uterotrophic assay, a rodent 20-day pubertal female assay with enhanced thyroid endpoints, a rodent 5–7-day Hershberger assay, a frog metamorphosis assay, and a fish gonadal recrudescence assay. Tier 2 tests include a two-generation reproductive toxicity study or a less comprehensive reproductive toxicity assay. In addition, EDSTAC recommends an avian reproduction toxicity assay, a fish life cycle toxicity assay, an opossum shrimp (Mysidacea) or other invertebrate life cycle toxicity assay, and an amphibian development and reproduction assay. Before these or any other proposed tests are put forth as guidelines, it is essential to standardize their protocols and validate the assay.

As the EDSTAC was completed in August of 1998, it was vital to establish another committee to oversee the standardization and validation of these screens and tests. The EPA has formed the Endocrine Disruptors Standardization and Validation Task Force whose members represent government, industry, and nongovernment organizations.

## Exposure Assessment

Exposure assessment involves determining the source and level of exposure to a chemical. Exposure may occur in the workplace, under conditions of use, and/or through residues in food or water. For production workers, exposure may be limited by Threshold Limit Values (TLVs) set by the American Conference of Governmental Industrial Hygienists, or Permissible Internal Exposure Limits (PIELs), set by the Occupational Safety and Health Administration (OSHA). The public is more likely to encounter pesticide exposure through ingestion of trace amounts in food (tolerance level) or water (Health Advisory or Maximum Contaminant Level), through use of home and garden products, or during crop or roadside applications.

In the past, estimates of dietary exposure to pesticides are often based on unrealistic assumptions about human exposure (Graham and Gray, 1995). Instead of establishing exposure estimates on measured amounts of pesticides in food as eaten, the EPA often use Theoretical Maximum Residue Concentrations(TMRCs) (EPA, 1992). The TMRC assumes that a pesticide is applied on all acres of each crop for which the pesticide is registered. In addition, it is assumed that the pesticide is found in those crops at its tolerance or half the Limit of Quantitation (LOQ) if no residues are found in the food crop.

The Food Quality Protection Act (Public Law 104-170, 1996) has mandated that the EPA consider the aggregate exposure, i.e., exposure arising from multiple sources (i.e., diet, water and nonoccupational sources) and pathways (i.e., oral, dermal, inhalation). A three-tier assessment approach is proposed. Tier 1 uses default assumptions and single point (deterministic) estimates of exposure, hazard and risk employing procedures routinely used by the EPA and the crop protection industry (EPA, 1999b). In Tier 2, combinations of deterministic and probabilistic (distributional) data are used, while Tier 3 assessments rely predominantly on distributional data.

### *Exposure from ingestion*

Procedures for estimating exposure to pesticides from dietary sources have been developed by the EPA (Petersen and Chaisson, 1988), and these are collectively known as the Dietary Residue Exposure System (DRES). DRES has provided the basis for past tolerance-setting decisions. Acute and chronic dietary exposure to a pesticide is calculated by assuming pesticide residues on food exist at tolerance levels, or at realistic anticipated residue levels. Refinements to the analysis at higher tiers may take into account information on market share, food processing factors, and studies that define the transfer of residues in fed commodities to milk, meat, and eggs.

New guidance has been developed by the EPA to assess the magnitude of pesticide exposure from food ingested during a single day (EPA, 1996e). The single highest residue value, the average residue, the 95th percentile of a residue distribution, or the entire distribution of residues may be used to derive the distribution for pesticide exposure. This is done using a sample population identified in the USDA continuing food intake survey for individuals (USDA, 1982).

An assessment of pesticide exposure from drinking water is less well developed. According to guidance established in the Primary Drinking Water Standard (EPA, 1991), the EPA has set the Maximum Contaminant Levels (MCLs) or Maximum Contaminant Level Guidelines (MCLGs) for selected pesticides. Traditionally, 20 per cent of the acceptable daily intake of a pesticide is allocated to drinking water based on a daily water consumption of 2 l for adults. Under FQPA, this general rule of thumb has been reconsidered such that Tier 1 analyses rely on the predicted concentration of pesticide residues in groundwater (Barrett, 1997) or surface water (Parker *et al.*, 1996). Refinements of these models have been proposed (Chen *et al.*, 1998), and it has been suggested that higher tiered analysis be based upon water monitoring data (Sielken *et al.*, 1998). More recently through aggregate exposure assessment, the EPA (EPA, 1999c) has introduced the concept of the Drinking Water Level of Comparison (DWLOC). This is the concentration of pesticide in drinking water based upon the acute, short-term, intermediate-term, or chronic hazard indicator.

### Dermal exposure

The Food Quality Protection Act mandates that exposure from nonoccupational sources be combined with exposure from ingestion. It is expected that a fraction of residential exposure will result from dermal exposure secondary to indoor or outdoor residential pesticide application. The Outdoor Residential Exposure Task Force (Breckenridge *et al.*, 1999) has been organized to develop a database that will include exposure data for individuals applying pesticides to turf. Subsequently, exposure related to reentry to the treated area as well as exposure for bystanders that may enter the treated area at various time intervals post-application will be assessed. A similar task force has been commissioned by industry to develop data that can be used to characterize pesticide exposure resulting from pesticide use in and around the home.

### Inhalation exposure

In most cases, the inhalation of pesticides is a minor route of exposure. Residential treatments are usually applied outside the home and the significant dilution in the atmosphere results in minimal opportunity for significant inhalation exposure. The exception to this is the use of pesticides in confined spaces such as termite treatment and other fogging uses.

### Multi-tier assessment

Probabilistic methods require extensive hazard and exposure data that are often not readily available. In order to efficiently use scientific resources, the simpler Tier 1 screening method currently can be used for a preliminary assessment. By conducting sensitivity analyses, exposure, hazard, and dose scaling factors that make significant contributions to risk can be identified, and research can be more effectively prioritized. Table 9.8 provides a list of factors commonly used in assessing the risk from exposure to pesticides arising from diet, water and nonoccupational pathways. A more detailed listing of such factors and the values commonly assigned to them can be found in *The Exposure Factors Handbook* (EPA, 1997b) developed and published by the EPA and *The Exposure Factors Sourcebook* (AIHC, 1994). Tier 1 analyses of the risk due to exposure to a pesticide from diet, water, and nonoccupational sources can be calculated by using constants for the parameters listed in Table 9.9.

This table does not list all parameters that might be utilized in a comprehensive risk assessment, only a representative few that are frequently encountered. Sensitivity analyses can be conducted using the Monte Carlo simulation method to determine the effect on the calculated risk when these parameters take on different values or distributional characteristics. Thus, the parameters that are thought to significantly impact risk can be identified and data can be collected for a higher tier analysis if the chemical fails to pass the Tier 1 screen.

The most appropriate approach to aggregate exposure to a given pesticide as required under the FQPA is still being defined.

## Risk assessment

Although the components for risk assessment used throughout the world embrace those defined by the Office of Science and Technology (OSTP, 1985), the goals defined by the different regulatory agencies globally may differ significantly.

### *Children and infants*

In addition, FQPA establishes a new safety standard and new procedures for the EPA's pesticide tolerance-setting activities (EPA, 1996f). Under the new Section 408 of FFDCA, the EPA can establish, revise, or leave in effect a tolerance (the legal limit for a pesticide chemical residue in or on a food) only if it is determined to be "safe". Section 408 defines safe to mean that there is a reasonable certainty that no harm will result from aggregate exposure to the pesticide chemical residue, including all anticipated dietary exposures and all other exposures for which there is reliable information. Section 408 requires the EPA to consider infants and children to ensure with reasonable certainty that no harm will result to infants and children from aggregate exposure to pesticide chemical residue. An additional 10-fold margin of safety for the pesticide chemical residue and other sources of exposure can be applied to account for potential pre- and post-natal toxicity and/or completeness of data with respect to exposure and toxicity to infants and children.

### *Carcinogen risk: weight of evidence approach*

A weight of evidence approach is described in the EPA cancer classification scheme (McGaughy, 1986). In a weight-of-evidence analysis, the evidence of oncogenicity in humans comes from long-term animal studies, and epidemiology (studies of man in exposed populations). Results from these studies are supplemented with available

*Table 9.9* Typical exposure, dose and risk assessment factors (Breckenridge *et al.*, 1999).

| *Exposure factors* | *Tier 1 (default value)* | *Tier 2* | *Tier 3* |
|---|---|---|---|
| *Exposure factors for dietary and water* | | | |
| Residue level | Diet: ($\sum$ tolerance); water: (MCLG) | | Residue distribution |
| Market share | Constant (100%) | | Variable (year, region) |
| Food intake | Mean (DRES) | ⇒ | Distribution |
| Water intake | Constant (2 l/day) | | Distribution |
| Exposure duration | Continuous (daily) | | Population linked distribution |
| Population linked | Constant (100%) | | Link to population |
| *Exposure factors for residential and lawn uses* | | | |
| Dislodgeable residue | Constant (100%) | | Distribution |
| Turf | | | |
| Residential | | | |
| Penetration factors | | | Distribution experimentally determined |
| Clothing/type | Constant (80%) | | |
| Dermal | Constant (100%) | | |
| Use pattern | | ⇒ | Distribution |
| Duration | Constant (daily) | | |
| Frequency | Constant (daily) | | |
| Reentry interval | Constant (specific) | | |
| Population linked | Constant (100%) | | Link to Population |
| *Dose factors* | | | |
| Body weight | Constant (70 kg) | | Distribution |
| Body surface area | Constant (21,110 $cm^2$) | | Weight-dependent variable |
| Respiration rate | Constant (29 l/min) | ⇒ | Distribution |
| Dose scaling factor | Constant $(B. Wt)^{3/4}$ | | Physiologically-based |
| Metabolism | No Metabolism | | Pharmacokinetic model |
| *Risk factors* | | | |
| Benchmark dose | Constant (chemical specific) | ⇒ | Distribution (chemical specific) |
| RfD | | | |
| $ED_{10}$, $LED_{10}$ | | | |
| Relative potency | Constant (chemical specific) | | Distribution (chemical specific) |
| $Q_1$* | | | |
| TEFs | | | |
| *Additivity factors* | | | |
| Conditional probability | | | |
| Multi-source | 1.0 | ⇒ | $< 1.0$ |
| Multi-chemical | | | |
| *Safety factor vs. percentiles of probability distributions* | | | |
| Health standard | Use safety factor (e.g., 10, 100, 1,000) | ⇒ | Use percentile of distribution (e.g., 95th percentile) |

information from other sources that include mutagenicity and other short-term tests (for genetic effects), metabolic or kinetic studies, and other relevant toxicological studies as presented in Table 9.10.

The EPA has essentially assumed that animal tumorigens are human carcinogens (McGaughy, 1986). This approach has been taken as a default assumption almost without regard to the quality of the conduct of the study, the level of test material administered, or the mechanism by which the tumor response is manifest (Sumner and Stevens, 1994). In this regard, all animal carcinogens are treated as though they have no threshold or there is a real risk at all exposure levels. The EPA has published a list identifying crop protection chemicals as known, probable, or possible human carcinogens without appropriate consideration of the mechanism by which the tumors occur or their relevance to man (Moolenaar, 1994).

However, most of the other OECD and EEC countries have assumed there are genotoxic agents that are carcinogenic in animal tests and are likely human carcinogens (UK Department of Health, 1991; EEC, 1993). It may be appropriate to regulate these chemicals as if there is a linear dose response extrapolated through zero. On the other hand, nongenotoxic agents are generally characterized as having a nonlinear dose response such that at some dose the cancer risk is zero. It is appropriate to regulate exposure to these chemicals using a suitable safety or uncertainty factor, i.e., the MOE approach.

### *Proposed cancer risk assessment guidelines*

The draft cancer risk assessment guidelines (EPA, 1996c,1999a) utilize the Bradford Hill criteria for causality to evaluate the biological plausibility of a proposed mode of action for tumor development (Hill, 1965). The proposed mode of action must identify a sequence of important events that lead up to the expression of a tumor response. The strength, consistency and the specificity of the association between chemical exposure and these events must be evaluated. The dose-response and the time-to-response associations must be consistent. For example, the tumor incidence is expected to increase with dose and those doses that cause tumors must trigger the essential events. Furthermore, essential events that are defined as precedent events must consistently occur earlier than those that occur later in the causal sequence. Overall, the proposed mode of action must be biologically plausible and equally plausible alternative explanations must be evaluated and discounted as being unlikely.

This framework can also be used to evaluate if the proposed mode of action is likely to be operative in humans and if the response is likely to be linearly related to dose throughout the entire dose-response range in animals or humans. These essential

*Table 9.10* US Environmental Protection Agency's weight-of-evidence approach (McGaughy, 1986)

| |
|---|
| In a weight-of-evidence analysis, the evidence of oncogenicity in humans comes from two sources |
| Long-term animal studies, and |
| Epidemiology (studies of man in exposed populations). |
| Results from these studies are supplemented with available information from other sources |
| Mutagenicity and other short-term tests (for genetic effects) |
| Metabolic or kinetic studies (looking at what the body does with the chemical), and |
| Other relevant toxicological studies. |

*Table 9.11* Comparison of the EPA old to proposed new scheme for classification of carcinogens

| *Old classification (McGaughy, 1984)* | *New proposed scheme (EPA, 1999a)* | *Criteria for classification* | *Evaluation* | |
|---|---|---|---|---|
| | | | *Uncertainty factor* | *LMS** |
| A – human | Known human carcinogen | Sufficient evidence in man | | X |
| B – probable human | Likely human carcinogen | B1 – limited evidence in man; sufficient evidence in animal (two species with tumors) | | X |
| | | B2 – inadequate human evidence; sufficient animal evidence | | X |
| C – possible human | Suggestive human carcinogen | No evidence in man; limited evidence in animals | X or | x |
| D – not classifiable | Inadequate data | Inadequate animal or human evidence | X | |
| E – not a human carcinogen | Not likely human carcinogen | Sufficient animal testing with no evidence of carcinogenicity and human experience | X | |

questions must be addressed before an informed decision about the human carcinogenic potential of the chemical.

Proposed guidelines suggest changes to address and accommodate new information on carcinogenesis, and hopefully advance cancer risk assessment (EPA, 1996c). Hazard assessment emphasizes analysis of all biological data. Agent's Mode of Action is emphasized to reduce the uncertainty in describing the likelihood of harm and in determining the dose response approach. Hazard characterization is added to integrate the data analysis of all relevant studies into a weight of evidence conclusion of hazard, to develop a conclusion regarding the mode of action in leading to tumor development. The conditions under which the hazard may be expressed such as route, pattern, duration, and magnitude of exposure must also be delineated. The proposed weight of evidence narrative, replacing the current alphanumeric classification, is intended to lie out a summary of the key evidence, describe the mode of action, characterize the conditions of hazard expression, and recommend an appropriate dose-response approach. The overall conclusion as to the likelihood of human carcinogenicity is given by route of exposure.

The weight of evidence descriptors for classifying human carcinogenic potential (EPA, 1999a) are contrasted against the six alphanumeric categories (A, B1, B2, C, D, E) in the 1986 cancer guidelines (McGaughy, 1986) in Table 9.11.

Although a biologically based extrapolation model should be the approach for quantifying risk, the proposed guidelines appear to default back to the LMS model.

Dose-response assessment is a two-step process. In the first step, response data are modeled in the range of observation. The second step is to determine the point of departure or range of extrapolation below the range of observation. Curve fitting in the observed range would be used to determine the effective dose corresponding to the lower 95 per cent limit on a dose associated with 10 per cent response (LED10) either with linear or nonlinear models (EPA, 1996c). The LED10 would then be used as a point of departure for extrapolation to the origin as the linear default or for a Margin of Exposure (MOE) discussion as the nonlinear default. The LED10 may become the standard point of departure, and may replace the NOEL. Despite the fact that these guidelines have not been finalized, the weight of evidence narrative descriptors and default methods of quantitation are being applied to new products in the registration process.

## References

AIHC. *Exposure Factors Sourcebook.* Washington, DC: American Industrial Health Council, 1994.

Aspelin AL. *Pesticide Industry Sales and Usage 1992 and 1993 market estimates*: US Environmental Protection Agency, Washington, DC, 1994, 733-K-94-001, 33 pp.

Aspelin AL. *Pesticide Industry Sales and Usage, 1994 and 1995 Market Estimates Reports.* Economic Analysis Branch, Biological and Economic Analysis Division, US Environmental Protection Agency, Washington, DC, 733-R-97-002, 1996.

Avery D. Environmental agriculture: 60 years of inspiration. *Natl Agric Chem Assoc Farm Chem Mag* 1993;1.

Ballantine LG. An overview of the US pesticide registration guidelines. *Agric Newslett* 1992;3(2):1–6.

Barrett M. *Initial Tier Screening of Pesticides for Ground Water Concentration Using the SCI -GROW Model.* US Environmental Protection Agency, December 3, 1997.

Breckenridge CB, Sielken Jr R L, Stevens JT. Aggregate and cumulative exposure and risk assessment. In: Ragsdale NN, Seiber JN, editors. *Pesticides: Managing Risks and Optimizing Benefits*. Washington, DC: American Chemical Society, 1999. pp. 38–67.

BIBRA. *Screening for Safety: Pesticides*. London: Liebling Stewart Design Associates Limited, 1988. pp. 3–4.

Chen W, Hertl P, Tierney D. *A Simple Regression Model for Predicting Surface Water Concentrations Resulting from Agricultural Field Runoff and Erosion*. Greensboro, NC: Novartis Crop Protection, Inc., 1998.

Conner Jr JD, Ebner LS, Landfair SW, O'Connor III C, Weinstein KW, Jovanovich AP. *Pesticide Regulations Handbook*, 3rd ed. New York: Executive Enterprises, 1991. pp. 1–2.

Cremlyn R. *Pesticides: Preparation and Mode of Action*. New York: John Wiley and Sons, 1978, p. 1.

Crofton K. Personal communication in regard to the status of the protocol for the developmental neurotoxicity study, 2000.

Dearfield KL Mutagenicity. *Addendum 9, EPA 540/09-91-122, Pesticide Assessment Guidelines, Subdivision F, Hazard Evaluation: Human and Domestic Animals, Series 84*. National Technical Information Service, Springfield, VA, pp. 1–11.

Ecobichon DJ. Toxic effects of pesticides. In: Doull J., Klaassen CD, Amdur MO, editors. *Casarett and Doull's Toxicology: The Basic Science of Poisons*, 4th ed. New York: Macmillan Publishing Co., Inc. 1993. pp. 565–621.

EPA. Proposed guidelines for registering pesticides in the US. *Fed Reg* 1978a;43(163):37336.

EPA. Toxic Substances Control Act; notification of substantial risk under Section 8(e); notice. *Fed Reg* 1978;43:11110 (FRL-849-2).

EPA. Good laboratory practice standard for health effects. *Fed Reg* 1979;44(91):27362.

EPA. *Pesticide Assessment Guidelines, Subdivision F. Hazard Evaluation: Human and Domestic Animals*. Washington, DC: Environmental Protection Agency, 1982a. No. 540/9-82-025. Available from NTIS, Springfield, VA.

EPA. *Pesticide Assessment Guidelines, Subdivision E. Hazard Evaluation: Wildlife and Aquatic Organisms*. Washington, DC: Environmental Protection Agency, 1982b. No. 540/09-87-198. Available from NTIS, Springfield, VA.

EPA. 40 CFR, Part 158: data requirements for pesticide re-registration. *Fed Reg* 1984;49(207):42856.

EPA.Toxic Substances Control Act test guidelines: final rules. *Fed Reg* 1985;50:39428–39429.

EPA. *FIFRA Accelerated Reregistration Phase 3 Technical Guidance, Appendix D. Office of Pesticides and Toxic Substances*. Washington, DC: Environmental Protection Agency, 1989. No. 540/09-90-078. Available from NTIS, Springfield, VA.

EPA. National Primary Drinking Water Regulations: Final Rule. Part II. *Fed Reg* 1991;3525–3757.

EPA. Health Effect Test Guidelines, 1996a. http://www.epa.gov/docs/OPPTS_Harmonized/870_Health_Effects_Test_Guidelines/Master/870mast.pdf (accessed 12/99).

EPA. Ecological Effects Test Guidelines, 1996b. http://www.epa.gov/docs/OPPTS_Harmonized/850_Ecological_Effects_Test_Guidelines/Master/850.master.pdf (accessed 12/99).

EPA. *Proposed Guidelines for Carcinogen Risk Assessment. Fed Reg* 1996c;17960–18011.

EPA. Endocrine disruptors; notice of public meeting. *Fed Reg* 1996d;61(230);60280–60281.

EPA *Final Office Policy for Performing Acute Dietary Exposure Assessment*. Office of Pesticide Programs, US Environmental Protection Agency, Washington, DC, 1996e.

EPA . Pesticides; policy issues related to the Food Quality Protection Act. *Fed Reg* 1996f;64(130):37001–37009.

EPA. Reporting requirements for risk/benefit information; final rule. *Fed Reg* 1997a;62(182):49369–49395.

EPA. *Exposure Factors Handbook: General Factors*. Washington, DC: National Center For Environmental Assessment, 1997b. EPA/600/P-95/002Fa.

EPA. Endocrine Disruptor Screening and Testing Advisory Committee (EDSTAC) final report. EPA/743/R-98/003 1998a. www.epa.gov/opptintr/opptendo/finalrpt.htm (accessed 11/99).

EPA. Endocrine Disruptor Screening Program. *Fed Reg* 1998b;63(154):42852–42855.

EPA. *Proposed Guidelines for Carcinogen Risk Assessment*. Washington, DC: Office of Research and Development, 1999a. EPA/600/p-92/003c.

EPA. Guidance for performing aggregate exposure and risk assessments. Pesticides; policy issues related to the Food Quality Protection Act. *Fed Reg* 1999b;64(217):61343–61346.

EPA. Estimating the drinking water component of a dietary exposure assessment. Pesticides; policy issues related to the Food Quality Protection Act. *Fed Reg* 1999c;64(217):61346–61348.

EEC. Commission Directive 93/67/EEC: 1993. *Laying Down the Principles for Assessment of Risks to Man and the Environment of Substances Notified in accordance with Council Directive 67/548/ EEC*. July 20, 1993.

EEC. Commission Directive 94/79/EC of December 1994 amending Council Directive 91/414/ EEC concerning the placing of plant protection products on the market. *Official Journal of the European Communities,* December 31, 1994. p. No.L354/16.

Farber TM. Pesticide Assessment Guidelines, Subdivision F, *Position Document: Selection of a Maximum Tolerated Dose (MTD) in Oncogenicity Studies.* Toxicology Branch, Hazard Evaluation Division, Office of Pesticides Programs, US Environmental Protection Agency. NTIS PB88-116736, 1987.

Foran J., the ILSI Risk Science Working Group on Dose Selection. Principles for the selection of doses in chronic rodent bioassays. *Environ Health Perspect* 1999;105(1):18–20.

Gianessi LP. *A National Pesticide Useage Data Base*. Washington, DC: Resources for the Future, 1986. pp. 1–14.

Gray GM, Graham JD. Risk versus risk: tradeoffs. In: Graham JD, Wiener JB, editors. *Protecting Health and the Environment.* Harvard University Press: Cambridge, MA. 1995. pp. 173–192.

Hayes Jr WJ. Introduction. In: Hayes Jr. WJ, Laws Jr. ER, editors. *Handbook of Pesticide Toxicology. Volume 1. General Principles.* San Diego, CA: Academic Press, Inc., 1991. pp. 1–37.

Hill B. The environment and disease: association or causation? *Proc R Soc Med* 1965;58:295–300.

JMPR. Evaluations: 1979. *Joint Meeting of the FAO Panel of Experts on Pesticide Residues in Food and the Environment and the WHO Expert Group on Pesticide Residues*. Sponsored jointly by FAO and WHO. Geneva, 3–12 December 1979. http://www.inchem.org/documents/jmpr/jmpmono/v079pr01.htm.

Klaassen CD, Doull J. Evaluation of safety Chapter 3. In: Klaassen CD, Doull J, Amdur MO, editors. *Casarett and Doull's Toxicology: The Basic Science of Poisons*. New York: MacMillan Publishing Co., Inc., 1980. p. 12.

McGaughy R. Guidelines for carcinogen risk assessment. *Fed. Reg*. 1986;51(185):33992–34003.

Mellanby K. *The DDT Story*. Farnham: British Crop Protection Council, 1992. pp. 6–7.

MAFF. Good Laboratory Practices. Noh San, Notification No. 3850. Japan: Agricultural Bureau, August 10, 1984.

MAFF. Notification of the Director-General. *Requirements for Safety Evaluation of Agricultural Chemicals*. Japan: Agricultural Production Bureau, Ministry of Agriculture, Forestry and Fisheries, 59 NohSan No.4200, 1985.

Moolenaar RJ. Default assumptions in carcinogenic risk assessment used by regulatory agencies. *Regulat Toxicol Pharmacol* 1994:20:S135–S141.

NRC. *Pesticides in the Diets of Infants and Children*. Washington, DC: National Academy Press, 1993. pp. 1–386.

NRDC. *ENVL417-Environmental Law. Markups added for course purposes*. 824 F.2D 1211, 263 U.S.APP.D.C. 231, 1987. http://web1.manhattan.edu/wmatysti/omgt/nrdcvepa.htm (accessed 1/2000).

OECD. *OECD Guideline for Testing of Chemicals, Section 4, Health Effects*, adopted May 12, 1981.

OECD. *Good Laboratory Practice in the Testing of Chemicals* – Final report of the OECD Expert Group on Good Laboratory Practice, 1989. ISBN 92-64-12367-9

Office of Science and Technology Policy (OSTP). Chemical carcinogens: review of the science and its associated principles *Fed. Reg*. 1985;50(50):10372–10442.

Parker RD, Nelson HP, Jones RD, Mostaghimi S. *Development for Screening Level Estimation of Pesticide Exposure in the Aquatic Environment*. Washington, DC: US EPA, 1996.

Petersen B, Chaisson CF. Pesticides and residues in food. *Food Technol* 1988;42(7):59–64.

Public Law 104-170. *Food Quality Protection Act of 1996*. 104 Congress, 2nd Session. Report 104-669, Part 2. Washington, DC: Government Printing Office, 1996. pp. 1–89.

Simmons SW. The use of DDT insecticides in human medicine. In: Miller P, editor. *DDT: The Insecticide Dichlorodiphenyl-Trichloroethane and Its Significance, Vol. 2*. Basel: Birkhaeriger, 1959. pp. 251–502.

Sielken RL, Bretzlaff RS, Valdez-Flores J. *Probabilistic Risk Assessment Using Atrazine and Simazine as a Model. Triazine Herbicides: Risk Assessment*. Washington, DC: American Chemical Society Symposium Series # 683, 1998.

Spindler M. *DDT:* Health aspects in relation to man and risk/benefit assessment based thereupon. *Resid Rev* 1990;90:1–34.

Stetter J. Trends in the future development of pest and weed control – an industrial point of view. *Regulat Toxicol Pharmacol* 1993;17:346–370.

Stevens JT. Risk assessment of pesticides. In: Sipes IG, McQueen CA, Gandolfi AJ, editors. *Comprehensive Toxicology. Volume 2. Toxicological Testing and Evaluation*. Oxford: Elsevier Science Ltd., 1997. pp. 17–26.

Stevens JT, Sumner DD, Luempert L. Agricultural chemicals: the impact of regulations under FIFRA on science and economics. In: Chenzelis C, Holson J, Gad S, editors. *Primer on Regulatory Toxicology*. New York: Raven Press, 1995. pp. 133–163.

Sumner DD, Stevens JT. Pharmacokinetic factors influencing risk assessment: saturation of biochemical processes and cofactor depletion. *Environ Health Perspect* 1994;102(Suppl 11): 3–22.

Thornaton R, Banks D, Bissing D, Carlson GA, Clydesdale F, Fukuto TR, Kennedy G, Merrill RA, Muir WR, Pesyna GM, Sharp D, Swanson E, Taylor MR, Tschirley FH , Upton FH, Van Ravenswaay E, Waggoner P. *Regulating Pesticides in Food: The Delaney Paradox*. Washington, DC: National Academy Press, 1987. pp. 1–272.

UK Department of Health. *Guidelines for the Evaluation of Chemicals for Carcinogenicity*. London: Committee on Carcinogenicity of Chemicals in Food, Consumer Products and the Environment, UK Department of Health, 1991. p. 1.

US Congress. Federal Insecticide, Fungicide and Rodenticide Act (FIFRA) Pub. L. No. 80-104, 61 Stat. 163, 1947. p. 1.

US Congress. Food additive amendments to the Federal Food, Drug, and Cosmetic Act (FFDCA) 409, Pub. L. No. 85-929, 72 Stat., 1785, 1958. p. 1.

US Congress. Federal Environmental Pesticide Control Act (FEPCA). Pub. L. No. 92-516, 86 Stat. 973, 1972.

USDA. National Food Consumption Survey 1977–78. *Foods commonly eaten by individuals: amount per day and per eating occasion*. Hyattsville, MD: USDA, 1982.

Weisburger JH. Chemical carcinogens. In: Casarett LJ, Duoll J, editors. *Toxicology, The Basic Science of Poisons*. New York: MacMillan Publishing Co. Inc., 1975. p. 333.

WHO. *WHO Expert Committee on Malaria. Thirteenth Report*. WHO Tech Rep Serv No. 357. Geneva: World Health Organization, 1967.

Chapter 10

# Industrial chemicals

## Regulation of new and existing chemicals (The Toxic Substances Control Act and similar worldwide chemical control laws)

*Richard C. Kraska*

### Introduction

In the major developed countries, laws to regulate chemicals used as drugs, food additives and pesticides were developed first due to public health and environmental concerns of these uses. Eventually these laws required pre-market clearances of new chemicals for these uses. As time went on, laws were implemented so that virtually all chemicals had some level of regulation. Chapters 10 and 11 deal with the laws that affect the industrial chemical industry. This chapter deals with the so-called chemical control laws which affect the introduction of new chemicals into commerce as well as attempt to manage newly discovered risks of existing chemicals. The next chapter deals with laws concerning worker safety which govern the communication of the hazard information and which establish safe exposure levels for chemicals.

The major chemical control laws include provisions for premanufacture or premarketing clearance for new chemicals. In addition, these laws contain various provisions for reporting information and controlling the risk of all chemicals, including preexisting chemicals that were in commerce before notifications for new chemicals were required. Legal, procedural and public policy analysis of these statutes is a fascinating subject covered adequately elsewhere (CEQ, 1971; EPA, 1996, 1997; OECD, 1997a, Bergeson *et al.*, 2000). This chapter will focus on the health and environmental issues with an emphasis on recent trends in these issues because they are of primary interest to toxicologists. In addition, the issues of interest to toxicologists are very complex and have many exceptions depending on the nature of various chemicals. Toxicologists need to consult the regulations, guidance documents and the experience of other practitioners in order to truly become an expert in the requirements of chemical control laws.

#### *General overview*

The first of these laws was passed in certain countries that experienced toxic incidents or concern about public and environmental health associated with the use and disposal of industrial chemicals. As time went on, other countries enacted similar laws, attempting to customize laws to national concerns and correct perceived defects in the laws they sought to emulate and improve upon.

Each of these laws defines a regulated community that usually consists of manufacturers and importers of chemicals. The laws give regulatory and enforcement authority to one or more national federal agencies. Since there were decades of industrial activity to reconcile, the laws commonly empower the agencies to require the regulated industry to report various information about manufacture and use of the chemicals as well as unpublished health and safety information about chemicals and processes.

It has been especially difficult in the past for practitioners to get information about requirements in other countries. This improved somewhat with time but has more recently greatly improved with information now available on the Internet. A list of useful web sites is given at the end of this chapter. Available guidelines and other publications will be mentioned at the appropriate points in this chapter. Another good source of information is the Organization for Economic Cooperation and Development (OECD). This organization has provided a platform for international discussion of a variety of issues associated with the regulation of chemicals and the work products of many work groups contribute significantly to the information available on many issues.

### *United States*

Congress enacted the Toxic Substances Control Act (TSCA) in 1976 (USC, 1976). The purpose of TSCA was to fill a regulatory gap by giving broad control to the Environmental Protection Agency (EPA) over industrial chemicals not regulated by other statutes. In the US, there was public concern over the risks from chemicals such as kepone, vinyl chloride, heavy metals and Polychlorinated Biphenyls (PCBs); also the President's Council on Environmental Quality (CEQ) identified a "high priority need for a program of testing and control of toxic substances" because the existing statutory mechanisms for protecting the environment against chemical hazards were neither cohesive nor adequate (CEQ, 1971).

Under TSCA, the EPA can impose wide-ranging requirements on importers and manufacturers of chemicals to test chemicals, to control the way chemicals are manufactured and used, and to report certain information and activities to the agency. TSCA is somewhat of a misnomer; the law gives authority to the EPA for all chemicals that are not regulated by other laws, such as pesticides and food additives. The term toxic is not even defined in TSCA or in EPA regulations.

The law is divided into a number of sections. Procedures that require premanufacturing clearances and other requirements for new chemicals are outlined in Section 5. Other sections deal with existing chemicals. Regulations promulgated by the EPA under TSCA are listed in Title 40 of the Code of Federal Regulations (CFR) in parts 700–799. The main sections of the law and corresponding parts of 40 CFR are listed in Table 10.1. Compared to its international counterparts, TSCA contains the most detailed reporting requirements on existing chemicals as well as specific provisions to require manufacturers, importers and, in some cases, processors of existing chemicals to conduct needed toxicology testing.

The EPA is divided into a number of different offices, each of which has responsibility for different laws that address different aspects of the environment: air pollution, water pollution, waste disposal, pesticides and toxic substances. Currently, the Office of Pollution Prevention and Toxics (OPPT) of the EPA enforces TSCA. The organization of these offices has changed from time to time (EPA, 1993a). There is little interchange between the offices or even divisions within an office that administer different laws. The

*Table 10.1* Major sections of TSCA

| *Section number* | *Subject* | *40 CFR reference* |
|---|---|---|
| 4 | Chemical testing | Parts 700–799 |
| | Good laboratory practices | Part 799 |
| 5 | New chemicals | Part 720 |
| | PMN exemptions | Part 723 |
| 5(a) | Significant new use rules | Part 721 |
| 6, 7 | Existing chemicals control | Part 750 |
| 8(a) | Reporting concerning chemical use and manufacture | Parts 704, 712 |
| 8(b) | Inventory reporting rules | Part 710 |
| 8(c) | Adverse reactions (allegations) reporting | Part 717 |
| 8(d) | Health and safety data reporting | Part 716 |
| 8(e) | Substantial risk reporting | |
| 12 | Export rules | Part 707 |
| 13 | Import rules | Part 707 |

scientists in the divisions of OPPT who are responsible for pesticides use many different tools and standards than the scientists who are responsible for industrial chemicals under TSCA.

A toxicology staff needs to play a key role in the TSCA compliance efforts of a company that manufactures or processes industrial chemicals. This chapter will focus on those sections of TSCA that are usually the responsibility of toxicologists. TSCA compliance, however, requires many activities that usually need the attention of regulatory affairs specialists and scientists with expertise in chemical nomenclature, manufacture, sales and marketing activities. These requirements will receive less attention in this chapter. The reader is referred to the text of the regulations and other information from the EPA for details.

One key to compliance is keeping up with new information on TSCA. Besides new regulations, the EPA frequently makes new policy statements and guidance documents available. A former key publication, the *Chemicals in Progress Bulletin* has been consolidated into a new publication entitled *Chemicals in our Community*. The EPA publishes this quarterly and subscriptions are free. It contains summaries of regulatory and compliance activities related to TSCA. Over the last few years, accessibility to EPA guidance documents and summaries of EPA programs has improved dramatically with the introduction of the EPA's web site and this is now the key source of updated information on TSCA.

## *Europe*

The Dangerous Substances Directive (DSD) became effective in 1967 (European Commission, 1967; 67/548/EEC). Many revisions have taken place since then in the form of eight amendments and twenty-five adaptations. Unlike the laws in the US, this directive is a more comprehensive authority to implement oversight on not only the manufacture and control of risks of new and existing chemicals, but also to mandate labeling and hazard communication requirements for both industrial chemicals and a wide range of consumer products other than drugs, food additives and pesticides. The labeling and hazard communication provisions of the DSD will be addressed in Chapter 11. Because the law is administered by agencies of all the member states of the EU, practitioners may encounter variations in the way the law is interpreted and administered in the individual member countries.

## *Japan*

Japanese laws and regulations provide much difficulty to companies based in Western nations. Japanese requirements for control of industrial chemicals have many important differences from the requirements in Europe and the US. The law that affects new chemical notification is the law concerning Examination and Regulation of the Manufacture, etc. of Chemical Substances and is administered by the Ministry of International Trade and Industry (MITI) and the Ministry of Health and Welfare (MHW). The law was established in 1973 to prevent environmental pollution and hazards to human health by chemical substances used for various purposes. The impetus for its enactment was the problem of environmental pollution caused by PCB in the late 1960s. New chemicals must also be notified under the Industry Safety and Health Law of 1977. This law is administered by the Ministry of Labor (MOL) and aims to protect health in the workplace. The MOL is primarily concerned with the prevention of occupational cancer.

The best source of information on Japanese laws and regulations in the English language is the *Handbook of Existing & New Chemical Substances* (The Chemical Daily Co, Ltd., 1999). The handbook contains full text of the laws, reporting forms, testing guidelines, review criteria as well the lists of Existing Chemical Substances, New Chemical Substances as well as the lists of Specified Chemical Substances.

## *Canada*

The Canadian Environmental Protection Act (CEPA) was passed in 1988 and significantly amended in 1999. The law is a comprehensive environmental statute rather than just a simple chemical control law. The law includes many provisions that are similar to TSCA but also contains some requirements similar to the DSD. CEPA established a chemical inventory of substances called the Domestic Substances List (DSL). The law contains authority to regulate both new and existing chemicals. The act also named eleven prohibited substances. In the recent amendment to the law, the term toxic was defined as

> a substance… entering the environment or (that) may enter the environment in a quantity or concentration having or that may have an immediate or long-term harmful effect on the environment, or its biological diversity, or on its human life or health….

The chemical control provisions of CEPA are primarily administered by Environment Canada with help from Health Canada. Environment Canada has a web site, publishes guidelines for compliance, and holds periodic educational workshops.

### *Australia*

The National Industrial Chemicals Notification and Assessment Scheme (NICNAS) was established in 1989 with the passage of the Industrial Chemicals Notification and Assessment Act to protect the public and the environment from the harmful effects of industrial chemicals. NICNAS is administered by three agencies, the National Occupational Health and Safety Commission (NOHSC, also known as Worksafe Australia), Environment Australia and the Therapeutic Goods Administration of the Department of Health and Family Services. NOHSC has a web site and a number of guidelines and publications on assessment activities are available.

### *Korea*

The Toxic Chemicals Control Act (TCCA) was enacted on August 1, 1990 to control chemical substances that are hazardous to human health or the environment. The law is administered by the Ministry of the Environment (MOE).

### *Philippines*

The Philippines chemical control law is the 1990 Toxic Substances and Hazardous Waste Control Act (the Philippines Republic Act 6969) which covers import, manufacture, processing, handling, storage, transport, sale, distribution, use and disposal of chemical substances and mixtures. The Act is administered by the Department of Environment and Natural Resources (DENR). DENR has a web site, however, as of this writing, it has no information on the chemical control law.

## Requirements for new chemicals

### *Overview*

Before any of the authority for requiring applications or notifications from industry to clear new chemicals for manufacture or marketing was implemented, each regulatory authority was required to establish a list or inventory of chemicals already in use. Usually this required several years of reporting by the regulated industry followed by compiling, publishing and correcting the list by the regulatory agency. Many chemical control laws established their inventories using the nomenclature system of the Chemical Abstract Service (CAS) and the CAS registry numbers assigned by this service.

The intricacies of the listing process and the various exemptions and notification procedures for various classes of new chemicals is usually of high interest to regulatory chemists and attorneys, but beyond the scope of this chapter. Consult the regulations in each country for these details. Another good source of great detail on the subject is the report of an OECD workshop on sharing chemical assessments between regulatory authorities (OECD, 1997a) and the New Industrial Chemicals Information Directory (OECD, 2000).

The testing and evaluation procedures are of more interest to toxicologists and will be discussed in greater detail. It is important to note that the notification procedure for new chemicals is more detailed for nonpolymeric chemicals. Testing requirements for the various laws are summarized in Table 10.2. Because of the lower concern for health and environmental effects of most polymers, there usually exists an abbreviated notification procedure or an outright exemption for some or all polymers under each scheme. Many countries have adopted the OECD base set tests for new chemicals (OECD, 1981) and the OECD guidelines for Good Laboratory Practices (GLP, OECD, 1997b).

Typically, these laws only allow a fixed period of 45–90 days for review of notifications by the government agency. Some of these laws do not even require a formal "positive approval" from the government agency and the submitter is free to proceed with manufacture or marketing if there is no response from the agency regarding a notification.

*Table 10.2* Testing requirements for new chemicals notifications for major countries[a]

| *Test type* | *OECD protocol* | *US*[b] | *EU* | *Japan* | *Canada* | *Australia* |
|---|---|---|---|---|---|---|
| *Mammalian studies* | | | | | | |
| Acute oral | 401 | SR[c] | R | R | R | R |
| Acute dermal | 402 | NR | R | NR | R | R |
| Eye irritation | 405 | NR | R | NR | R | R |
| Skin irritation | 404 | NR | R | NR | R | R |
| Dermal sensitization | 406 | NR | R | R | R | R |
| Oral repeated dose | 407 | SR[c] | R | R[d] | R | R |
| Chronic toxicity | 452 | SR | NR | SR[e] | R | R |
| Reproductive toxicity | 415 | SR | FR | NR | NR | NR |
| *Genotoxicity* | | | | | | |
| Bacterial mutagencity (*in vitro*) | 471 | SR[c] | R | R | R | R |
| Mammalian cytogenetics (*in vitro*) | 473 | SR[c] | SR | SR | SR | SR |
| Germ cell cytogenetics | 478 | NR | NR | NR | SR | R[e] |
| Mouse micronucleus[f] | 474 | SR[c] | SR | SR | SR | SR |
| *Environmental* | | | | | | |
| Acute fish | 203 | SR[g] | R | R | R | R |
| Acute daphnia | 202 | SR[g] | R | NR | R | R |
| Acute algae | 201 | SR[g] | R | NR | R | R |
| Chronic daphnia | 202 | NR | NR | NR | NR | R |
| Biodegradation | 301 or 302 | NR | R | R[h] | R | R |
| Fish bioaccumulation | 305 | NR | NR | R[i] | NR | NR |

[a] R = required in most instances; SR = sometimes required under certain circumstances; NR = not required in most instances; FR = future requirement likely.

[b] Although no formal testing requirements exist for new chemicals under TSCA, the EPA has authority to require virtually any test if serious questions of health and environmental safety arise. 40 CFR 720.50 requires that any available data on the health and environmental effects of the notified substances be submitted.

[c] Required when exposure based concerns for human health are triggered.

[d] Japanese guidelines contain various additions to an OECD 407 such as additional tissues to be examined. Urinalysis and 14-day recovery groups for control and high dose groups also strongly recommended.

[e] Mouse micronucleus is usually allowed as a substitute for a germ cell cytogenetics assay.

[f] Mouse micronucleus may be required to resolve a positive *in vitro* cytogenetics result.

[g] Required when exposure based concern for environmental effects are triggered. Check with the EPA for special customized protocols.

[h] Requires use of Japanese sludge sample and identification of metabolites of degradation.

[i] Japanese reviewers prefer "cold" chemical analysis methods as opposed to radiotracer studies.

### *United States*

The TSCA Chemical Substances Inventory was developed according to the procedures Section 8(b) and initially consisted of those substances in commerce in the US between January 1, 1975 and June 1, 1979. As new chemicals are notified to the EPA and are commercialized, they are added to the TSCA Chemical Substances Inventory.

A new chemical or polymer is notified through an application process called a Premanufacturing Notice (PMN). The notification must be submitted on the prescribed PMN form which calls for a complete chemical identity, impurities, use description, manufacturing locations and process descriptions, and worker, customer and environmental exposure data. Requirements for PMNs in the US differ from those in many other countries in that no specific toxicity testing is required to be conducted before notification. By regulation (40 CFR 720.50), however, all available toxicity data known to the notifier as well as any data related to health and environmental impact of the chemicals must be submitted. A detailed guideline for PMN submitters is available on the EPA web site (EPA, 1997).

In order for the EPA to make a decision within the statutory 90-day review period without benefit of a standard data set, the EPA relies heavily on structure activity relationship (SAR) determinations on the chemical to judge whether it can be manufactured and used safely. The EPA develops structure activity relationships from known publicly available data and proprietary data submitted by other notifiers or reporters of information under other sections of TSCA (Wagner *et al.*, 1995). This review is conducted under a very regimented procedure that is tightly scheduled in the 90-day review period as described in Figure 10.1. Over the last several years, the EPA has added several computer models for exposure modeling and for SAR determinations. The EPA has begun to hold workshops for scientists that work for regulated companies so that they can use the models to anticipate the results of the EPA review. The list of models currently used by the EPA is given in Table 10.3 (EPA, 1998) and details on how the exposure models are used are in the 1997 guidance document.

In cooperation with European regulators, the EPA participated in a joint study comparing their SAR projections against data submitted under the European notification scheme. A report is available which compares the results of the two approaches (EPA/EC, 1994). The EPA has used this study to help improve the SAR models that they have developed.

The 1997 guidance describes eleven possible outcomes of PMN review. The most common outcome is an early "drop" from the review process. This occurs for about 80% of the PMNs. Further review exonerates the majority of the remaining PMNs from concern and all these chemicals are allowed to be marketed without further controls. About 5 per cent of the cases progress to "standard review".

The EPA uses a variety of methods to control chemicals that make it to standard review. The EPA will sometimes issue a "letter of concern" to the submitter asking for some specific voluntary control in the way the chemical is manufactured or used. The EPA has a great deal of flexibility under Section 5(e) of TSCA to impose controls or require the submitter to submit further information. Commonly, additional testing data is required via a formal written agreement called a consent order with the submitter. Consent orders can be used to impose controls on manufacturing or use of the chemical substance. Depending on the need for notifying other possible manufacturers,

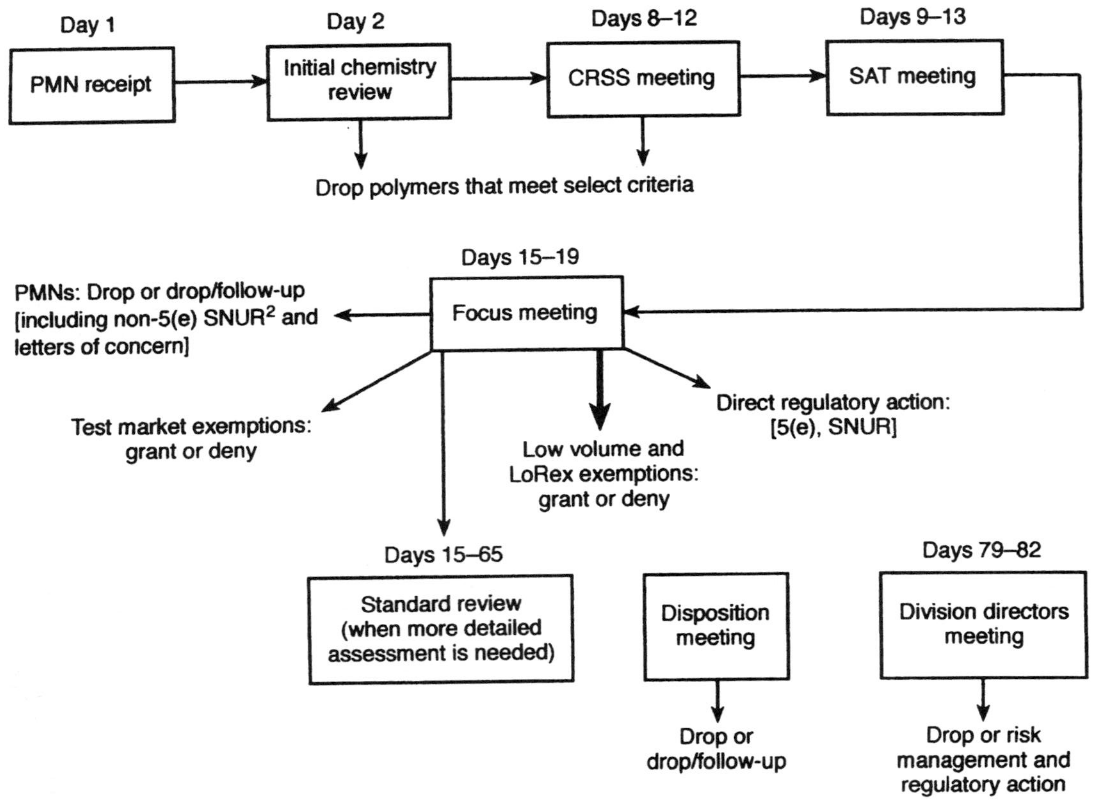


*Figure 10.1* EPA/OPPT review process for PMNs on new chemicals.

*Table 10.3* Models used by scientists at OPPT to review estimate exposure, environmental fate and hazards of chemical substances

| *Model* | *Predicts* |
|---|---|
| *Environmental fate and exposure models* | |
| KOWWIN | Octanol water partition coefficient (atom fragment method) |
| AOPWIN | Atmospheric half life |
| HENRYWIN | Henry's law constant (air/water partition coefficient) |
| MPBPWIN | Melting point, boiling point and vapor pressure |
| BIOWIN | Rate of biodegradation |
| PCKOCWIN | Soil and sediment adsorption |
| WSKOWIN | Octanol water partition coefficient and water solubility |
| HYDROWIN | Rate of hydrolysis |
| BCFWIN | Bioconcentration factor |
| WVOLWIN | Rate of volatilization from surface waters |
| STPWIN | Removal by waste water treatment plant |
| LEV3EPI | Fugacity, partitioning between soil and water |
| *Hazard modeling* | |
| ECOSAR | Toxicity to aquatic species |
| ONCOLOGIC | Potential for carcinogenicity |

the EPA may also issue a Significant New Use Rule (SNUR). In the rare instance that the chemical cannot be manufactured and used safely, the EPA can prohibit manufacture and sale under Section 5(f).

The EPA has formalized many of the procedures for requiring test data or implementing SNURs after reviewing a notice. A "risk based" determination can be based on high probability of a suspected hazard. Notifiers can anticipate these by reviewing the EPA's report on suspected health and environmental concerns of various chemical categories (EPA, 1993b). This report is periodically updated on the EPA's web site.

The EPA introduced an "exposure based" program to prescribe testing when high human or environmental exposure of the notified chemical is coupled with lack of human health or environmental effects data, respectively (EPA, 1991a). The exposure criteria are listed in Table 10.4 and the testing requirements are listed in Table 10.2. The EPA implements the risk and exposure procedures in a manner that is difficult for notifiers to predict. Not every chemical that meets the exposure criteria is required to be tested. The EPA will not require testing if an adequate SAR determination is attainable. The lack of predictability of when the EPA will require testing is due to the complex nature of their SAR methods and the fact that the EPA is privy to proprietary data submitted by other notifiers and reporters. Practitioners can learn to anticipate these requirements for particular chemical categories if their company notifies a series of similar chemicals over time.

The various offices and programs of the EPA are currently focusing more of their efforts to anticipate chemicals that exhibit environmental persistence and bioaccumulative and toxic properties (PBT chemicals). This is being done proactively to help prevent the proliferation of chemicals that may have properties similar to PCBs, poly brominated biphenyls (PBBs) and dioxins. The EPA has implemented a set of criteria to trigger testing for new chemicals that might be suspect to have the properties of a PBT chemical (Federal Register, 1999a). Criteria for testing under this policy are listed in Table 10.5.

*Table 10.4* The EPA exposure based criteria for PMNs. If information in a PMN indicates that the production volume is exceeded and also one of the following criteria are met and there is insufficient data on similar chemicals to make a judgment on safety, the EPA may require the studies indicated in Table 10.2.

| *Exposure parameter* | *TSCA 5(e) exposure-based policy criterion* |
|---|---|
| Production volume | 100,000 kg/year |
| Significant or substantial human exposure: high number of workers exposed | ≥ 1,000 workers |
| Significant or substantial human exposure: acute worker exposure, inhalation | 100 workers exposed to ≥ 10 mg/day |
| Significant or substantial human exposure: chronic worker exposure, inhalation | ≥ 100 workers exposed to 1–10 mg/day for ≥ 100 days/year |
| Significant or substantial human exposure: chronic worker exposure, dermal | ≥ 250 workers exposed by routine dermal contact for ≥ 100 days/year |
| Significant or substantial human exposure: consumer | Presence in consumer product where exposures are likely |
| Significant human exposure: ambient general population | ≥ 70 mg/year exposure via drinking water, air, or groundwater |
| Substantial human exposure: ambient general population | ≥ 10,000 kg/year release to environmental media |
| Substantial environmental release | ≥ 1,000 kg/year total release to surface water calculated after wastewater treatment |

SNUR procedures are listed in 40 CFR Part 721, Subpart A. Subpart B lists scores of various boilerplate descriptions of conditions that the EPA typically uses to define significant new uses. Chemicals with SNURs are listed in Subpart E. These restrictions define the conditions under which the use of the chemical would be considered a significant new use as defined by the statute. The major categories of SNUR restrictions include occupational exposure controls, pollution prevention measures and marketing restrictions. If a company wishes to manufacture or use a chemical as defined in the SNUR, a Significant New Use Notification (SNUN) would be required to be

*Table 10.5* New chemicals program PBT: category criteria and process

| | *TSCA Section 5(e) action* | |
|---|---|---|
| | *5(e) Order Pending Testing Significant New Use Rule (SNUR)*[a] | *5(e) Ban Pending Testing*[b] |
| Persistence (transformation half-life) | > 2 months | > 6 months |
| Bioaccumulation (fish BCF or BAF)[c] | > 1,000 | > 5,000 |
| Toxicity | Develop toxicity data where necessary[d] | Develop toxicity data where necessary[d] |

[a] Exposure/release controls included in order; testing required.

[b] Deny commercialization; testing results may justify removing chemical from “high risk concern”.

[c] Chemicals must also meet criteria for MW ( < 1,000) and cross-sectional diameter ( < 20A, or $< 20 \times 10^{-8}$ cm).

[d] Based upon various factors, including concerns for persistence, bioaccumulation, other physical/chemical factors, and toxicity based on existing data.

submitted. This is done using the same form for PMNs. Practitioners should be aware that a significant amount of new data is normally required by the EPA to convince them that the new use is safe.

Once a consent order is signed, or the 90-day review period has expired without the EPA contacting the notifier about extending the review period, the chemical can be imported or manufactured for commercial use. Within 30 days of its first manufacture or import, notifiers must submit a Notice of Commencement to the EPA.

Polymers can be notified on the same PMN form. Manufacturers of polymers that meet the structural and compositional exemption criteria in 40 CFR 723.250 can elect to submit annual reports on exempt polymers rather than notifying them individually.

Due to the flexible nature of TSCA as a "gap filling" statute, the EPA has developed detailed procedures for the regulation of new products of biotechnology for industrial use (Federal Register, 1997).

## *European Union*

The DSD, Directive 67/548/EC (European Commission, 1967) provides the basis for the harmonized classification of packaging and labeling of chemicals in the EU. In 1979, the Council of Ministers of the European Community adopted the 6th Amendment to the DSD, or Directive 79/831/EC (European Commission, 1979), which introduced a notification system and mandatory testing requirements for new chemicals in addition to requirements for classification and labeling of dangerous chemicals. All member states had 2 years to adopt these harmonized procedures. This 6th amendment was superseded by the 7th amendment, Directive 92/32/EC (European Commission, 1992), which had to be implemented by October 31, 1993. The 7th amendment defines the current requirements for introduction of new chemicals to markets in the EU.

A new chemical is defined as one that is not on the European Inventory of Existing Chemical Substances (EINECS). EINECS was compiled from industry nomination of chemicals that were placed on the EU market in the 10-year period between January 1, 1971 and September 18, 1981. It is a closed list to which no additions are permitted. Polymers are "considered as notified" and are exempt from notification provided they do not contain more than 2 per cent of a new monomer that is not on EINECS.

Once notified, new chemicals are listed on the European List of Notified Chemical Substances (ELINCS). ELINCS chemicals are identified by trade name until such time as the substance is added to Annex 1 of the DSD. This lists the chemical substances that have mandatory hazard classifications under the DSD. Listing in ELINCS does not mean that the substance no longer has notification requirements. A notification is required from each new manufacturer of the chemical although data sharing with a previous notifier is encouraged.

The notification requirements for a new chemical substance are laid down in several annexes to the Directive. Annex VIIA is a full notification for chemicals placed on the market in quantities greater than 1 metric ton/year. If less than 1 metric ton/year, consult Annexes VIIB or C and, if greater than 10 metric tons/year, consult Annex VIII, Levels 1 and 2. If a polymer needs to be notified then consult Annex VIID. In any case, there is an annex that will describe the appropriate situation and the corresponding notification requirements. A typical chemical notification dossier, Annex VIIA, (for

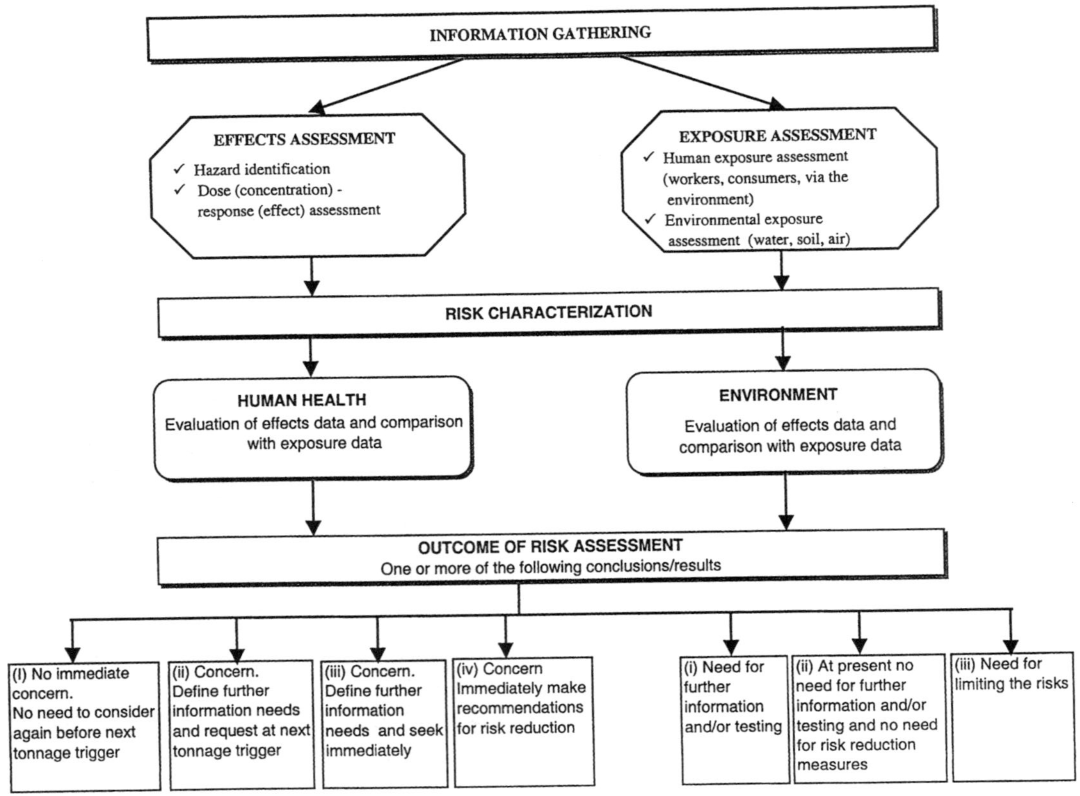


*Figure 10.2* A schematic outline of the major steps and possible outcomes of the EU risk assessments for new and exiting chemicals under adaptation EC/793/93 of the Dangerous Substances Directive.

a chemical placed on the market at 1–10 metric tons/year), contains spectral, physiochemical, toxicological and ecotoxicological data, together with a summary of information which must be submitted as an electronic file in Standard Notification Information Format (SNIF). Table 10.2 indicates the tests required for this type of a notification.

The EU was the first to adopt the base set of tests recommended by the OECD for screening for the hazards of new chemicals (OECD, 1981). All tests are to be performed by current OECD test protocols and the laboratory performing the tests must abide by GLP. These details are all spelled out in Annex V. Again, since national agencies in the various member states are charged with reviewing the notifications, subtle differences in requirements, preferences and review predisposition become evident with repeated notification experience. Data from laboratories around the world that meet GLP requirements are generally accepted although subtle nuances between national agencies are observed. For instance in many member states, dermal sensitization by the maximization protocol is preferred over the topical assay even though both are acceptable alternatives in an OECD 406 study (OECD, 1993).

In addition, the notifier supplies a classification and labeling proposal for the substance and a Material Safety Data Sheet (MSDS). Council Regulation No. 93/67 (European Commission, 1993a) mandates a risk assessment approach in the review of new chemicals. The notifier may provide a preliminary risk assessment for the substance, although the ultimate responsibility of the risk assessment as required by the 7th Amendment rests with the competent authority. Possible outcomes of the risk assessment process are given in Figure 10.2. More details on the risk assessment approach can be found in the extensive technical guidance document (European Commission, 1996).

Notifiers of new chemicals must report annual production or import volume each year. Once an annual volume or cumulative volume exceeds a trigger volume, additional testing must be negotiated with the authorities. The tests that must be considered are listed in Schedule 3 of Annex VIII.

## *Japan*

The Chemical Substances Control Law was established in 1973 to prevent environmental pollution and hazards to human health by chemical substances. The impetus for its enactment was the problem of environmental pollution caused by PCBs in the late 1960s. It provides for a classification of new chemical substances that have similar properties to PCB (low biodegradability, high bioaccumulation and chronic toxicity) as Class I Specified Chemical Substances and, in fact, virtually prohibits the manufacture and import of such substances.

The law was amended in 1986, which introduced the system for assigning Designated Chemical Substances and Class II Specified Chemical Substances. This originated out of the necessity to regulate the substances having the properties of low bioaccumulation, but low biodegradability and chronic toxicity, depending on the degree of persistence in the environment.

The main objective of the law is to protect humans from exposure to dangerous substances in the environment and especially from dangerous substances that could enter the food chain. Their approach to new chemical control is somewhat different from the rest of the world. They place great emphasis on biodegradation and bioaccumulation. The biodegradation test is required to identify any unique metabolites resulting from biological action. Depending on the results of the biodegradation study, ecotoxicity and toxicity studies may be required on the environmental degradants and not necessarily on the parent compound. For this reason, as well as other possible nuances in interpretation of the results, practitioners usually perform the required tests in a stepwise fashion and discuss the results with scientists from the Japanese agencies before going on to the next test. The basic testing requirements are outlined in Table 10.2, but the reader should take notice of the many footnotes in the table.

Although the tests for MITI/MHW/MOL notifications are based on OECD test methods, one should consult the Japanese test guidelines because often the tests are more stringent and may require, for example, unique Japanese test media or species. In practice, most practitioners have found that a greater probability of a successful notification is obtained when the environmental tests are done by Japanese laboratories. Data from laboratories outside of Japan are usually accepted for health effects studies as long as the laboratory has successfully undergone the Japanese certification process for GLP.

The test data is reviewed by MITI and MHW. Separate notification must be made to the MOL. Their review focuses on the protection of workers from new chemicals and they are particularly concerned about the introduction of new carcinogens. The only data required to be submitted to the MOL is a bacterial mutagenicity test.

If a notified chemical is biodegradable, it is classified as a "safe" chemical and no further testing is necessary. However, very few synthetic chemicals meet the biodegradation criteria. Next the substance or its nonbiodegradable metabolite is assessed for bioaccumulation. If the material is not bioaccumulative, it undergoes a set of toxicity studies: a 28-day subacute oral toxicity in rat, Ames mutagenicity and *in vitro* chromosome aberration tests. If the material does bioaccumulate, then a more detailed set of toxicity testing is required. This entails a great many discussions with MITI and could become an expensive and long notification process. In the end, these substances may be deemed "safe" or may become Class I or II Specified Chemical Substances or Designated Chemical Substances. In the Japanese notification process, consultations with MITI are expected and advisable. Rarely can a notifier be successful by simply submitting test results and a form to the Japanese authorities without prior consultation. The major decision points in the review of submitted data are listed in Figure 10.3.

All notified chemicals are eventually added to the list of new substances. Initially, only the notifier can manufacture or import this substance until the substance is published in the Official Gazette; this usually occurs 1–3 years after the new chemical dossier is examined and approved. The lists of existing and new chemical substances do not use CAS nomenclature but rather consists of some 20,000 entries which are often more generically described rather than listing specific substances. The generic listings make it possible that an entirely new chemical may be adequately described by a preexisting listing so that no notification would be required. The lists are difficult to use and practitioners should consult the Explanation and Examples of Classification Based on Chemicals Structure (The Chemical Daily Co, Ltd., 1999).

There is also a polymer notification scheme. The polymer is evaluated for photo, thermal and hydrolytic stability, solubility in water under acidic and alkaline condi-

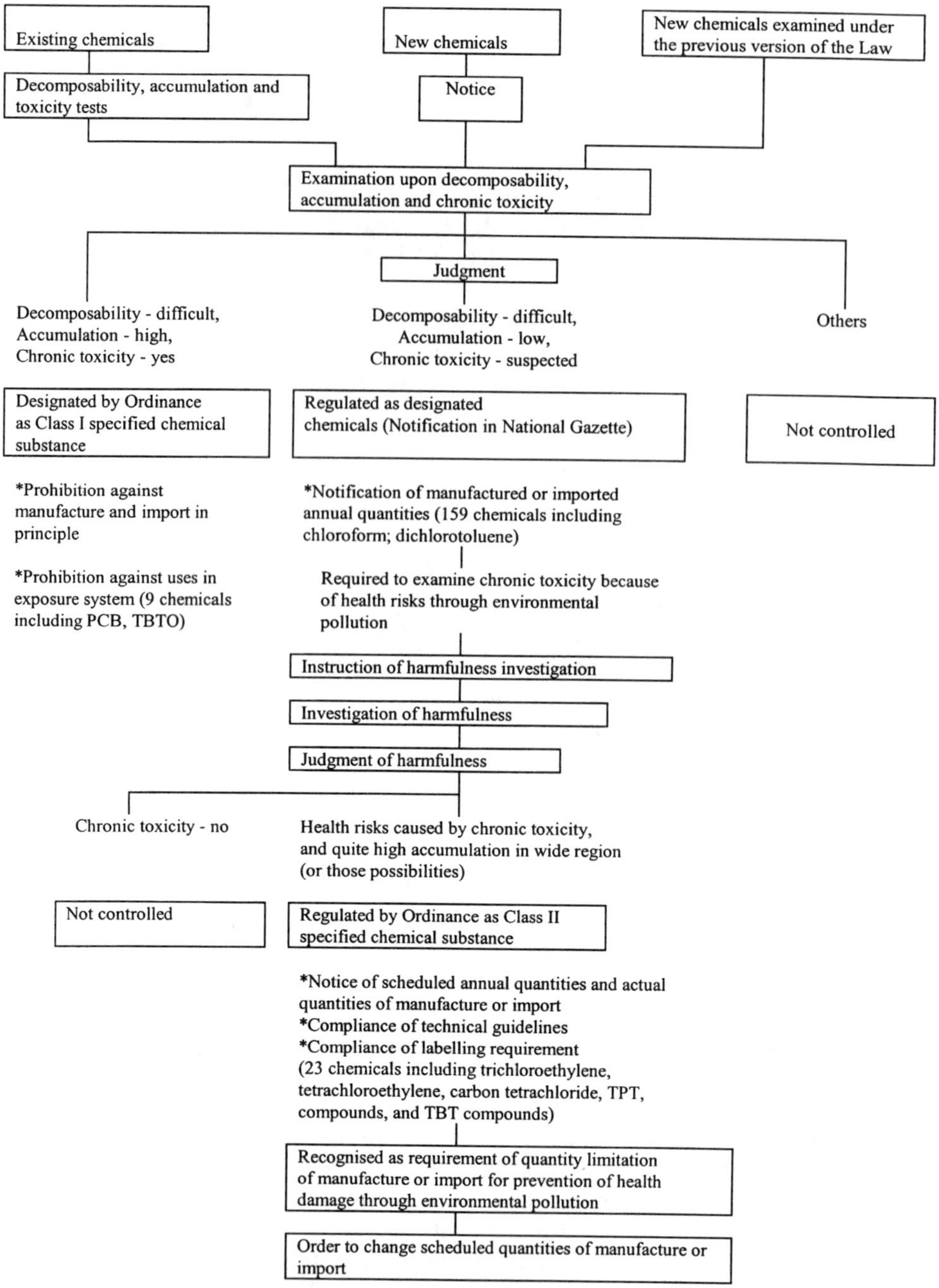


*Figure 10.3* A flow chart indicating data needs and logic of assessments and possible outcomes under the Japanese system administered by MITI and MHW (OECD, 1997). Systematic chart of the law concerning examination and regulation of manufacture, etc. of chemical substances (those in parentheses designated as of November 1995)

tions, solubility in various solvents, structural characteristics; molecular weight distribution and proportion of oligomers. If it passes this evaluation it is deemed "safe", but if not, it is considered nonpolymeric and must be fully tested.

## *Canada*

CEPA is the primary legislative instrument in Canada for environmental protection. Part II of the Act concerns the introduction, by import or manufacture, of new substances into Canada through a requirement for a pre-import or pre-manufacture notification and assessment. The legislation came into force in July, 1994. A number of procedural inefficiencies were addressed by the CEPA amendment of 1999.

The Canadian DSL is their inventory of chemicals that were in commercial use in Canada between January 1, 1984 and December 31, 1986. As new chemicals are notified they are placed on the DSL. Canada also created an NonDomestic Substances List (NDSL). The original NDSL consisted of chemicals on the 1985 TSCA inventory that were not on the DSL. Chemicals listed on the NDSL could be placed on the DSL with reduced notification requirements.

The Department of the Environment (Environment Canada) and the Department of National Health and Welfare (Health Canada) assess new chemical notifications. The assessment will result in:

1 A determination that the substance is not suspected of being toxic; or
2 A suspicion that the substance is toxic, which may require: (1) controls on, or prohibition of, import and manufacture or (2) prohibition pending submission and assessment of additional information; or
3 Limiting the purpose for which a substance may be used to permit the waiver of information requirements.

The notification requirements are tiered in a unique fashion under the Canadian notification system. The information requirements depend on the chemical class (e.g., polymer, chemical, biotechnology product), volume of import or manufacture and proposed use (e.g., research and development, etc.). The Guidelines for the Notification and Testing of New Substances: Chemical and Polymers (Environment Canada, 1993) is a critical reference tool for Canadian notifications. The appropriate flow charts and the appropriate notification schedule in the guidelines must be ascertained before filing a notification. The full data package for a nonpolymeric chemical (Schedule III), when required, is very similar to a EU test package based on OECD protocols done under GLP procedures (Table 10.2). However, data can be supplied in three forms: actual test data, surrogate data (data either calculated or based on structural analogs), or requests for waivers of information requirements if testing is not possible or relevant due to the properties of the chemical. Once a schedule is filed and reviewed, the quantity of chemical manufactured or imported into Canada is tracked until the next schedule is filed. Eventually an ultimate schedule is filed and once the trigger volume is exceeded, the material is listed on the DSL.

## *Australia*

In Australia, an industrial chemical is defined as one that is not an agricultural or veterinary chemical, a therapeutic good, or a food or food additive. It is interesting to note, that unlike the status in the US, chemicals used in cosmetics are considered industrial chemicals.

The chemical inventory was created from the chemicals that were commercially in use in Australia from December 1, 1977 to July 16, 1990. New chemicals are added to the inventory 5 years after a notification is approved. During the 5-year interim period, the notifier alone has the right to manufacture or import the new chemical.

There are notification requirements for chemicals and polymers. Assessments should be completed in 90 days. Manufacture cannot begin until an assessment certification is given. It seems to take several weeks to months to obtain the assessment certificate before manufacture or import is allowed. A Notice of Commencement is required upon first manufacture or import.

The notification for a chemical requires an information set very similar to that of the EU. All the tests are required to be performed according to OECD guidelines under GLP procedures. See Table 10.2 for details. The notification scheme also provides for a certain amount of flexibility in the data requirements. A waiver may be requested (for a fee) if the test required can be shown to be irrelevant, unnecessary or economically prohibitive. As with most notification schemes in other countries, there are reduced notifications or exemptions for small volumes of chemicals, site-limited chemicals and substances used in various quantities for research and development. A Handbook for Notifiers can be ordered through the NOHSC web site (NICNAS, 1995).

Polymers are notified as new synthetic polymers with number-average molecular weights of less than 1,000, as new synthetic polymers with number-average molecular weights of more than 1,000 or as polymers of low concern. If the polymer molecular weight is under 1,000, then the notification resembles a chemical notification. The other polymer notifications do not require toxicity tests but do require characterization for molecular weight distribution, residual monomers, impurities and stability.

## Korea

The Korean TCCA became effective on February 8, 1991. It has been modified several times since and this has significantly simplified and streamlined the import and notification procedures. The Korean MOE administers the law.

The original Korean Existing Chemical Inventory included chemical substances manufactured or imported into Korea prior to February 8, 1991. All new chemical substances must be notified to the MOE at least 90 days before the first manufacture or import. The notification information includes technical and commercial information and some details on use and disposal. There are required studies on acute toxicity in rats or mice, mutagenicity studies (Ames and chromosomal aberration), and an *in vivo* mouse micronucleus test to confirm mutagenic potential if either of the mutagenic studies are positive. A biodegradation study, review of hydrolysis, photolysis and physical and chemical properties indicating persistence, and a review of bioaccumulation potential are required. These studies may be from the published literature or unpublished studies according to OECD or other acceptable protocols on the exact chemical substance or on an acceptable surrogate substance. An abstract in Korean is required for all foreign language test reports. The notification requirements are reduced for substances that are reported on two foreign inventories before 1991. The usual technical, commercial, use and disposal information is required; but only an acute toxicity and Ames test are required.

The notification requirements for polymers (as defined by OECD) are simplified. A determination of number average molecular weight, weight per cent of residual monomers and oligomers with molecular weights below 1,000, certain physical properties (i.e., melting point, solubility in common solvents), as well as information on intended use and some manufacturing details are required.

### *Philippines*

Title II of the Toxic Substances, Hazardous Waste and Nuclear Waste Control Act, deals with toxic substances. DENR is charged with protecting the public health and the environment from unreasonable risks posed by these substances.

DENR compiles, maintains and updates an inventory of chemical substances known as the Philippines Inventory of Chemicals and Chemical Substances (PICCS). The PICCS is composed of those chemicals manufactured, used or imported in the Philippines prior to December 31, 1993. As new chemicals are notified and reviewed, they are eventually added to the PICCS inventory. The PICCS is updated once every 5 years. During a 5-year interim period, only the notifier of the chemical may sell commercially.

Manufacturers and importers of new chemicals are required to notify the DENR of their intent to manufacture or import the new chemical. The DENR is responsible for assessing the potential risk posed to the public and health and the environment by the new chemical substance. This notification requires the submission of a Pre-Manufacturing and Pre-Importation Notification (PMPIN) form.

There are two kinds of PMPIN forms. The abbreviated form is used when a new chemical is in commerce with no controls in a country with a similar review process as the Philippines, and when the notifier believes there is sufficient information that clearly exhibits that the chemical will not pose an unreasonable risk. The instructions for the form call for a short description of potential effects of the chemical. All available toxicological and environmental information should be addressed. Depending on the nature of the chemical, a good quality material safety data sheet may be an acceptable summary. If DENR is not satisfied with the first submission, they may require more comprehensive information on a more detailed PMPIN form including additional testing if they determine insufficient information has been submitted to assess the safety of the chemical.

Once DENR reviews the information and determines that the chemical will not pose an unreasonable risk, it will issue a clearance to import or manufacture the new chemical. A Notice of Commencing Import and Manufacture is required upon first import or manufacture.

Experience with the Philippine notification system is limited. PMPINS have been submitted to the Philippines since 1994 but DENR has only recently (mid-1999) appointed staff to review notifications. Little experience has been reported in the industry about the adequacy of the data submitted for these new chemicals and how they are being reviewed by the agency. There is little additional guidance available on the notification of chemicals and polymers in the Philippines.

### *International harmonization*

The OECD has been active in developing consensus on a number of processes associated with chemical regulation. Their recommendations for the mutual acceptance of data (OECD, 1981) for a minimum data set for the assessment of a new chemical (OECD, 1982) were published almost two decades ago. They have also published internationally recognized standards for GLPs (OECD, 1997b) and guidelines for health and environmental testing (OECD, 1993).

Their more recent activities have arisen from the United Nations Conference on Environment and Development held in Rio de Janeiro in June 1992. Agenda 21 from this conference describes a comprehensive program in the areas of environmental protection, climate change and sustainable development. Chapter 19 of Agenda 21 addresses the management of toxic chemicals. OECD activities have been directed to the development of recommendations for chemical management systems that are sound and globally harmonized.

The OECD held a conference in 1996 to explore means to increase international cooperation among the existing authorities that review new chemical notifications by these various laws. As a first step, this conference discussed means of sharing data among government agencies obtained from notifications for new chemicals. The background information in the report of the conference is a good source of information for the practitioner seeking more details of the various review and administrative processes under each law (OECD, 1997a). The conference identified potential common ground and goals of the regulatory systems of the various countries and possible means to share data. The US and Canada are sharing assessments under a pilot program called the Four Corners Agreement.

Harmonization of the notifications in these countries would take statutory changes by the legislative bodies of these countries. This would certainly take a long time to accomplish. New laws from other countries add to an already complex situation. Hopefully, any country contemplating instituting a new chemical control law will consider the guidance from OECD to avoid adding any more confusion.

## Requirements for existing chemicals

### *United States*

#### *Chemical testing requirements*

The EPA has many options under TSCA to assess and control the risks of existing chemicals. Under Section 4 of TSCA, the EPA can require manufacturers and importers of chemical substances to conduct health and environmental testing under certain criteria listed in Section 4(a) of TSCA. The exact legal interpretation of these criteria has been somewhat controversial, but basically testing can be required when the EPA determines that an unreasonable risk may exist from current uses (hazard finding) or if there is substantial human or environmental exposure (exposure finding). In essence, although Congress realized that new information could become known which would make further testing desirable and prudent, it did not give the EPA the authority to require testing without good cause.

The procedure for developing test rules is shown in Figure 10.4. Section 4 established the Interagency Testing Committee (ITC), an independent expert panel made up of scientists from various regulatory and research agencies. The law specifies that the ITC should issue a report annually to designate chemicals that would appear to be candidates for test rules, based on statutory criteria. New toxicology data and exposure patterns are taken into account. The EPA studies the recommendation and obtains additional information from industry through Section 8: reporting rules (Walker, 1993).

The EPA further investigates the production, use, and existing data on the chemicals listed by the ITC. If sufficient reason appears to exist to support a test rule, the EPA publishes a notice of proposed rulemaking that discusses the need for the test rule and a list of tests that are thought to be needed to more fully evaluate the risks of the chemical. The manufacturers and importers can challenge the basis for the rule and also dispute the lists of tests by submitting comments on the notice. Under the rule-making procedures of the Administrative Procedures Act, the EPA is required to take this information into account before publishing a final rule. The final rule will indicate the deadline for the submission of test reports. If the industry is still unsatisfied with the test rule, suit can be brought against the EPA. Many lawsuits have been brought against the EPA on Section 4 matters.

The tests must be done according to guidelines published under TSCA. The EPA has established dozens of test guidelines for health and environmental effects and environmental fate. As part of a deregulation effort, the EPA has removed these guidelines from 40 CFR Part 798 and now makes these available on their web site with cross reference to OECD methods. The tests need to be conducted according to GLP regulations (40 CFR Part 799).

The EPA has noted the need for testing besides the recommendations of the ITC. Some US laws lack authority to require regulated industries to conduct testing that various regulatory agencies may feel are necessary to determine public health risks of various industrial activity. OPPT has proposed testing initiatives under Section 4 authority for data needed by OSHA for determining the potential for skin absorption of certain chemicals (Federal Register, 1999b) and for the EPA's Office of Clean Air for further information on certain hazardous air pollutants (Federal Register, 1996). Test rules are expected to be developed for several chemicals for needed data identified by the Agency for Toxic Substances and Disease Registry (ATSDR), the National Toxicology Program (NTP), and under the Comprehensive Environmental Response, Compensation, and Liability Act (CERCLA) (Federal Register, 1999c). Usually these initiatives encounter delay in the regulatory process. The EPA has more recently encouraged voluntary test programs. The EPA tracks all of these testing programs on various chemicals through a Master Testing List. More recent additions of chemical categories added to the Master Testing List illustrate the breadth of programs that is being considered. Categories added since 1996 are listed in Table 10.6.

At present, the EPA is expending much of its resources on two major testing initiatives. One relates to recent concern about toxicity of various chemicals to humans and wildlife through mechanisms of action that may affect endocrine systems. This concern has lead to the passage of the Food Quality Protection Act and amendments to the Safe Drinking Water Act that require the EPA to develop screening tests and implement a program to screen various chemicals for biological activity which target endocrine

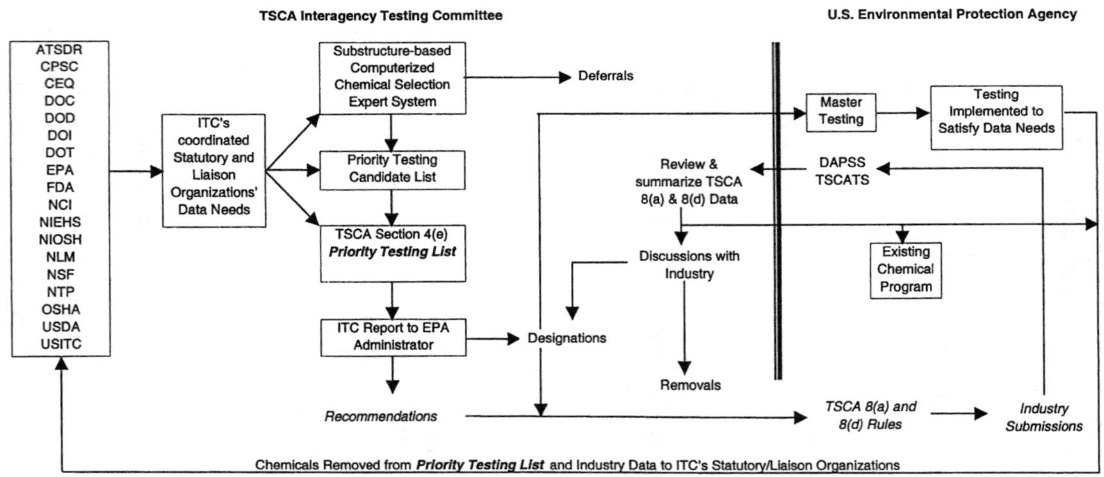


*Figure 10.4* Schematic of process to develop Section 4 test rules.

*Table 10.6* Categories of chemicals and exposure situations added to the EPA's master test list in 1996

| |
|---|
| Persistent bioaccumulators |
| New chemicals program "Chemical Categories" |
| EPCRA Section 313 ("Tri Screening") |
| Clean Air Act Section 112 "Air Toxics" (hazardous air pollutants) |
| SARA Section 104 "Priority Data Needs" |
| Respirable fibers |
| Indoor air source characterization – carpet/carpet-related products |
| Indoor air source characterization – interior architectural coatings |
| Polychlorinated dioxins/furies in wood pulp/paper mill sludge |
| Endocrine disrupters (new category) |
| Machining fluid products/chemicals (new category) |
| Paint stripping products use cluster (new category) |
| Oxygenated fuel additives (new category) |

systems. Although the main focus of this program will be on pesticides, these new laws give the EPA the legal authority to address chemicals that have been found to contaminate drinking water sources. As of this writing, the screening tests are still under development. It is unclear to what extent the EPA will require screening of industrial chemicals (Federal Register, 1998a,b).

The second new program is much further along. In 1998, the EPA requested voluntary testing of High Production Volume (HPV) chemicals. The EPA defines HPV chemicals as those that are manufactured at an annual volume in excess of one million pounds. The EPA is focusing on a list of approximately 2,800 nonpolymeric organic chemicals that exceed this production limit. The EPA requested that companies volunteer for this program by December 1, 1999. In addition, the EPA has worked with various industry and trade associations and foreign regulatory authorities to make this testing effort more of an international effort. This effort was quite successful since EPA only needed to address approximately 40 chemicals in its proposed test rule that lacked volunteers (Federal Register, 2000).

The HPV testing program is modeled after the voluntary OECD Screening Information Data Set (SIDS) Program. The SIDS battery was developed by regulatory toxicologists as the minimum data needed to screen HPV chemicals for health and environmental concerns. Approximately 400 chemicals have been addressed in the SIDS program, however, only about 100 have completed the entire process culminating in peer review (OECD, 1981).

Whether every test in the SIDS battery needs to be conducted on each chemical on the list is a controversial subject. The SIDS program and the EPA have published guidance for using category approaches for structurally related chemicals (EPA, 1999a,b). This guidance attempts to define how scientific judgment can be used to determine whether data on similar chemicals is sufficient to estimate the hazard potency of structurally similar chemicals. Most consortiums were formed to address various classes of chemicals. Several test programs were submitted in the latter half of the year 2000. It will be interesting to see how aggressive the test proposals will be in the use of the category approach to minimize test costs. But it is clear that the HPV testing program will be the most expensive test program ever through the use of TSCA Section 4 authority.

The EPA and industry have allied to make all the information generated in the HPV program available through the Internet. Industry participants will make robust summaries of preexisting and newly generated data in a publicly available database. Testing proposals will also be made public. It is expected that these proposals will be scrutinized not only by the EPA, but also by public interests groups especially the Environmental Defense Fund (EDF) and People for Ethical Treatment of Animals (PETA). The point of view of these public interest groups is diametrically opposed. EDF called for an increased level of testing by industry to support the safety of their chemicals in a report entitled *Toxic Ignorance* (Environmental Defense Fund, 1997). PETA on the other hand, believes that the testing required is unnecessary and will cause unnecessary pain and suffering to laboratory animals. The proper level of testing promises to become an interesting public debate (EPA, 1999b). It is also expected that, as more data emerges on the HPV chemicals, increased awareness about the hazards of chemicals will result in further discussion about the safety of these chemicals to health and the environment.

### *Sections 6 and 7: existing chemicals control*

If new test data or information indicate that control of a chemical is needed to protect health or the environment, TSCA gives the EPA the authority to take a wide variety of actions under Sections 6 and 7.

This authority includes limiting or banning the manufacture and use of chemicals for just cause depending on the seriousness of the effect and the appropriateness of the action. When TSCA went into effect, it was originally thought that the EPA would often take action to limit the use of chemicals as a result of data submitted under Section 4 and other reporting rules. However, little action has been taken under Sections 6 and 7. The EPA has taken final action only on PCBs, fully halogenated chlorofluorocarbons and asbestos.

The lack of EPA action is due to a variety of causes. Often, industry takes voluntary action to move onto substitute chemicals when a serious problem arises. It is also difficult to determine if the actual exposure to certain chemicals warrants regulatory action. The EPA also has found that it is difficult to take action under Sections 6 and 7 due to the formal rulemaking procedures in TSCA that require the EPA to identify the most cost-effective regulatory approach, consider the benefit of the chemical, and assess the availability of substitutes for the chemical.

In 1991, parts of the EPA ban and limits on asbestos were overturned due to a court ruling in which the EPA failed to properly assess the economic burden of the rule as well as assess alternate solutions (Corrosion Fittings vs. The EPA, 1991). Critics of industry and the EPA say that the success of this lawsuit indicates that TSCA is too weak to effectively limit chemicals. However, since the asbestos regulations were deliberated internally at the EPA for 10 years before they were published, a little criticism of the agency on its inability to establish a successful review framework to use Section 6 authority would seem to be justified to a certain extent.

Recently, the EPA has tried to develop criteria and procedures to manage and prioritize the review of existing chemicals. A phased risk management has been developed (EPA, 1992). Chemicals are first assessed at a cursory level called RM1. The EPA asks industry to voluntarily supply exposure information to help assess the risks. After RM1 is complete, later phases pursue the investigation in greater detail (RM2 and post

RM2 stages). After RM1 review, the EPA sometimes asks for voluntary controls or for exposure studies for data that will be used in the further review. The EPA is also concentrating on more limited action on the highest risk activities for a chemical; such as the use of neurotoxic and carcinogenic acrylamide monomer in sewer-grouting operations (Federal Register, 1991) and the use of nitrosamine-forming nitrites in metalworking fluids (Federal Register, 1993). The EPA has also started a number of cooperative, voluntary programs to reduce the risk of chemicals in certain industries.

### *Section 8: reporting rules*

There are a wide variety of reporting rules in Section 8 of TSCA. The purpose of these rules is for the EPA to gather information from industry on the uses and hazards of chemicals. Section 8(a) rules gather information on the use and manufacture of chemicals. Section 8(b) is the authority that the EPA used to require reporting to compose the original TSCA Inventory of Chemical Substances. Section 8(c), 8(d) and 8(e) deal with the reporting on the health and environmental effects of chemicals. Because the latter three sections are of more interest to toxicologists, they will be covered in more detail here.

### *Section 8(c): adverse reaction reporting*

Various US laws require manufacturers to keep records on and report adverse reactions to products. Probably the most elaborate example is the requirement of the Food, Drug and Cosmetic Act on drug manufacturers (see Chapter 2).

Similar requirements are delineated in Section 8(c) of TSCA. Manufacturers, importers and processors of chemicals are required to keep records of allegations of health and environmental effects. These records must be kept for at least 5 years, and in the case of employee health effects, for at least 30 years. The regulations governing this section require recording the allegation of an effect on health or the environment regardless of proof.

Industrial toxicologists responsible for Section 8(c) compliance need to make sure that internal company reporting mechanisms for incidents involving employees, plant neighbors, and customer complaints keep them informed of all allegations. These records are required to be kept in a central location at the company. The EPA, when investigating a particular chemical can require companies to submit copies of the allegations on the chemical to the agency. The EPA has rarely invoked this reporting requirement.

### *Section 8(d): health and safety data reporting*

The legislative history of TSCA acknowledges that industry conducts many voluntary health and safety studies, but these are rarely published in the scientific literature. It is important that the EPA have access to all health and safety data when the investigation of a chemical reaches a certain point. The EPA routinely adds to the list of reportable substances in 40 CFR 716.120 after receiving an ITC report.

When a new Section 8(d) rule becomes effective, manufacturers, importers and processors of any newly listed chemical have 60 days to report all health and safety studies in their files. In some cases, only a list of studies is required, but if the report is readily available to the company, a copy of the full report is usually required. The EPA recently made a number of improvements to the procedural rule for reporting to make the reporting more efficient, including shortening the period for continued reporting of new studies from 10 years to 1 year (Federal Register, 1998c). Exceptions to this are possible for up to 2 years on a chemical-by-chemical basis. Sunset dates are included for every chemical listed in 40 CFR 716.120. Federal Register notices on ITC reports now routinely request that industry submit available unpublished data before the Section 8(d) rule is published to help expedite the review of newly listed chemicals.

The regulatory definition of a health and safety study is complex and occupies approximately a half page in the Code of Federal regulations (40 CFR 716.3). It includes not only laboratory toxicology and environmental effect and fate studies, but also certain industrial hygiene studies, environmental monitoring studies and computer modeling studies. As with all TSCA regulations, reporting rules and procedures are complex and 40 CFR Part 716 should be consulted to assure proper and complete reporting.

### *Section 8(e): substantial risk reporting*

Section 8(e) of TSCA requires manufacturers, importers and processors of chemicals to immediately submit to the EPA any information that may indicate a substantial risk to health and the environment. These reporting requirements assure that the EPA is immediately notified of studies or of incidents that may need quick remedial regulatory action. The EPA first published guidelines on the reporting requirements in 1978 (Federal Register, 1978). A summary of the guidelines is given in Table 10.7.

During the first decade of the existence of TSCA, the EPA took surprisingly little action based on substantial risk reports that were submitted. Industry submitted hundreds of reports based on the vague 1978 guidelines. Although the EPA action was largely absent, there was much interchange of information among companies that manufactured, purchased and used a particular chemical after submittal of a substantial risk report. The submitting company found that it is necessary due to the public nature of the reports to reassure customers about the safety of the chemical or updated safe handling recommendations.

After a number of inspections of corporate records, the EPA felt that not all substantial risk information had been reported. The EPA issued new, more specific guidance in 1991 (EPA, 1991b). These guidelines were issued in conjunction with a voluntary Compliance Audit Program (CAP) in which much of the regulated industry participated.

Highlights of the 1991 guidance are outlined in Tables 10.8–10.10. Unfortunately, the new guidelines encourage the reporting of information that many toxicologists consider trivial. Some examples of overly conservative guidance include evidence of neurotoxicity at virtually any dose or exposure route and organ toxicity observed at high doses in repeated dose studies.

The volume of reporting has increased due to the new guidelines. Over 7,000 reports were submitted between 1991 and 1993 under CAP. Although this increases the amount of information available to EPA scientists and the public, the net effect of the new guidelines as judged by this author is counterproductive, since substantial risk reports no longer have the same significance. Due to the trivial nature of most new

*Table 10.7* 1978 Substantial risk reporting guidance: March 16, 1978 Federal Register

V. What constitutes substantial risk
The agency considers effects for which substantial risk information must be reported to include the following:

*a. Human health effects*
1. Any instance of cancer, birth defects, mutagenicity, death, or serious or prolonged incapacitation, including the loss of or inability to use a normal bodily function with a consequent relatively serious impairment of normal activities, if one (or a few) chemical(s) is strongly implicated
2. Any pattern of effects or evidence which reasonably supports the conclusion that the chemical substance or mixture can produce cancer, mutation, birth defects or toxic effects resulting in death, or serious prolonged incapacitation

*b. Environmental effects*
1. Widespread and previously unsuspected distribution in environmental media, as indicated in studies (excluding materials contained within appropriate disposal facilities)
2. Pronounced bioaccumulation. Measurements and indicators of pronounced heretofore unknown to the Administrator (including bioaccumulation in fish beyond 5,000 times water concentration in a 30-day exposure or having an *n*-octanol/water partition coefficient greater than 25,000) should be reported when coupled with potential for widespread exposure and any nontrivial adverse effect
3. Any nontrivial adverse effect heretofore unknown to the Administrator, associated with a chemical known to have bioaccumulated to a prolonged degree or to be widespread in environmental media
4. Ecologically significant changes in species' interrelationships; that is, changes in population behavior, growth, survival, etc., that in turn affect other species' behavior, growth, or survival. Examples include
i. Excessive stimulation of primary producers (algae, macrophytes) in aquatic ecosystems, e.g., resulting in nutrient enrichment, or eutrophication, of aquatic ecosystems
ii. Interference with critical biogeochemical cycles, such as the nitrogen cycle
5. Facile transformation or degradation to a chemical having an unacceptable risk as defined above

*c. Emergency incidents of environmental contamination* – any environmental contamination by a chemical substance or mixture to which any of the above adverse effects has been described and which because of the pattern, extent, and amount of contamination:
1. Seriously threatens humans with cancer, birth defects, mutation, death or serious or prolonged incapacitation, or
2. Seriously threatens nonhuman organisms with large-scale or ecologically significant population destruction

VI. Nature and sources of information which "reasonably supports the conclusion" of substantial risk
1. Designed, controlled studies
i. In vivo experiments and tests
ii. In vitro experiments and tests
iii. Epidemiological studies
iv. Environmental monitoring studies
2. Reports concerning and studies of undersigned, uncontrolled circumstances
i. Medical and health surveys
ii. Clinical studies
iii. Reports concerning and evidence of effects in consumers, workers, or the environment

reports, companies in industries that use chemicals have become increasingly more complacent about the substantial risk reports of their suppliers.

As of this writing, the EPA is still working on revising the guidance for companies to

*Table 10.8* 1991 substantial risk reporting guidance: neurotoxicity observations in general toxicology studies

I. Not serious
 1. Effects only seen in moribund animals or in only one or a few isolated cases in non-moribund animals.
 2. Effects which are transient in nature, rather than intermittent or continuous

II. Probably not serious, but may be supportive evidence of neurotoxicity if observed with more serious effects
 1. Lethargy
 2. Salivation

III. Serious effects, if not judged nonserious by I
 1. Paralysis
 2. Convulsions
 3. Ataxia

report environmental accidents (Federal Register, 1999d). The only recent action under Section 8(e) is defining the nature of endocrine disruption effects that require reporting (Federal Register, 1998a–c). It is uncertain whether the EPA will conduct another comprehensive review of the reporting criteria in the future.

### *Public databases maintained by the EPA for TSCA data*

The EPA compiles data submitted under Sections 4 and 8 reporting rules in a database called Toxic Substances Control Act Test Submissions (TSCATS). This database is extremely useful to toxicologists in locating data not published in the scientific literature. The database can be accessed directly through the Internet via the EPA's web site.

## *European Union*

In 1993, The European Council instituted a comprehensive review of existing chemicals under the Dangerous Substances Directive (European Commission, 1993b). This directive calls for a risk assessment approach on priority chemicals. Reporting of use and health information by European chemical companies was required in three phases. Tiers of production volume established the phases with highest volume chemicals reported first. The regulation obliges industry to submit all readily available data on HPV.

*Table 10.9* 1991 substantial risk reporting guidance: Table 10.1 – factors to consider in determining reportability of lethality information under TSCA Section 8 (e)

| *$LD_{50}$ oral dose (mg/kg)* | *$LD_{50}$ dermal dose (mg/kg)* | *4-h $LC_{50}$ inhalation dose (ppm/mg/l)* | *Consider exposure/ other factors?* |
|---|---|---|---|
| 5 | 20 | $<$ 50 (0.5) | No (EXTREMELY TOXIC) |
| $>$ 5–50 | $>$ 20–200 | $>$ 50 ( $>$ 0.5) to 200 (2) | Only to some reasonable degree (HIGHLY TOXIC) |
| $>$ 50 | $>$ 200 | $>$ 200 ( $>$ 2) | Yes (MODERATELY TOXIC) |

*Table 10.10* 1991 substantial risk reporting guidance

Summaries called status reports were issued on each 8(e) report received before 1991. As part of the 1991 guidance, the EPA cataloged the most important status reports by category. These categories are illustrative of the breadth of information subject to 8(e) reporting.
SECTION 8(e) GUIDANCE/POLICY REFLECTED IN STATUS REPORTS

I. Toxicological/exposure findings
 A. Acute toxicity (animal)
 B. Acute toxicity (human)
 C. Subacute toxicity (animal)
 D. Immunotoxicity (animal)
 E. Neurotoxicity (animal)
 F. Neurotoxicity (human)
 G. Oncogenicity (animal)
 H. Oncogenicity (human)
 I. Reproductive/developmental (animal)
 J. Reproductive/developmental (human)
 K. Genotoxicity (*in vitro*)
 L. Genotoxicity (*in vivo*)
 M. Aquatic toxicity/bioconcentration
 N. Emergency incidents of environmental contamination
 O. General/nonemergency environmental contamination
II. General reporting issues
 A. Intracorporate reporting procedures
 B. Subject persons
 C. Subject chemicals
 D. Research and development chemicals
 E. Drug export
 F. Pesticide export
 G. Previous manufacture/import/process/distribution
 H. Obtaining information
 I. Pre-1977 information
 J. Actual knowledge by the EPA
 K. Published scientific literature
 L. Information obtained from other federal agencies
 M. Information corroborating well-established effects
 N. Relationship to other TSCA reporting requirements
 O. Relationship to other the EPA administered authorities
 P. Relationship to authorities not administered by the EPA
 Q. Section 8(e) reporting procedures

The International Uniform Chemicals Information Database (IUCLID) database is a repository of the reported information and a tool for setting priorities for further risk assessment. European chemical companies and regulatory authorities are expected to cooperatively participate in the worldwide HPV chemicals testing effort through the International Council of Chemical Associations (ICCA). Article 10 of the directive mandates that the real or potential risk for man and environment of priority substances is to be assessed using principles laid down in Commission Regulation (EC) No. 1488/94 (European Union, 1994) on risk assessment for existing substances. The risk assessments are carried out by competent authorities designated by the responsible member states to act as rapporteurs. An extensive technical guidance document was published on the risk assessment process in 1996 (European

Commission, 1996). A schematic of the risk assessment process and range of outcomes on particular chemicals is given in Figure 10.2.

Four priority lists of substances have been listed since 1995 under existing substances regulation 793/93/EEC. Each list consists of 30–50 substances. A rapporteur from a member country is appointed. Action has been slow on the priority lists so far. Many of the chemicals on the first and second lists have not proceeded beyond the preliminary discussion stage. In addition, the European Commission has published a document outlining an action plan for endocrine disrupters (European Commission, 1999) as well as studying regulatory actions for Persistent Organic Pollutants (POPs).

Although no formal authority exists in the DSD to mandate testing on existing substances, it would be possible to seek voluntary testing on any of the chemicals on the priority lists. The DSD differs from TSCA in that it does not contain reporting requirements for new risk information or the recording of adverse reaction reports.

Although EU laws and directives were meant to supersede national laws, this has happened very slowly. The Scandinavian countries and Germany have been more persistent in retaining their national initiatives on industrial chemicals than other member states. Many times this activity will catalyze additional new regulations in the EU similar to the way various states such as California spur reform in the US.

### Japan

The Japanese law also gives the implementing agencies authority over existing chemicals. The same criteria and procedures used to classify chemicals as designated substances are used (see Figure 10.3). There are currently nine Class I and twenty-three Class II Specified Substances listed which consist mainly of PCBs and other chlorinated hydrocarbons and pesticides as well as a number of organotin compounds. Class I compounds are banned practically speaking and Class II compounds can only be marketed under tight controls. Currently, those new chemicals that become Designated Substances as a result of a review of data presented by an intended manufacturer are merely being tracked. However, they can be subject to restrictions if concern about any of these chemicals increase. As this chapter went to press, approximately 10 designated substances were being reviewed for more stringent controls.

The Japan Environment Agency is actively studying the endocrine disrupter issue. A list of sixty-seven suspected endocrine disrupters has been published and an environmental monitoring program is in place. It is likely that any risk management actions to protect public health and the environment will be coordinated with international efforts (Japan Environment Agency, 1998).

### Canada

Because CEPA is a relatively new law, regulatory programs on existing chemicals have not progressed to a great extent in Canada. Much control is exerted through hazard communication authority (see Chapter 11). Substantial risk reporting similar to TSCA is required under Section 17 of CEPA. Guidelines have been published for submission of these reports (Environment Canada, 1994).

### Australia

During 1997 and 1998, a number of legislative and streamlining activities were introduced into the Existing Chemicals Assessment Program. These introduced greater flexibility in the declaration of Priority Existing Chemicals (PECs) and facilitated preliminary assessments. Assessments of only six existing chemicals were published in prior years. Candidate chemicals can be nominated by industry, the public or government. After a review of forty-one candidate chemicals, eleven were declared as priority substances in 1978. This process of declaring PECs is careful to focus on chemicals that are extensively used in Australia and avoids chemicals that are subject to review in other countries to conserve resources. Additional information and input from industry is solicited during the process. A preliminary or full quantitative risk assessment is done as part of the review of PECs (NOHSC, 1999).

Recently there has been more evidence of activity on a wider variety of generic existing chemical topics such as endocrine disruption and the HPV issue on the NOHSC web site.

### Philippines

The Philippines has a Priority Chemical List (PCL). It currently lists twenty-eight chemicals. As DENR reviews new chemicals, some may be added to this PCL if determined to pose unreasonable risk to public health, workplace and the environment. These chemicals are subject to Chemical Control Orders (CCOs) that can prohibit, limit or regulate the use, manufacture, import, export, transport, processing, storage, possession and wholesale of the chemicals.

### Outlook for the future

It is likely that future regulatory programs on the control of industrial chemicals will arise from international discussions, if not cooperative programs, among the various regulatory agencies around the world. These discussions, which arose from meetings scheduled largely to discuss larger environmental issues such as global warming have become more important and frequent over the last decade. At least three major issues will be the focus of first decade of the new century.

1 Endocrine toxicity. Although the US regulatory agencies are focusing on validating screening tests before moving on with a regulatory program, the European Commission may begin regulating certain suspect chemicals before definitive data is obtained. Whether the dangers of the current level of endocrine modulators in the environment are judged to be sufficiently serious to justify regulatory action as new data is generated will be an interesting debate.
2 High production volume chemical testing. The international effort to conduct more testing will result in a great quantity of data that can be scrutinized by regulators, environmentalists and the general public. Debates on the safety of these chemicals, the need for further testing on these chemicals and the proper level of testing to screen new chemicals in the future will likely arise.
3 PBT/POP chemicals. Regulatory programs are beginning to emerge in the US and Europe that will begin to focus on existing chemicals that may be problematic

because of their environmental persistence. A new protocol for examining new chemicals for PBT properties in the US is being initiated.

These and other issues will be on the forefront of activity. As always, political agendas and socioeconomic events will also play an important role in defining the future of chemical control regulations.

## Acknowledgments

I would like to thank my colleagues at Lubrizol for their technical advice and comments, particularly John Blickensderfer, Tracy Harris, Joe Kostusyk, Chris Sgarlata, David Sharp, Steve Signs, Mick Wragg and Ping Zhu. Special thanks go to Meera Raghuram for help with the figures as well as technical assistance and to Linda Toth and Jennifer Hanna for many hours of editorial assistance.

## Web sites of interest

| | |
|---|---|
| *Governments* | |
| Australia (NOHSC) | www.nohsc.gov.au |
| Canada (Environment Canada) | www.ec.gc.ca |
| Japan (MITI) | www.miti.gd.jp |
| Korea | www.infokorea.com |
| Philippines (DENR) | www.denr.gov.ph |
| EPA | www.epa.gov |
| US Government Printing Office | www.gpo.gov |
| *Other organizations* | |
| Chemical Abstracts Services (CAS) | www.cas.org |
| Chemical Manufacturers Associations (CMA) (now the American Chemistry Council) | www.cmahq.com |
| Environmental Defense Fund (EDF) | www.edf.org |
| Office of Economic Cooperation and Development (OECD) | www.oecd.org |
| International Council of Chemical Associations (ICCA) | www.icca-chem.org |
| United Nations Environmental Programs (UNEP) | www.unep.org |

## References

Bergeson LL, Campbell LM, Rothenberg L. TSCA and the future of chemical regulation. *EPA Admin L Rep* 2000;15(4):290–310.

The Chemical Daily Co., Ltd., 1999. *Handbook of Existing & New Chemical Substances*, 8th edition. Tokyo: The Chemical Daily Co., Ltd.

CEQ. *Toxic Substances, Reprinted in Staff House Committee on Interstate and Foreign Commerce*, 94th Congress, 2nd Session, Legislative History of the Toxic Substances Control Act (TSCA Legislative History) at 760, 1971.

Environmental Defense Fund. *Toxic Ignorance*, 1997.

Environment Canada. *Guidelines for the Notification and Testing of New Substances: Chemicals and Polymers*, 1993. ISBN 0-662-20542-1.

Environment Canada. *Guidelines for Submissions under Section 17 of the Canadian Environmental Protection Act*, 1994.

EPA/EC. Joint project on the evaluation of (quantitative) structure activity relationships. 1994.

EPA, Overview EPA Exposure based Policy, EPA 743-R-94-001, 1991a.

EPA. TSCA Section 8(e) Reporting guide. June 1991b.

EPA. Office of Pollution Prevention and Toxics. OPPT's RM2 process identifies risk management options for existing chemicals. Chem Prog Bull 1992;13(2):16–23.

EPA. Office of Pollution Prevention and Toxics. OPPT restructures functions, programs. Chem Prog Bull 1993a:14(1).

EPA. Office of Pollution Prevention and Toxics. Categories provide guidelines for premanufacture notice submitter. Chem Prog Bull 1993b;14(2):35.

EPA. Office of Pollution Prevention and Toxics. The Toxic Substances Control Act at Twenty. *Chem Environ* 1996;4:1–28.

EPA. Office of Pollution Prevention and Toxics. *Chemistry Assistance Manual for Premanufacture Notification Submitters*. EPA 744-R-97-003, March 1997.

EPA. Office of Pollution Prevention and Toxics. Pollution Prevention (P2) Assessment Framework. November, 1998 Draft.

EPA. *Development of Chemical Categories in the HPV Challenge Program*, 1999a.

EPA. *HPV Chemical Human Health Testing: Animal Welfare Issues and Approaches*, 1999b.

European Commission. Council Directive 67/548/EC on the Approximation of the Laws, Regulations and Administrative Provisions Relating to the Classification, Packaging and Labeling of Dangerous Substances, 1967.

European Commission. Council Directive 79/831/EC Amending for the Sixth Time Directive 67/548/EC on the Approximation of the Laws, Regulations and Administrative Provisions Relating to the Classification, Packaging and Labeling of Dangerous Substances, 1979.

European Commission. Council Directive 92/32/EC Amending for the Seventh Time Directive 67/548/EC on the Approximation of the Laws, Regulations and Administrative Provisions Relating to the Classification, Packaging and Labeling of Dangerous Substances, 1992.

European Commission. Commission Directive 93/67/EC on Rick Assessment for New Notified Substances, 1993a.

European Commission. Council regulation (EC) No. 793/93. on The Evaluation and Control of Existing Substances, 1993b.

European Commission. Commission Regulation (EC) No. 1488/94 on Risk Assessment for Existing Substances, 1994.

European Commission. Technical Guidance in Support of Commission Directive 93/67/EEC on Risk Assessment for New Notified Substances and Commission Regulation EC No 1488/94 on Risk Assessment for Existing Substances, 1996.

European Commission. Communication from the Commission to the Council and European Parliament. Community Strategy for Endocrine Disrupters, 1999.

Federal Register. Environmental Protection Agency. Toxic Substances Control Act. Statement of interpretation and enforcement of policy; notification of substantial risk. *Fed Reg* 1978;43;11110.

Federal Register. Environmental Protection Agency. Toxic Substances Control Act. Notice of proposed rulemaking: acrylamide and methylol acrylamide. *Fed Reg* 1991;56:49863.

Federal Register. Environmental Protection Agency. TSCA Section 8(e); notice of clarification and solicitation of public comment. *Fed Reg* 1993;58:37735.

Federal Register. Toxic Substances Control Act. Proposed test rule for Hazardous Air Pollutants (HAPS). *Fed Reg* 1996;61:33178.

Federal Register. Toxic Substances Control Act. Final rule: microbial products of biotechnology. *Fed Reg* 1997;62:17909.

Federal Register. Notice: outline of screening program. *Fed Reg* 1998a;63:42852.

Federal Register. Notice: proposed screening program and request for comment. *Fed Reg* 1998b;63:71542.

Federal Register. Direct final rule: revisions to reporting regulations under TSCA Section 8(d). *Fed Reg* 1998c;63:15765.

Federal Register. Environmental Protection Agency, Policy Statement: category for persistent, bioaccumulative, and toxic new chemicals. *Fed Reg* 1999a;64:60194.

Federal Register. Proposed test rule for *in vitro* dermal absorption rate testing of certain chemicals of interest to occupational safety and health administration; proposed rule. *Fed Reg* 1999b;64:31073.

Federal Register. Semi-annual regulatory agenda: test rule: ATSDR substances. *Fed Reg* 1999c;64:21997.

Federal Register. Semi-annual regulatory agenda: TSCA Section 8(e) policy: notice of clarification. *Fed Reg* 1999d;64:22006.

Federal Register. Testing of certain high production volume chemicals; proposed rule and notice. *Fed Reg* 2000;65:81657.

Japan Environment Agency. *Strategic Programs on Environmental Endocrine Disruptors*, 1998.

NOHSC. *Annual Report on NICNAS*, 1999.

NICNAS. *Handbook for Notifiers*, 1995.

OECD. Decision of the Council concerning the mutual acceptance of data of chemicals [C(81)30(Final)], 1981.

OECD. Decision of the Council concerning the minimum pre-marketing set of data in the assessment of chemicals, 1982.

OECD. *Guidelines for the Testing of Chemicals*. Paris Cedex: OECD Publications, 1993.

OECD. Report of the OECD Workshop on sharing information about new industrial chemicals assessment OCDE/GD(97)33, 1997a.

OECD. *OECD Principles for GLPs*. Paris Cedex: OECD Publications 1997b.

OECD. *New Industrial Chemicals Information Directory*. Paris Cedex: OECD Publications, 2000

USC. Title 15, USC 2601, Public Law No. 94-469, 1976.

Wagner PM, Nabholz JV, Kent RJ. The new chemicals process at the environmental Protection Agency (EPA): structure–activity relationships for hazard identification and risk assessment. *Toxicol Lett* 1995;79:67–73.

Walker JD. *The TSCA Interagency Testing Committee, 1977–1992: Creation, Structure Functions and Contributions. Environmental Toxicology and Risk Assessment: Second Volume*. ASTM STP 1216, 1993. Philadelphia, PA: American Society for Testing and Materials.

Chapter 11

# Industrial chemicals

## Hazard communication, exposure limits, labeling and other workplace and transportation requirements under OSHA, DOT, and similar authorities around the world

*Richard C. Kraska and Debora L. Hooper*

### Introduction

Although the chemical control laws discussed in the last chapter certainly impose a major regulatory burden on the industrial chemical industry, this chapter will discuss other authorities that control more of the day to day activities in the operations where industrial chemicals are manufactured, used and transported. The most wide-ranging requirement on these operations is hazard communication regulations. These regulations usually require, if not covered by other regulations, that adequate training and hazard information be provided to the workers that actually handle chemicals in the workplace. Underpinning all hazard communication regulations around the world is the requirement that Material Safety Data Sheets (MSDSs) on hazardous chemicals are provided in the workplace and that hazardous chemicals are adequately labeled. These regulations also define, in elaborate detail, what makes a chemical hazardous and define many rules on how this information should be communicated on MSDSs and labels. Hazard communication requirements usually work in tandem with separate complementary laws or regulations that define safe exposure levels for the most hazardous of chemicals and mandate other practices to assure a safe workplace as well as provide for special labels for chemicals that have hazards that are important to communicate for chemicals while being transported.

The US has been one of the world leaders for legislation and regulation on matters of workplace safety. The Occupational Safety and Health, (OSH) Act of 1970 arose as a result of two decades of increasing concern about hazards in the workplace. In 1969, a congressional report estimated that 14,000 deaths and two million injuries occur each year from workplace accidents. The OSH Act established the Occupational Safety and Health Administration (OSHA) in the Department of Labor (DOL) and the National Institute of Occupational Safety and Health (NIOSH) in the former Department of Health, Education, and Welfare (DHEW) (Government Institutes, 1991).

Regulatory authority was given to OSHA to establish occupational standards and conduct inspections to assure that employers protect their workers from various hazards. The original writers of the law realized that OSHA would not be able to write a standard to cover every conceivable hazard. Therefore, the law also includes a "general duty" clause to assure that employers take reasonable steps to provide a safe workplace. NIOSH was established to conduct research into occupational safety and health concerns and to recommend various standards to OSHA (Mintz, 1984; Rothstein, 1990).

Prior to the passage of the OSH Act, the American National Standards Institute (ANSI) published voluntary standards to control occupational hazards. In the early 1970s, OSHA quickly adopted many standards for various workplace tasks to control workplace hazards (Government Institutes, 1991).

One of the first actions of OSHA regarding chemical hazards was the 1971 publication of Permissible Exposure Limits (PELs) for 400 chemicals. For the next decade, OSHA established generic standards for personal protective equipment and engineering controls to help control hazards due to chemicals in the workplace. OSHA also established very specific standards for about three dozen chemicals, mostly carcinogens, to control hazards (Mintz, 1984). These rulemaking procedures were very laborious and painstaking because, under the Administrative Procedures Act, federal agencies like OSHA need to consider the comments of the public on proposed regulations before issuing final regulations. Two opposing factions, labor and business entities, usually submit voluminous and opposing comments on OSHA-proposed regulations. More than 2,000 pages of OSHA regulations have been published over the years, and OSHA has found it difficult to keep up with the need for new and updated PELs.

It was soon realized that OSHA could not possibly establish standards for every hazardous chemical; yet hazard information on each hazardous chemical used in the workplace was needed by every worker coming into contact with the chemical. In 1974, NIOSH recommended label regulations, and a Standards Advisory Committee was established to develop a standard for hazard communication (DHEW, 1974; DOL, 1975). The Committee issued a report on June 6, 1975, recommending a comprehensive program of labels and other information reinforced by training programs.

The pace of voluntary action began to increase in the late 1970s. ANSI published voluntary hazard warning and precautionary labeling guidelines in 1976. OSHA published a recommendation for a form to be used (OSHA Form 20) as an MSDS. An MSDS is a comprehensive document that should include all known information which allows safe use of a chemical including a description of hazards, precautionary practices and protective equipment, first aid, and information to assist in spill clean-up and fire-fighting. Although providing MSDSs was only optional at the time, most major chemical companies began providing MSDSs because of increasing customer requests. Many chemical companies began to find that customers would not accept new chemicals without an MSDS or similar documents being made available. After a decade of regulatory and legislative negotiating and lobbying, the OSHA Hazard Communication Standard became effective in 1985 as a regulation under the existing OSH Act without need of new legislation.

The regulatory history for control of workplace hazards has been similar in the major industrialized nations around the globe. Laws and regulations regarding hazard communication, transportation requirements and the development of safe exposure limits for chemicals that have been developed over the last 30 years will be described below.

## Hazard communication: background and basic requirements for labels and MSDSs

Hazard communication regulations usually specify a comprehensive set of requirements such as worker training, availability of information for workers such as MSDSs, container labels and lists of hazardous chemicals in the workplace. The regulations also specify in great detail the minimum content of MSDSs and labels and the definitions of health and safety hazards. The main role of toxicologists employed by companies that need to write MSDSs and labels is evaluating the health hazards of the chemicals manufactured and sold as products. This is a complex task for companies that sell their products internationally since the requirements and definition of hazards can be quite different in various countries. The health hazard definitions of the various major international hazard communication requirements will be presented in detail, since this will be of the greatest interest to toxicologists.

Efforts are underway to harmonize the differences in international hazard communication requirements. A significant step was taken in 1994, led by the ANSI, to develop a basic format for section headings and section order in MSDSs. There was a great deal of international cooperation in the development of this standard. MSDSs that meet this format are virtually accepted worldwide. MSDS content and hazard definitions are still a contentious issue on the international front so toxicologists and other members of companies' MSDS writing teams need to pay close attention to these differences to assure their company's MSDSs are acceptable in the countries where their chemicals are manufactured and their products are sold. The ANSI standard document is an excellent reference for more information on the detailed requirements on MSDSs in each country. The sixteen sections specified by the ANSI standards are given in Table 11.1.

The Organization of Economic Cooperation and Development (OECD) has undertaken the task of harmonizing the international hazard classifications and definitions. Hopefully this effort will lead to internationally accepted hazard classifications for workplace and transport labels that will be adopted by the regulatory or legislative bodies of the industrial world. This work is still in progress and will be summarized below.

*Table 11.1* Section headings and order specified by the ANSI Standard for MSDSs (ANSI, 1993, 1998)

| | |
|---|---|
| Section 1 | Product and company identification |
| Section 2 | Composition/information on ingredients |
| Section 3 | Hazards identification |
| Section 4 | First aid measures |
| Section 5 | Fire fighting measures |
| Section 6 | Accidental release measures |
| Section 7 | Handling and storage |
| Section 8 | Exposure controls/personal protection |
| Section 9 | Physical and chemical properties |
| Section 10 | Stability and reactivity |
| Section 11 | Toxicological information |
| Section 12 | Ecological information |
| Section 13 | Disposal considerations |
| Section 14 | Transport information |
| Section 15 | Regulatory information |
| Section 16 | Other information |

### *United States*

Under 29 CFR 1910.1200, manufacturers and importers of certain hazardous industrial chemicals have the responsibility for creating MSDSs and labels for each hazardous product. Exempt products include food, pesticides, drugs and consumer products. Another exemption exists for products or articles where exposure to hazardous chemicals does not result from occupational use. Therefore, the regulation basically covers chemical type products – non-articles in the form of hazardous liquids, gases, and granular solids. The main industry covered is the chemical industry, but most other industries that are users of chemicals to formulate products such as paints, inks, cleaners and adhesives are also covered. Although these industries are usually considered processors of chemicals under the Toxic Substances Control Act (TSCA) (see previous chapter), they are product manufacturers under Hazard Communication Standard (HCS) and have hazard communication responsibilities.

Although this requirement only affects hazardous chemicals or products that release hazardous chemicals, many industrial users of various products purchased from other companies have difficulty determining whether a product is non-hazardous or exempt from MSDS requirements. Therefore, most companies that supply products to other industrial users find that these customers insist on receiving an MSDS for every product. Therefore, MSDSs have been written for many unusual products, such as metal ingots, building materials, saws and other tools and equipment as well as non-hazardous products such as food ingredients.

The health and safety hazard definitions are listed in Appendix A of 29 CFR 1910.1200 (Table 11.2). Definitions of some terms such as *compressed gas, oxidizer, water reactive, sensitizer*, and *corrosive* are rather generic. Others, such as *pyrophoric, combustible, flammable, skin irritant, eye irritant, toxic*, and *highly toxic* are more technically exact or at least semi-quantitative. Definitions of some terms, such as *carcinogenicity*, resemble legal jargon rather than scientific definitions. Some definitions are clearer than others, but these definitions form the crux of the hazard evaluation that manufacturers must perform on all of their products.

These hazard definitions are used to assess chemicals and chemical products for communication on MSDSs and labels. The information on hazards that appears in these two documents has notable distinctions consistent with the purposes of the two documents. MSDSs are multi-page documents with complete hazard information. They are reference documents intended for reading by the workers before using a particular chemical and also to be referred to whenever any worker has a hazard or handling question. Labels are shorter documents that are intended to relay the most important information. The label, which is required to be on every container of a hazardous chemical, reminds the worker of important hazards every time the worker sees it.

The OSHA HCS is very much a performance standard where basic definitions and requirements are listed, but not every conceivable situation is covered by a prescriptive "how to" regulation. Manufacturers are expected to use good judgment coupled with their own knowledge of how products are used and the severity and nature of the hazards presented by the products to provide good information to their workers and customers. Manufacturers are given great latitude on how to compose their MSDSs and labels. The regulation requires only chemical or product name, hazard warnings, and hazardous ingredients to appear on the label. However, most of US industry follows the standard first developed by ANSI in 1976 and last revised in 2000 that recommends that labels include important handling, precautionary and first aid information. The ANSI standard also recommends the use of various signal words (Danger, Warning, Caution) with their list of recommended hazard warning statements depending on the severity of the hazard (ANSI, 1976, 2000).

The HCS regulation establishes an important distinction on the severity of hazards to be communicated on labels and MSDSs. MSDSs are to report all hazards; labels are to note "appropriate hazards". While there is an intentional distinction in the regulatory language between "all" and "appropriate" hazards, OSHA has occasionally lost sight of the distinction, which has caused some confusion in the regulated community. This was probably due to a misunderstanding in the agency that led to an overly conservative interpretation of the regulations. However, the intent of the regulations is clear and the distinction has recently been confirmed in the courts (Secretary of Labor vs. American Cyanamid, 1992).

OSHA clearly confirms the distinction between *all* and *appropriate* in a guidance document passage on carcinogenicity. In this document (OSHA, 1986), OSHA makes it clear that, based on the certainty or strength of the evidence, carcinogenic warnings need not appear on the label. The document refers to the classification system of the International Agency for Research on Cancer (IARC, 1987). Cancer warnings need to be communicated for chemicals that are considered Group 1 (human) or Group 2A (probable human) on the MSDS and on labels. Group 2B (possible human) carcinogens do not require a label warning, but need to be reported on the MSDS.

The regulations do require that MSDSs and labels be consistent with each other. Part of the working definition of *consistency*, therefore, is that all hazards that appear on the label should be noted also on the MSDS, but because of the "appropriate" hazard language, not all hazards that are explained on the MSDS need to appear on the label. This is an appropriate standard that is consistent with the intent of labels.

Labels are intended to convey the most important information in as clear and concise a form as possible. Due to the litigious climate in the US with regard to legal liabilities, many companies feel compelled to give a great deal of information on the label as well as legal disclaimers. Unfortunately, wordy labels do not convey information easily and may even discourage the product users from reading the label at all. The labeling standard established and periodically improved by ANSI is widely used by US companies and establishes a reasonable standard beyond OSHA requirements while avoiding over-communication (ANSI, 2000). OSHA only requires hazard warning statements and the name of the hazardous chemicals on the label. The ANSI standard suggests appropriate signal words (Danger, Warning, Caution) to be used with the hazard warning statement as well as precautionary statements and brief first aid information. The ANSI document includes example labels as a guide for practitioners.

The HCS requires that hazard assessments be complete (Code of Federal Regulations, 1999a; 29 CFR 1910.1200 Appendix B) and be kept current. In implementing a performance standard, OSHA did not attempt to define the hazards of any chemicals because of the dynamic nature of hazard information and the realization that agency resources and the US regulatory machinery could not keep mandatory hazard classifications up to date. Recognizing the US industrial culture of "supplier responsibility", companies that buy chemicals for use in their products need not conduct complete

*Table 11.2* Health hazard definitions under the OSHA Hazard Communication Standard (HCS) listed in 29 CFR 1910.1200 Appendix A

Carcinogen

A chemical is considered to be a carcinogen if:

1 It has been evaluated by the International Agency for Research on Cancer (IARC), and found to be a carcinogen or potential carcinogen
2 It is listed as a carcinogen or potential carcinogen in the Annual Reports on Carcinogens published by the National Toxicology Program (NTP) (last edition); or
3 It has been evaluated by the International Agency for Research on Cancer (IARC),3 It has been evaluated by the International Agency for Research on Cancer (IARC),It is regulated by OSHA as a carcinogen

Corrosive

A chemical that causes visible destruction of, or irreversible alternations in, living tissue by chemical action at the site of contact. For example, a chemical is considered to be corrosive if, when tested on the intact skin of albino rabbits by the method described by the US Department of Transportation in Appendix A to 49 CFR Part 173, it destroys or changes irreversibly the structure of the tissue at the site of contact following an exposure period of 4 h. This term shall not refer to action on inanimate surfaces.

Highly toxic

A chemical falling within any of the following categories:

1 A chemical that has a median lethal dose ($LD_{50}$) of 50 mg or less per kg of body weight when administered orally to albino rats weighing between 200 and 300 g each.
2 A chemical that has a median lethal dose ($LD_{50}$) of 200 mg or less per kg of body weight when administered by continuous contact for 24 h (or less if death occurs within 24 h) with the bare skin of albino rabbits weighing between 2 and 3 kg each.
3 A chemical that has a median lethal concentration ($LC_{50}$) in air of 200 ppm by volume or less of gas or vapor, or 2 mg/l or less of mist, fume, or dust, when administered by continuous inhalation for 1 h (or less if death occurs within 1 h) to albino rats weighing between 200 and 300 g each.

Irritant

A chemical, which is not corrosive, but which causes a reversible inflammatory effect on living tissue by chemical action at the site of contact. A chemical is a skin irritant, if when tested on the intact skin of albino rabbits by the methods of 16 CFR 1500.41 for 4 h exposure or by other appropriate techniques, it results in an empirical score of 5 or more. A chemical is any eye irritant if so determined under the procedure listed in 16 CFR 1500.42 or other appropriate techniques.

Sensitizer

A chemical that causes a substantial proportion of exposed people or animals to develop an allergic reaction in normal tissue after repeated exposure to the chemical.

Table 11.2 (*continued*)

Toxic

A chemical falling within any of the following categories:

1 A chemical that has a median lethal dose ($LC_{50}$) or more than 50 mg/kg but not more than 500 mg/kg of body weight when administered orally to albino rats weighing between 200 and 300 g each.
2 A chemical that has a median lethal dose ($LD_{50}$) of more than 200 mg/kg but not more than 1,000 mg/kg of body weight when administered by continuous contact for 25 h (or less if death occurs within 24 h) with the bare skin of albino rabbits weighing between 2 and 3 kg each.
3 A chemical that has a median lethal concentration ($LD_{50}$) in air of more than 200 ppm but not more than 2,000 ppm by volume of gas or vapor or more than 2 mg/l but not more than 20 mg/l of mist, fume, or dust, when administered by continuous inhalation for 1 h (or less if death occurs within 1 h) to albino rats weighing between 200 and 300 g each.

Target organ effects

The following is a target organ categorization of effects that may occur, including examples of signs and symptoms and chemicals that have been found to cause such effects. These examples are presented to illustrate the range and diversity of effects and hazards found in the workplace, and the broad scope employers must consider in this area, but are not intended to be all-inclusive.

| *Effect* | *Definition* | *Signs and symptoms* | *Chemicals* |
|---|---|---|---|
| Hepatotoxins | Chemicals which produce liver damage | Jaundice; liver enlargement | Carbon tetrachloride; nitrosamines |
| Nephrotoxins | Chemicals which produce kidney damage | Edema; proteinuria | Halogenated hydrocarbons; uranium |
| Neurotoxins | Chemicals which produce their primary toxic effect on the nervous system | Narcosis; behavioral changes; decrease in motor functions | Mercury; carbon disulfide |
| Agents which act on the blood or hematopoietic system | Decrease hemoglobin function, deprive the body tissue of oxygen | Cyanosis; loss of consciousness | Carbon monoxide; cyanides |
| Agents which damage the lung | Chemicals which irritate or damage the pulmonary tissue | Cough; tightness in chest; shortness of breath | Silica; asbestos |
| Reproductive toxins | Chemicals which affect the reproductive capabilities including chromosome damage (mutations) and effects on fetuses (teratogenesis) | Birth defects; sterility | Lead, DBCP |
| Cutaneous | Chemicals which affect the dermal layer of the body | Defatting of the skin; rashes, irritation | Ketones; chlorinated compounds |
| Eye hazards | Chemicals which affect the eye or visual capacity | Conjunctivitis; corneal damage | Organic solvents; acids |

hazard evaluations on purchased chemicals. The OSHA HCS explicitly allows reliance on the hazard evaluation of the supplier.

If new information significantly changes the hazard assessment or recommendations, the HCS requires that MSDSs and labels be changed within 90 days. This is an important part of the HCS that is easy to overlook. New information can affect large product lines almost on a daily basis. New information that needs to be considered in the hazard assessment includes new toxicology test results, new information on product composition, new scientific literature, new government regulations and test results. It is important that the hazard assessment team establish a routine procedure to gather and review new information. This provides formidable compliance challenges for the hazard communication staff of many companies.

Other requirements of HCS, such as worker training and MSDS distribution, are usually not the responsibility of the toxicologist and will not be explained in detail here. Basic requirements are listed in Table 11.3. Many of the latter requirements are usually the responsibility of industrial hygienists. These are reviewed quite well in *Patty's Industrial Hygiene* (Heidemann and Simkins, 1991). Much information is now available on OSHA's web site (www.osha.gov). This includes guidance materials, regulatory agendas, and letters of interpretation.

### Europe

Regulations on hazardous chemicals were first established in 1967 by the European Commission under the Dangerous Substances Directive (DSD; European Commission, 1967). Since 1967, the directive has undergone eight amendments and twenty-seven adaptations for technical progress. The last two appearing in the year 2000 (European Commission 2000a,b). Hazard definitions and labeling requirements are laid out by the DSD. In 1988, the Dangerous Preparations Directive (DPD; European Commission, 1988; last amendment, European Commission, 1999) expanded labeling requirements to mixtures of chemicals. Subsequently, MSDSs became a formal requirement with the MSDS directive (European Commission, 1991).

Although these directives would also be considered performance standards since responsible parties must perform a good deal of data interpretation and regulatory interpretation, the requirements are somewhat more prescriptive than in the OSHA HCS. These details are contained in the various annexes to the DSD.

*Table 11.3* Summary of requirements of OSHA Hazard Communication Standard (29 CFR1910.1200)

1. Conduct hazard determination
2. Prepare and conduct a hazard communication program which is documented in writing which includes:
   - List of hazardous chemicals in the workplace
   - Method to inform hazards of non-routine tasks
   - Method to inform contract employees of hazards
   - System for container labeling
   - Provide MSDSs that are updated with new information within 90 days
   - Information on location of hazards and access methods to training materials and MSDSs
   - Training of all employees
3. Dissemination of information to health professionals in medial emergencies

Annex I lists official hazard classifications of approximately 1,000 chemicals. These listings prescribe label symbols, risk and safety phrases (see Annexes II, III, IV) and threshold percentages for classifying mixtures under the DPD (see discussion on mixtures below).

Annex II lists the official pictorial and letter symbols to be used on labels for certain hazards. The pictorial symbols are not reproduced here. The letter symbols for health and environmental hazards are:

| | |
|---|---|
| T + | Highly Toxic |
| T | Toxic |
| Xn | Harmful |
| Xi | Irritant |
| C | Corrosive |
| N | Dangerous to the Environment |

Annex III describes the complete list of risk or R phrases to be used on labels when the chemical meets a hazard definition. The R phrases are comparable to the hazard warning phrases in the ANSI standard. Official wording in all EU languages is listed in the annex so that translation is not an issue. A complete list of these phrases is given in Table 11.4. The Annex also specifies that certain combinations of hazards result in combined R phrases with specified wording.

Annex IV describes official wording in all EU languages of safety or S phrases to be used on labels. These are comparable to the precautionary statements recommended by ANSI that give short advice on various practices to help guide safe handling of the product. A complete list of S phrases is given in Table 11.4.

Annex V provides the detailed hazard definitions for each R phrase. In addition it mandates which symbols and S phrases should be used in conjunction with each R phrase.

Annex VI lists approved test methods to determine the various hazard classifications. These are largely harmonized with OECD guidelines.

The text of the R and S phrases is the only warning and precautionary language allowed on a label; therefore, there is less flexibility allowed in the EU with regard to labeling which may cause difficulties for companies who wish to achieve consistency with US labels. However, European labels do have the advantage of being simple and easy to read. The regulations had specified that no more than four risk and four safety phrases be used on the label, but this has recently been increased to six each to accommodate a number of new phrases recently added to describe environmental hazards (European Commission, 1999).

## *Canada*

In Canada, the Workplace Hazardous Materials Information System (WHMIS) was established in 1987 to administer provisions in the Hazardous Products Act and other federal authorities to assure that information about the hazards of materials produced or sold in Canada is provided to workers who use these chemicals in the workplace. Health Canada, one of the more important federal agencies in WHMIS published regulations in the *Canadian Gazette* dictating requirements and hazard criteria for material safety data sheets and labels for hazardous chemicals (Controlled Products Regulations, 1988). These requirements for MSDSs in Canada go somewhat beyond

*Table 11.4* Risk (R) and safety (S) phrases specified in the EU by the Dangerous Substances Directive

R 1 – Explosive when dry
R 2 – Risk of explosion by shock, friction, fire or other sources of ignition
R 3 – Extreme risk of explosion by shock, friction, fire or other sources of ignition
R 4 – Forms very sensitive explosive metallic compounds
R 5 – Heating may cause explosion
R 6 – Explosive with or without contact with air
R 7 – May cause fire
R 8 – Contact with combustible material may cause fire
R 9 – Explosive when mixed with combustible material
R 10 – Flammable
R 11 – Highly flammable
R 12 – Extremely flammable
R 14 – Reacts violently with water
R 15 – Contact with water liberates highly flammable gases
R 16 – Explosive when mixed with oxidizing substances
R 17 – Spontaneously flammable in air
R 19 – May form explosive peroxides
R 20 – Harmful by inhalation
R 21 – Harmful in contact with skin
R 22 – Harmful if swallowed
R 23 – Toxic by inhalation
R 24 – Toxic in contact with skin
R 25 – Toxic if swallowed
R 26 – Very toxic by inhalation
R 27 – Very toxic in contact with skin
R 28 – Very toxic if swallowed
R 29 – Contact with water liberates toxic gas
R 30 – Can become highly flammable in use
R 31 – Contact with acids liberates toxic gas
R 32 – Contact with acids liberates very toxic gas
R 33 – Danger of cumulative effects
R 34 – Causes burns
R 35 – Causes severe burns
R 36 – Irritating to eyes
R 37 – Irritating to respiratory system
R 38 – Irritating to skin
R 39 – Danger of very serious irreversible effects
R 40 – Possible risk of irreversible effects
R 41 – Risk of serious damage to eye
R 42 – May cause sensitization by inhalation
R 43 – May cause sensitization by skin contact
R 44 – Risk of explosion if heated under confinement
R 45 – May cause cancer
R 46 – May cause heritable genetic damage
R 48 – Danger of serious damage to health by prolonged exposure
R 49 – May cause cancer by inhalation
R 50 – Very toxic to aquatic organisms
R 51 – Toxic to aquatic organisms
R 52 – Harmful to aquatic organisms
R 53 – May cause long-term adverse effects in aquatic environment
R 54 – Toxic to flora
R 55 – Toxic to fauna
R 56 – Toxic to soil organisms
R 57 – Toxic to bees

Table 11.4 (*continued*)

R 58 – May cause long-term adverse effects in the environment
R 59 – Dangerous for the ozone layer
R 60 – May impair fertility
R 61 – May cause harm to the unborn child
R 62 – Possible risk of impaired fertility
R 63 – Possible risk of harm to the unborn child.
R 64 – May cause harm to breastfed babies
R 65 – Harmful: may cause lung damage if swallowed
R 66 – Repeated exposure may cause skin dryness or cracking
R 67 – Vapors may cause drowsiness and dizziness

S 1 – Keep locked up
S 2 – Keep out of reach of children
S 3 – Keep in a cool place
S 4 – Keep away from living quarters
S 5 – Keep contents under...
S 6 – Keep under...
S 7 – Keep container tightly closed
S 8 – Keep container dry
S 9 – Keep container in a well ventilated place
S 12 – Do not keep the container sealed
S 13 – Keep away from food, drink, and animal feeding stuffs
S 14 – Keep away from...
S 15 – Keep away from heat
S 16 – Keep away from sources of ignition – NO SMOKING
S 17 – Keep away from combustible material
S 18 – Handle and open container with care
S 20 – When using do not eat or drink
S 21 – When using do not smoke
S 22 – Do not breathe dust
S 24 – Avoid contact with skin
S 25 – Avoid contact with eyes
S 26 – In case of contact with eyes, rinse immediately with plenty of water and seek medical advice
S 27 – Take off immediately all contaminated clothing
S 28 – After contact with skin, wash immediately with plenty of water
S 29 – Do not empty into drains
S 30 – Never add water to this product
S 33 – Take precautionary measures against static discharges
S 35 – This material and its container must be disposed of in a safe way
S 36 – Wear suitable protective clothing
S 37 – Wear suitable gloves
S 38 – In case of insufficient ventilation, wear suitable respiratory equipment
S 40 – To clean the floor and all objects contaminated by this chemical, use...
S 43 – In case of fire, use...
S 45 – In case of accident or if you feel unwell, seek medical advice immediately (show label where possible)
S 46 – If swallowed seek medical advice immediately and show this container label
S 47 – Keep at temperature not exceeding... C.
S 48 – Keep wetted with...
S 49 – Keep only in the original container
S 50 – Do not mix with...
S 51 – Use only in well ventilated areas
S 52 – Not recommended for interior use on large surface areas
S 53 – Avoid exposure - obtain special instructions before use

Table 11.4 (*continued*)

| |
|---|
| S 56 – Do not discharge into drains or the environment, dispose to an authorized waste collection point |
| S 57 – Use appropriate containment to avoid environmental contamination |
| S 62 – If swallowed, do not induce vomiting: seek medical advice immediately and show this container or label |

the typical requirements in other countries in that particular toxicological data, such as the oral $LD_{50}$ of all components of the product, must be listed on the MSDS. There is also an ingredient disclosure list that contains more than 1,700 chemical substances. Health Canada published a guidance document in 1997 entitled *Guidelines for the Disclosure of Toxicological Information on a Material Safety Data Sheet in 1997* (Health Canada, 1997). Justice Canada has an excellent web site that links texts of statutes with the current regulations in force

### *Japan*

Several Japanese laws contribute to the requirements for MSDSs and labels for chemicals. However the Japanese requirements are less defined than in other countries. Guidelines were issued in 1993 by the Ministry of Health and Welfare (MHW) and Ministry of International Trade and Industry (MITI) to clarify the requirements (MHW/MITI, 1993; OECD, 1998).

### *Australia*

In 1986, the Australian National Occupational Health and Safety Commission published voluntary guidance for MSDSs. The commission agreed in 1992 to convert this guidance into a national code of practice supporting its National Model Regulations for the control of workplace substances. Under these model regulations, manufacturers and importers of workplace hazardous substances are required to produce MSDSs and supply them to purchasers. The regulations require the same risk and safety phrases used in the EU for MSDSs and labels.

### *Harmonization efforts on MSDS and labels*

ANSI has developed a standard for a sixteen-section MSDS. This standard was intended to provide a format that specifies an order of information presented which meets all international requirements. OECD is working on a Globally Harmonized System (GHS) to standardize labeling phraseology and hazard criteria. Progress on hazard criteria will be discussed in more detail below. Once the OECD recommendation is final, a lengthy period of debate to adopt the recommendation into national legislations and regulations is expected.

## Definition of hazardous substances

The various international hazard communication regulations define a hazardous substance as any chemical or mixture that meets the hazard definitions defined by the regulation. Many of these definitions will be discussed in detail below. Table 11.5 is also provided to assist the reader in locating the detailed hazard definitions in the HCS, DSD and WHMIS regulations. In addition, each regulation defines a "floor of hazardous chemicals", i.e., those which are defined as being automatically hazardous.

### United States

The hazard definitions under the OSHA HCS are listed in Appendix A of 29 CFR 1910.1200. The regulation also specifies that substances that have Permissible Exposure Limits (PELs) under OSHA or threshold limit values assigned by the American Conference of Government Industrial Hygienists (ACGIH) are considered hazardous under the regulation. It should be noted that OSHA does not prescribe any hazard classifications for these chemicals. That is left to the judgment and responsibility of the scientists conducting the hazard evaluation for the companies that manufacture and use these chemicals. In addition, carcinogens listed by the International Agency for Research on Cancer (IARC), the National Toxicology Program (NTP) or specifically regulated by OSHA as a carcinogen are also considered hazardous with mandatory cancer warnings.

### Europe

In contrast to OSHA's approach, the European Commission has established in the DSD a list of hazardous chemicals entitled Annex I of the directive. Annex I specifies an exact hazard classification which should be used for over a thousand chemical substances and percentage thresholds to be used in classifying mixtures in which an Annex I substances are contained. Annex I is periodically updated, and both the 25th and 26th adaptation for technical progress of the DSD made additions and changes to the list (European Commission, 1998, 2000a).

### Canada

Criteria for hazardous substances were published in 1988 (Controlled Products Regulations, 1988. These hazard definitions bear more similarity to the DSD than the OSHA HCS. The hazards were divided into various classes with divisions and subdivisions. The class, division and subdivision must appear on MSDSs and labels in Canada. The following classes refer to health effects:

Class D: poisonous and infectious material

1. Division 1: materials causing immediate and serious toxic effects
   - Subdivision A: very toxic material
   - Subdivision B: toxic material
2. Division 2: materials causing other toxic effects
   - Subdivision A: very toxic material
   - Subdivision B: toxic material

Class E: corrosive material

The WHMIS disclosure list is not considered a floor list of hazardous chemicals. The

*Table 11.5* Reference to hazard definitions. Paragraph or section numbers where major health and environmental hazard definitions are found in the OSHA HCS, the EU DSD and Canadian WHMIS requirements

| *Hazard* | *United States* | *Europe* | | *Canada* | |
|---|---|---|---|---|---|
| | *29 CFR 1910.1200 Appendix A reference*[a] | *R phrase* | *DSD reference* | *Hazard class* | *WHMIS reference* |
| *Health effects* | | | | | |
| Acute toxicity (lethality) | | | | | |
| Very toxic | 3 | 26, 27, 28 | 3.2.1 | D1A | 46 |
| Toxic | 6 | 22, 23, 24 | 3.2.2 | D1B | 49 |
| Harmful | N/A | 20, 21, 22 | 3.2.3 | N/A | N/A |
| Acute toxicity (symptoms) | 7 a, b, c, d, g, h | 39 | 3.2.1,3.2.2 | N/A | N/A |
| | | 48 | 3.2.2, 3.2.3, 3.2.4 | | |
| | | 40 | 3.2.3 | | |
| | | 65 | 3.2.3 | | |
| | | 67 | 3.2.8 | | |
| Skin corrosion | 2 | 34, 35 | 3.2.5 | E | 65 |
| Skin irritation | 4, 7g | 38, 66 | 3.2.6.1, 3.2.8 | D2B | 60(a) |
| Eye irritation | 2, 4, 7h | 36, 41 | 3.2.6.2 | D2B | 60(b) |
| Skin sensitization | 5 | 43 | 3.2.7.2 | D2B | 61 |
| Respiratory irritation | 2,4, 7e | 37 | 3.2.6.3 | N/A | N/A |
| Respiratory sensitization | 5 | 42 | 3.2.7.1 | D2A | 56 |
| Delayed effects | 7 a, b, c, d, g, h | 33 | 3.2.8 | D2A | 52 |
| | | | | D2B | 59 |
| Carcinogenicity | 1 | 40, 45, 49 | 4.2.1 | D2A | 54 |
| Teratogenicity | 7f | 61, 63 | 4.2.3 | D2A | 53 |
| Reproductive effects | 7f | 60, 61, 62, 63 | 4.2.3 | D2A | 55 |
| Mutagenicity | 7f | 40, 46 | 4.2.2 | D2A | 57 |
| | | | | D2B | 62 |
| *Environmental fate and effects*[b] | | | | | |
| Aquatic toxicity | N/A | 50, 51, 52 | 5.2.1 | N/A | N/A |
| Biodegradation (aquatic) | N/A | 53 | 5.2.1.2 | N/A | N/A |

Table 11.5 (*continued*)

| *Hazard* | *United States* | *Europe* | | *Canada* | |
|---|---|---|---|---|---|
| | *29 CFR 1910.1200 Appendix A reference*[a] | *R phrase* | *DSD reference* | *Hazard class* | *WHMIS reference* |
| Bioaccumulation | N/A | 53 | 5.2.1.2 | N/A | N/A |
| Plant toxicity | N/A | 54 | 5.2.2 | N/A | N/A |
| Terrestrial organisms | N/A | 55, 56, 57 | 5.2.2 | N/A | N/A |
| Biodegradation (soil) | N/A | 58 | 5.2.2 | N/A | N/A |
| Ozone layer | N/A | 59 | 5.2.2 | N/A | N/A |

[a] OSHA HCS states hazard lists are not meant to be a comprehensive lists. Hazards are to be defined broadly and may go beyond this list. See ANSI labeling guide for more complete lists of hazards used in the US.

[b] OSHA HCS and Canadian WHMIS have no authority on environmental effects.

only chemicals considered hazardous by definition are carcinogens that meet certain criteria under ACGIH and IARC. For more details see the carcinogenicity section below.

### *Japan*

Japanese requirements rely on many lists of chemicals under various laws that are specific for various physical and health hazards. Definitions for hazards exist but are more loosely defined than in other laws around the world (MHW/MITI, 1993; OECD, 1998).

### *Australia*

Hazard communication requirements in Australia are very similar to those of the EU. The National Occupational Health and Safety Commission (NOHSC, also known as Worksafe Australia) established the National Model for Regulations for the Control of Workplace Hazardous Substances (NOHSC, 1994a). NOHSC published a national code of practice for the preparation of MSDSs (NOHSC, 1994b). A hazardous substance is defined as any chemical listed in the List of Designated Hazardous Substances (NOHSC, 1994c) or any other substance that meets the criteria specified in Approved Criteria for Classifying Hazardous Substances (NOHSC, 1994d). This list will be updated from time to time. The 1994 list of designated substances is almost identical to the European Annex I, as it existed with the thirteenth adaptation of technical progress (European Commission, 1991). The hazard criteria are identical to the EU and the European system of risk (R) and safety (S) phrases and safety was also adopted. Due to these similarities, the Australian requirements will not be discussed further. The reader is referred to the Australian guidance documents for any differences with European requirements.

## Testing programs and hazard definitions

The OSHA HCS is merely a communication regulation. It requires an honest evaluation and representation of the hazards of which the manufacturer is aware on MSDS and labels. It does not require any testing per se. Lawyers have argued whether the general duty clause of the OSH Act requires manufacturers to test in order to gain knowledge of "foreseeable hazards". Today, prudent companies and their toxicologists and product safety specialists generally collect a minimum set of data so that the hazards of products or a class of products can be predicted.

Therefore, it is desirable for all manufacturers required to classify products under HCS hazards to have an active testing program for heath and other hazards. A variety of standard toxicology test protocols are available and may suffice, especially where the regulatory hazard definitions are vague on test methods. For less vague definitions such as skin irritation, test protocols should follow the testing guidelines as closely as possible. Where testing needs to help define hazards for a variety of jurisdictions, care should be exerted to incorporate all the important features and endpoints required.

It is not necessary that all tests be done at once or even before the product is on the market. If a toxicologist is hired by a chemical company that simply has no toxicology data (hopefully this is a rare circumstance today), the testing program would need to be developed beginning with simple tests and proceeding to more complex tests. In addition, it would be important to develop a priority list of products predicted to be most likely hazardous so that they could be tested first. It would make sense for a testing program to be developed so that acute hazards were covered first. Later, as needed due to new product uses, product sales volume, or suspicion based on occupation exposure, tests employing repeated-dose exposures and more sophisticated endpoints could be undertaken.

In actuality, the testing requirements for regulatory approvals such as PreManufacture Notifications (PMNs) under TSCA and other world-wide chemical inventory laws or food additive or food contact use approvals under FDA sometimes lead to the accumulation of greater than normal amounts of toxicological data on new products. It is important that the hazard evaluation system of any company ensures that all data are evaluated for hazard communication, regardless of the reason for accumulation. Although a test may have been conducted for a purpose of a regulatory approval under another law, all available data needs to be evaluated under hazard communication requirements for consideration of communication on the MSDS or label.

The hazard evaluation team must be aware of all information generated by the corporation that may indicate hazards, as the courts hold the corporation liable for all information known about a product by any employee in the corporation. Therefore, the MSDS writing team must anticipate sources of valuable information within the corporation. For example, business units have been known to conduct their own toxicology studies; manufacturing plant and occupational health staff investigate suspected hazards; and research organizations may accumulate new information on product composition.

Much of this chapter will discuss the types of toxicology testing which may need to be conducted in order to have a successful hazard identification program. Of all the fields of regulatory toxicology, test programs for the HCS are potentially very rewarding to the toxicologist because the extent of testing needed is not specified by the regulations. The toxicologist is largely free to develop a unique program according to his or her professional judgment.

## Product composition

One basic set of data that cannot be overlooked in order to conduct an adequate hazard assessment on a chemical product is chemical composition. This is simply good sense – one cannot accurately assess the hazards of a product if one does not have a good idea of what the product is chemically. This is especially true of chemical mixtures. The HCS has specific rules for assessing the hazards of mixtures based on their composition, and these will be discussed later in this chapter.

It should be noted that most so-called pure chemicals are really mixtures that have a primary economic constituent and many contaminants of varying prevalence. The percentage of the contaminant may be great enough for the hazards of the contaminant to be considered as part of the hazard assessment of the product. The MSDS team needs to gather chemical composition information before beginning the hazard assessment. Hazard assessments should be updated as new chemical composition information is gathered by quality assurance, research, or other departments. Similar information from suppliers of raw materials should also be monitored as part of the ongoing hazard assessment.

## Acute toxicity

### United States

The HCS has defined the terms *toxic* and *very toxic* based on both an oral and a dermal $LD_{50}$ and also a 1-h respiratory $LC_{50}$ for vapors and particulates. Criteria for these classifications are given in Table 11.6.

The ANSI labeling standard recommends a short statement of hazard, signal word, precautionary and first aid statements for each of these classifications. In addition, consideration needs to be given to target organs of acute exposure. It is common for MSDSs in the United States to warn of absorption by dermal contact, aspiration hazard of accidental swallowing, common signs of neurotoxicty (dizziness, nausea) by any route of exposure and possible lung injury by inhalation. Examples of these effects are given in Appendix A of the OSHA HCS. These are merely examples and it is worthwhile pointing out once again that OSHA intends that hazards be considered broadly so that all important information about hazardous products is communicated. These situations are also well covered for recommendations by the ANSI labeling standard document.

### Europe

The DSD divides acute toxicity into three categories of severity: harmful (R 20–22), toxic (R 23–25) and very toxic (R 26–28). Criteria based on $LD_{50}$ and $LC_{50}$, along with the required label symbol, are listed in Table 11.6. It should be noted that inhalation classifications are based on 4-h $LC_{50}$ values. The DSD also recommends various protocols to obtain an oral $LD_{50}$ using as few animals as possible.

In addition, the DSD requires the use of R 39 (danger of irreversible effects) when serious effects are noted in the toxic or very toxic dose range and R 40 (possible risk of irreversible effects) when effects are noted in the harmful range.

Some common acute hazards were only recently addressed by the DSD. Aspiration hazard (R 65) was not addressed until the 24th adaptation of the DSD. The 25th adaptation of the DSD has defined R 67 for vapors that may cause drowsiness and dizziness.

### Canada

Acute toxicity classifications in Canada are solely based on experimental evidence of lethality by the oral, dermal and inhalation routes. The criteria with corresponding hazard classes and divisions/subdivisions are listed in Table 11.6.

### Japan

Acute toxicity to humans is a recognized hazard in Japan but no rigorous definition is given. There are lists of acutely toxic chemicals described under several laws. The lists that need to be consulted are defined under the Poisonous and Deleterious Substances Control Law, Ships and Safety Act (Annex 4), Ordinance on the Prevention of Organic Solvent Poisoning, Ordinance of Prevention of Hazards (Group 3 substances), Ordinance on Prevention of Lead Poisoning, Ordinance on Prevention of Tetraalkyl Lead

*Table 11.6* Comparison of acute toxicity hazard criteria of hazard communication regulations from the major regulatory jurisdictions.

| *Route of exposure* | *Classification* | *United States* | *Europe* | | *Canada*[a] | |
|---|---|---|---|---|---|---|
| | | *Criteria LD/LC$_{50}$ range* | *Symbol/R-phrase* | *Criteria LD/LC$_{50}$ range* | *Hazard class* | *Criteria LD/LC$_{50}$ range* |
| Oral | Very or highly toxic | $<$ 50 mg/kg | T + /R 28 | $<$ 25 mg/kg | DIA | $<$ 50 mg/kg |
| | Toxic | 50–500 | T/R 25 | 25–200 | DIB | 50–500 |
| | Harmful | N/A | Xn/R 22 | 200–2000 | N/A | N/A |
| Dermal | Very or highly toxic | $<$ 200 mg/kg | T + /R 27 | $<$ 50 mg/kg | DIA | $<$ 200 mg/kg |
| | Toxic | 200—1000 | T/R 24 | 50–400 | DIB | 200–1000 |
| | Harmful | N/A | Xn/R 21 | 400–2000 | N/A | N/A |
| Inhalation (vapor/gas) | Very or highly toxic | $<$ 200 ppm; 1 h value | T + /R 26 | $<$ 0.5 mg/l; 4 h value | DIA | $<$ 2500 ppm (gas) $<$ 1500 ppm (vapor); 4 h value |
| | Toxic | 200—2000 | T/R 23 | 0.5–2.0 | DIB | 1500–2500 (vapor) |
| | Harmful | N/A | Xn/R 20 | 2–20 | N/A | N/A |
| Inhalation (particulate) | Very or highly toxic | $<$ 2 mg/l; 1 h. value | T + /R 26 | $<$ 0.25 mg/l; 4 h value | DIA | $<$ 0.5 mg/l; 4 h value |
| | Toxic | 2—20 | T/R 23 | 0.25–1.0 | DIB | 0.5–2.5 |
| | Harmful | N/A | Xn/R 20 | 1–5 | N/A | N/A |

[a] Canadian criteria for vapor also includes that saturated vapor concentration must exceed 0.4 times the LC$_{50}$ value.

Poisoning and substances designated by the Safety and Health Department of the Ministry of Labor (MHW/MITI, 1993; OECD, 1998).

### *Recommendations for acute testing*

As noted above, the acute toxicity classifications consider only the oral, dermal and inhaled route of administration. Technically under the regulations, data obtained by the subcutaneous, intradermal, intraperitoneal, or intravenous routes can be ignored. It would be prudent, however to confirm any suspiciously toxic result obtained by any route of exposure, however, using a more occupationally relevant route such as dermal or inhalation.

The acute toxicity-testing program, in any case, should be geared to gain information by the most occupationally relevant route of exposure based on knowledge of product manufacture and use. Dermal contact is the most common occupational exposure. It is imperative that chemical products which meet the toxic definition by dermal exposure need to be identified, because many workers may assume casual skin contact to be unimportant, especially if the product is nonirritating.

It is important that inhalation data be obtained for volatile materials or products that have significant amounts of volatile constituents or contaminants. In addition, if mists of a nonvolatile liquid product or dusts from a solid product are likely to cause significant occupational exposure based on product use, inhalation testing should be considered. Species and exposure methods (whole-body, nose only, etc.) are up to the judgment of the toxicologist; however, data obtained by a consistent methodology are useful to compare products.

Although oral exposure is not a routine problem in an occupational setting, acute oral toxicity data are useful for a variety of reasons:

1 It is a useful worst-case screen for dermal testing. If the material is not toxic orally, it is not likely to be toxic dermally. Other information, such as molecular weight, could be considered before determining whether dermal testing would be prudent.
2 Classification of products by "toxicity level" found in many standard texts (Klaassen, 1986) is a useful set of data for ranking the relative hazard of chemicals. An oral $LD_{50}$ value may be the first thing noted by industrial hygienists on an MSDS of a new product at a customer's work site in assessing the relative hazard of the material.
3 Since accidental oral exposure to industrial products does occur, acute oral data is a useful resource to assess appropriate emergency treatment.
4 Some toxicologists feel acute oral data is useful in prioritizing chemicals for inhalation testing or in estimating inhalation toxicity.

## Skin irritation

### *United States*

Skin corrosion and irritation criteria are summarized in Table 11.7. A single corrosion classification is defined in the OSHA HCS. Corrosion is defined as visible destruction of, or irreversible alterations, of living tissue by chemical action at the site of contact. The definition of skin corrosion and irritation is based on results of testing in rabbits

(Draize, 1944). Reportedly, OSHA will be officially endorsing the use of an *in vitro* test system for skin corrosivity (Ballard, 2000).

OSHA chose criteria that include conditions and evaluation criteria to avoid overestimating the potential for skin irritation. The definition of skin irritation (see Table 11.1) refers to the Federal Hazardous Substances Act (FHSA) protocol (Code of Federal Regulations, 1999b) but makes important modifications in the test procedure. A single 4-h semi-occlusive exposure on intact skin is specified as opposed to the 24-h occlusive exposure in the FHSA regulations. The logic here is that a 4-h exposure is more than enough to predict a problem in an occupational setting where reasonable hygiene is being practiced. The criteria for an irritant is a Primary Irritation Index (PII) of 5 where erythema and edema are scored by the method of Draize 24 and 72 h after exposure. Because of the sensitivity of the rabbit, the literature indicates that following these OSHA definitions will lead to minimal false-negative predictions (Phillips *et al.*, 1972; Steinberg *et al.*, 1982).

However, the OSHA hazard definitions specify minimum criteria. Many industrial toxicologists use more conservative criteria. The ANSI labeling standard recommends the use of the term moderate irritant in the PII range of 2–5 and severe irritant with a PII $> 5$. The toxicologist needs to be guided by professional experience, by the history of the chemical or product at the manufacturing site, and by customers. In many cases a "may cause" warning may be prudent when the PII is less than 5. On the other hand, using the PII of 5 may work well with product lines that seem to have an absence of accidental irritation associated with them (Campbell *et al.*, 1975).

The HCS is less clear on how to handle situations where data are available in other species, including human beings. The situation is more complex when conflicting data are available on the same product in different species. Consider the case where rabbit data on a product indicates a PII of greater than 5, while the results of human skin patch testing in a large volunteer panel indicate no irritation at all. Due to the performance nature of the HCS, it would seem to be justifiable in this case to cite only the human data on the MSDS. Caution should be exerted here because instances like these have not been fully discussed by OSHA or tested in the courts. It is important that adequate records and strong documentation is retained to support these decisions.

Non-allergic dermatitis from repeated or prolonged skin contact should be considered a target organ effect in Appendix A. The examples given in Appendix A are rashes and defatting of the skin. This is usually suspected based on workplace exposures rather than on laboratory animal data. The language associated with target organs indicates that OSHA wished hazards to be addressed broadly and flexibly to assure completeness of hazard evaluation beyond the few target examples listed.

### Europe

Two corrosive categories and one irritant category are specified for skin by the DSD. These criteria are listed in Table 11.7. Animal welfare considerations have resulted in a recommendation for the use of a three-animal assay rather than the six-animal assay used in various US protocols. The criteria for irritants differ whether data is available for a three- or six-animal assay. Human experimentation is discouraged by the European Commission and European customers may express displeasure about classifications that

*Table 11.7* Worldwide skin corrosion and irritation criteria

| *Hazard* | *United States* | *Europe* | | *Canada* | |
|---|---|---|---|---|---|
| | *Criteria* | *R phrase* | *Criteria* | *Hazard class* | *Criteria* |
| Severe corrosive | N/A | C/R 35 | Full thickness destruction as result of 3 min of contact or less | N/A | N/A |
| Corrosive | Visible destruction or irreversible alteration of living tissue by chemical action at site of contact | C/R 34 | Full thickness destruction as a result of 4 h of exposure or less (but more than 3 min) | E | Full thickness destruction as a result of 4-h exposure in an OECD 404 test |
| Irritant | Primary irritation index of 5 by a test method in 16 CFR 1500.41 using 4-h exposure with semi occlusive dressing | Xn/R 38 | Six- rabbit protocol: combined average erythema or edema observed at 24, 48, 72 h of 2 or greater<br>Three rabbit protocol: average erythema or edema of 2 or greater in at least two animals | D2B | Average erythema or edema of 2 or greater in an OECD 404 test |

refer to human data. The DSD also considers organic peroxides to be skin irritants unless evidence exists to the contrary. In the 27th adaptation to the DSD, the European Commission published guidelines on the use of the *in vitro* rat skin Transcutaneous Electrical Resistance (TER) assay as an alternate method for skin corrosivity (European Commission, 2000b).

The 25th adaptation of the DSD has finally addressed non-allergic dermatitis from prolonged or repeated contact that do not meet the criteria for R 38. The language for the new classification R 66 will be "repeated exposure may cause skin dryness or cracking". Unlike the OSHA HCS and Canadian WHMIS, the 25th adaptation also now states that evidence in a repeated dose dermal irritation study can trigger R 38. This departure from criteria which judge "primary irritation", that is resulting from a single, short exposure, will certainly lead to more confusion for companies who may wish to describe skin irritation in a consistent fashion on the MSDSs for customers in different countries.

### Canada

The Canadian definitions for skin corrosive and skin irritant and corresponding hazard classes and divisions/subdivisions are listed in Table 11.7. The regulation refers to data in a rabbit test by OECD guideline no. 404. The Canadian definition for corrosion is similar to the US definition while their definition for irritation is more similar to the European criteria.

### Japan

Corrosive substances are defined as those that cause irreversible skin damage. Specific skin corrosives are listed under the Ships and Safety Act (Annex Table 3), Ordinance on Industrial Safety and Health Law, and those designated by the Safety and Health Department of the Ministry of Labor. Irritants are defined as those causing erythema, blisters or edema in human skin. No prescribed lists exist for skin irritants (MHW/MITI, 1993; OECD, 1998).

### Recommendations for skin irritation testing

The challenge of obtaining relevant animal data for predicting the potential of a chemical to cause skin corrosion and irritation in humans has been the subject of an excellent overview by McCreesh and Steinberg (1983).

The rabbit is the most prevalent animal model used, not so much because of its relevance to human experience, but more because of the ease of handling and because of the extensiveness of historical data which can be used for comparison and relative ranking. As the authors point out, the rabbit, in general, over predicts the human experience. In general, rabbit tests are best at predicting innocuous and severely irritating or corrosive materials. Moderate irritants are less predictable. As public concern for animal welfare increases, many companies are validating various *in vitro* assay systems.

Part of the confusion caused by rabbit testing is due to the wide variety of experimental procedures well documented by the same authors. Length of exposure, method of occlusion, skin preparation, method of termination of exposure and scoring periods are important factors that have contributed to variation in the historical database. The availability of data by a wide variety of methods has been encouraged by testing needed to meet various regulatory requirements. FHSA, Department of Transportation (DOT), and Environmental Protection Agency (EPA) protocols are subtly different from one another (Steinberg *et al.*, 1982). Toxicologists who need to classify products under OSHA, DSD and WHMIS criteria will find that the OECD protocol will basically meet their needs because of the use of a 4-h semi-occluded exposure with observations scheduled at 24, 48 and 72 h (OECD, 1993).

## Skin sensitization

The scientific literature on skin sensitization is full of confusing and conflicting reports. The vague regulatory definitions do not offer much help. In essence, the absolute terms *non-sensitizer* and *sensitize*r may have little relevance. Many chemicals can cause allergic reactions in a small subset of the population. Differences between weak and strong sensitizers are related to the percentage of the population susceptible and the severity of exposure needed to elicit an allergic reaction. The regulations do allow some judgment so that documentation of a single allergic individual does not necessarily mandate that a chemical be classified as a sensitizer. The regulations also require that evidence of sensitization in workplace experience be considered.

The field of skin sensitization has been well reviewed. The toxicologist must determine what test method is appropriate for product exposure patterns. The most commonly used experimental systems are the maximization assay (Magnusson and Kligman , 1969) and the topical assay (Buehler, 1965).

Both assays use guinea pigs. The former includes an intradermal treatment with adjuvant and a topical induction exposure while the latter includes 3–9 topical inductions. It has been the experience of many in the field that the use of adjuvant methods over-predicts the sensitization potential in the workplace, probably due to the severe exposure regimen in the induction phase of the assay. Where results of topical human testing have been obtained, results from Buehler topical guinea pig assays usually compare better than the maximization assay (Gad and Chengelis, 1988; Robinson *et al.*, 1989). Protocols for test batteries using traditional guinea pig methods in conjunction with human testing in a weight of the evidence approach have been defined (Robinson *et al.*, 1989).

Because of the unreliability of these assays, the subjectivity in the visual assessment of effects and because the protocol requires a long experimental period with many test animals, there has been activity in developing alternative assays. The assay that has the most favorable experience reported is the Mouse Local Lymph Node Assay (LLNA, Kimber, 1996).

### *United States*

The OSHA HCS qualitatively defines a sensitizer as "a chemical that causes a substantial proportion of exposed people or animals to develop an allergic reaction in normal tissue after repeated exposure". Due to the unreliability of typical guinea pig assays, some companies have used experimental protocols with human volunteers to discern the inductive potential of chemicals that have exhibited equivocal or weak responses in guinea pig assays. Prudent scientific judgment based on weight of the evidence is commonly accepted in the US industrial community although OSHA guidance is lacking.

The LLNA in mice has been evaluated by the National Institute of Environmental Health and Sciences (NIEHS) Interagency Coordinating Committee on the Validation of Alternative Methods (ICCVAM) and by the NTP Interagency Center for the Evaluation of Alternative Toxicological Methods (NICEATM) and has referred it to the regulatory agencies for their consideration (Federal Register, 1999a). As of this writing, OSHA has not indicated any activity on consideration of this method.

### Europe

The DSD specifies R 43 "May cause sensitization by skin contact" with a symbol of Xn (harmful) "if practical experience shows the substances and preparations to be capable of inducing a sensitization reaction in a substantial number of persons by skin contact, or on the basis of a positive response in experimental animals". A positive response in animals is defined as a minimum of 30 per cent incidence with adjuvant test methods, and a 15 per cent response by topical methods. Authorities in the EU frown upon using experiments in human volunteers to clarify a weak positive response in animals. The LLNA is also gaining popularity in Europe (ECVAM, 2000).

### Canada

Skin sensitization is considered a class D hazard, Division 2, Subdivision B. The Canadian regulations refer to OECD test protocols and use response criteria of 30 or 15 per cent depending on the method as described for European regulations above. Classification is also required if evidence shows that skin sensitization occurs following exposure in the workplace. No guidance exists as to how prevalent human reaction needs to be in the workplace before classification is required.

### Japan

The MITI/MHW guidelines do not refer to skin sensitization (MHW/MITI, 1993; OECD, 1998).

## Eye corrosion and irritation

### United States

OSHA adopted the FHSA criteria and rabbit protocols for eye irritation (Code of Federal Regulations, 1999b; 16 CFR 1500.42). The FHSA criteria are fairly conservative and only a single irritant classification exists (Table 11.8). Two responding animals of an experimental group of six in any of the Draize parameters evaluated is sufficient to consider the test material an eye irritant. A total score of 4 under the Draize *et al.* (1944) scoring system of a possible 110 can be sufficient to consider the test material an irritant. The conservative basis of this hazard class has caused little controversy due to

*Table 11.8* Current eye corrosive and irritant criteria and classifications for US OSHA, Canadian WHMIS and EU DSD

| *Hazard* | *United States* | *Europe* | | *Canada* | |
|---|---|---|---|---|---|
| | *Criteria* | *R phrase* | *Criteria* | *Hazard class* | *Criteria* |
| Eye corrosive | OSHA definition for corrosive refers to destruction of any tissue | R 34, 35 | EU criteria does not address eye corrosion, but assumes that any skin corrosive will also meet R 41 criteria | | |
| Severe irritant | N/A | R 41 | Iritis > 2. Corneal opacity > 3. Irreversible effects; positive scores do not clear by day 21 | | |
| Irritant | Four or more positive of six animals. Corneal opacity > 1. Iritis > 1. Conjuctival redness > 2. Chemosis > 2 | R 36 | Two or more positive of three animals. Corneal opacity > 2. Iritis > 1. Conjuctival redness > 2.5. Chemosis > 2 | D2B | Mean score of six animals. Corneal opacity > 2. Iritis > 1. Conjuctival redness > 2.5. Chemosis > 2.5 |

the commonness of eye irritants and the routine use of eye protection in the workplace in the US. This conservative hazard definition is, however, a main reason that very few chemicals can be considered non-hazardous under HCS.

Eye corrosion is covered by the general definition of *corrosive.* However, there is no requirement under the HCS to describe the severity of eye irritation between corrosive and irritant. Due to the vast number of chemicals that meet the definition of eye irritant, there is the danger that over-warning will result in an indifferent worker attitude. The prudent toxicologist will prefer to give some indication of severity to emphasize the need for protective eyewear. Appendix A of the HCS does require that any irreversible damage observed in an ocular test needs to be noted as required under target organ effects. Any reasonable severity scale, consistently applied, will enhance accurate communication of relative hazard. The definition of a severe eye irritant in the EU DSD or the guidance in the ANSI labeling standard is a useful voluntary tool to be considered for communication in the US.

### *Europe*

There are two eye irritant classifications under the DSD (see Table 11.8). R 41 is a classification for severe or irreversible effects. The other classification for irritants is R 36. In order to discourage eye irritation testing of corrosive material in animals, the DSD automatically considers skin corrosives as eye hazards. Under the criteria for R 36, a significantly greater extent of inflammation is necessary than under OSHA criteria. Therefore, fewer chemicals are classified as eye irritants in Europe than in the US.

### *Canada*

A single irritant classification is described by Canadian regulations. The criteria are similar to the criteria of the EU, although a somewhat greater level of conjuctival chemosis is needed to be considered an eye irritant (Table 11.8).

### *Japan*

Eye irritation is loosely defined as clouding, abnormalities of the iris, conjunctivitis or conjunctival edema under the MHW/MITI guidelines (MHW/MITI, 1993; OECD, 1998).

### *Recommendations for eye testing programs*

Due to animal welfare considerations, there is much interest in new *in vitro* systems and much care is commonly exerted in traditional eye testing with rabbits. Many follow protocols that start with observations of a single animal. If corrosive or irreversible effects are observed, no further animals are tested. In order to meet DSD criteria for R-41, it is recommended that in studies where moderate to severe irritation is observed, that the studies be extended up to 14 days to determine if lesions will clear.

## Respiratory irritation and sensitization

The lack of a specific and quantitative definition for respiratory irritation and respiratory sensitization reflects the lack of validated models to measure this effect. Although experimental systems are available to study respiratory irritation (Bos *et al.*, 1992; Draize *et al.*, 1944), most toxicologists and rely on occupational experience for existing chemicals and laboratory experience from research chemists on new chemicals regarding the potential for respiratory irritation. Experimental procedures for respiratory sensitization are also available (Karol *et al.*, 1981) but have not been sufficiently validated for regulatory adoption. Most chemicals have not been screened by such procedures due to lack of perceived inhalation exposure, cost and variability of the assay, and the apparent lack of workplace problems.

Respiratory sensitization properties seem to be limited to only a few classes of chemicals, such as isocyanates, anhydrides and certain metal compounds. The toxicologist needs to consider exposure, occupational experience, and chemical similarity to known respiratory sensitizers in order to decide whether testing is prudent.

None of the worldwide hazard communication regulations recommend a specific protocol for laboratory animal studies for either of these respiratory hazards. It is interesting to note that none of the mandatory testing for any of the chemical control laws discussed in Chapter 10 address either of these two hazards.

### *United States*

Respiratory irritation classifications are fairly common under the HCS since many chemicals that have had Threshold Limit Value (TLVs) set by ACGIH are based on concentrations that irritate mucous membranes in the workplace. Again the general definition of corrosion applies if irreversible effects can occur. Although experimental criteria are defined for eye and skin irritants, respiratory irritation is covered by reference to non-corrosive inflammation of any tissue. Respiratory irritation is mentioned also in the examples of target organs in Appendix A.

Respiratory sensitization is defined empirically under the HCS. If a substance is found to cause an effect in a significant number of animals or humans, then the hazard category applies. No experimental protocols are specified in the regulation. Choice of experimental system is entirely up to the discretion of the toxicologist.

Knowledge about both of these respiratory hazards largely relies on workplace experience for classification.

### *Europe*

"Substances and preparations which cause serious irritation to the respiratory system, based on practical observation" are defined as respiratory irritants (Xn, R 37). Respiratory irritation data on experimental animals is not recognized by the DSD. In practice, only chemicals which have been listed by the regulatory authorities as R 37 in Annex I are found to be classified as respiratory irritants on European labels and MSDSs. There are far fewer chemicals labeled as R 37 in Europe than have a label warning for respiratory irritation in the US.

Respiratory sanitizers are classified as Xn/R 42. The definition refers to practical experience in the workplace. Isocyanates are considered as respiratory sensitizers by definition, unless there is evidence to the contrary. Little voluntary testing to screen unknown chemicals for respiratory sensitization seems to occur. Again there is a heavy reliance on Annex I to identify respiratory sensitizers.

### *Canada*

Respiratory tract sensitizers are defined by evidence in the workplace and classified as D2A.

Curiously, a definition for respiratory irritation is missing from Canadian regulations.

### *Japan*

Respiratory irritation and sensitization are not addressed by the MHW/MITI guidelines (MHW/MITI, 1993; OECD, 1998).

## Target organs

### *United States*

The language in Appendix A of the HCS reiterates OSHA's intent to allow much latitude in the identification of hazards. It was clearly OSHA's intent that important or unusual hazards not be missed by defining hazards too prescriptively. Target organs are defined by a set of examples that are not meant to be all-inclusive. These examples are given in Table 11.2.

Exposure length to the agent is deliberately not specified. Any effect resulting from acute, subacute, subchronic, or chronic exposure needs to be considered. It is important that all sources of data are considered, even if no studies were designed specifically to evaluate target organs. Examples of data that can be overlooked are:

1 Toxicology studies conducted for a regulatory approval, such as for a food additive petition or TSCA PMN.
2 Definitive organ effects observed in an acute study. Acute studies need to be evaluated beyond the recognition of a lethal or toxic dose.
3 Epidemiological studies and occupational health investigations.
4 Information on supplied chemicals that may appear on the MSDS or in the published literature.

Target organ data need to be critically evaluated to determine whether a true occupational hazard exists. As indicated earlier, scientific judgment must be used to differentiate between a complete hazard description for inclusion on the MSDS and those descriptions that are occupationally relevant or appropriate for a label warning. OSHA guidance concedes that evaluation of animal studies obviously requires subjective, professional judgment, especially in regard to whether particular observations indicate an adverse effect.

Factors to be legitimately considered for occupational relevance include:

1 The degree and specificity of the effect on the organ observed in the study. For instance, a mere increase in liver weight usually indicates a metabolic adaptive response rather than toxicity.
2 The relevance of the experimental route of exposure to the occupational exposure. Effects observed in oral studies should be evaluated depending on the experimental doses and the relevance to dermal absorption or inhalation exposure levels expected.

Hazard warnings must be specific. "Harmful if inhaled," for instance, is considered an inadequate warning by OSHA if a particular organ is known to be affected due to inhalation exposure (OSHA, 1990).

For additional guidance in interpreting repeated dose studies, toxicologists may wish to consider using the European or Canadian criteria shown in Table 11.6 that specify dose levels, study duration and seriousness of effect as voluntary guidance.

### *Europe*

Various R phrases apply to organ effects. The European criteria are much clearer about which apply to acute or repeated exposure. The phrases R 39 and R 40 apply to acute exposure (see section on acute toxicity above). The phrases R 33 and R 48 are reserved for delayed or cumulative effects (see Table 11.2). Effects needed to trigger R 48 are more serious than to trigger R 33. The detailed criteria for effect levels in animal studies necessary to trigger R 48 are listed in Table 11.9. In order to designate the route of exposure of concern, R 48 is combined with one or more statements (R 20–28). Again most European manufacturers rely on Annex I to designate substances that require R 33 or R 48.

### *Canada*

Canadian criteria for organ effects are also very detailed and listed in Table 11.9. They address only effects from repeated exposure.

### *Japan*

The MHW/MITI guidelines do not refer to target organ toxicity (MHW/MITI, 1993; OECD, 1998).

### *Recommendations for test programs*

Many manufacturers have little data from repeated-dose studies. Due to the cost and length of time needed to conduct these studies, prospective toxicology programs should be implemented in a phased approach. Priorities should be set on volume of manufacture, occupational exposure, product uses, volatility, and likelihood of systemic absorption. Route of exposure should mimic occupational experience. Care should be exerted before using protocols or procedures that have not been validated to evaluate endpoints such as neurotoxicity or immunotoxicity. Study of such endpoints will, however, be more common in the future due to regulatory interest under other laws such as those discussed in the previous chapter.

## Cancer

Many authors and organizations have suggested criteria for reviewing and classifying carcinogens and this remains a controversial subject (Ashby *et al.*, 1990; Federal Register, 1996). The major hazard communication regulations refer to carcinogen lists of various expert bodies for chemicals that are required to be subject to cancer warnings. The lists invoked by US, European and Canadian requirements are listed in Table 11.10.

Most companies do not generate carcinogenicity data on many of their chemicals. Data should be acquired as needed to support expanded product uses, to follow up on suspicions, or as dictated by priority-setting schemes.

The experimental route of exposure should be carefully considered in relation to use of the chemical, occupational exposure, and other relevant properties of the chemical. Exposure route can affect the carcinogenic potency observed (Pepellco, 1991. In designing a study for carcinogenicity to pursue an occupational concern, the toxicologist is theoretically free to design a protocol that best suites needs and does not need to be locked into a certain guideline. In view of costs of systemic carcinogenicity studies and the prevalence of dermal exposure in the workplace, skin-painting studies should be considered (EPA, 1988).

### *United States*

The definition of *carcinogen* found in the Appendix A of the HCS is more a legal definition than a scientific one. The definition incorporates all chemicals considered to be carcinogenic by the IARC (IARC, 2000) or the NTP (NTP, 2000) report on carcinogens, as well as any other chemical regulated by OSHA in other sections of Title 29 as a carcinogen. In addition, chemicals which have not been evaluated by these experts but which have sufficient animal or human data as described in Appendix B of the HCS should also be considered carcinogens.

*Table 11.9* Dosage criteria for subchronic studies under the EU DSD and Canadian WHMIS for classifying delayed effects similar to the target organ requirements under the OSHA HCS

| *Route of administration* | *Europe*[a] | | *Canada*[b] | |
|---|---|---|---|---|
| | *Classification* | *Criteria (LSEL)*[c] | *Classification* | *Criteria (LSEL)* |
| Oral | T/R 48 | < 5 mg/kg | D2A | < 10 mg/kg |
| Dermal | T/R 48 | < 10 mg/kg | D2A | < 20 mg/kg |
| Inhalation | T/R 48 | < 0.025 mg/l | D2A | < 0.01 mg/l |
| Oral | Xn/R 48 | 5–50 mg/kg | D2B | 10–100 mg/kg |
| Dermal | Xn/R 48 | 10–100 mg/kg | D2B | 20–200 mg/kg |
| Inhalation | Xn/R 48 | 0.025–0.25 mg/l | D2B | 0.01–0.1 mg/l |

[a] The criteria listed for the European DSD are for studies of 90 days or longer in duration. For 28-day studies commonly run for European PMN requirements, the DSD specifies that dose ranges 3-fold higher should be used to judge effects.

[b] The criteria listed by the Canadian WHMIS regulations are for studies of duration of 90 days or longer. No guidance is given for studies of shorter duration.

[c] LSEL means lowest effect level where severe effects are observed.

*Table 11.10* Carcinogen lists incorporated into the definition of a carcinogen under US OSHA HCS, EU DSD and Canadian WHMIS requirements

*United States*
International Agency for Research on Cancer (IARC).
Substances classified as group 1, 2A and 2B
National Toxicology Program (NTP) annual report on carcinogens
Substances otherwise regulated by OSHA as carcinogens

*Europe*
Substances classified as R 45, R 49 and some substances classified as R 40 in Annex I

*Canada*
American Conference of Government Industrial Hygienists (ACGIH)
Substances classified as group A1a, A1b and A2
International Agency for Research on Cancer (IARC)
Substances classified as group 1, 2A and 2B

Although the incorporation by reference of these "carcinogen lists" does add some clarity to which chemicals should be considered carcinogenic, this feature does add some difficulties for the following reasons:

1 IARC and NTP use different criteria for determining whether a chemical is carcinogenic.
2 The expert bodies do not limit their review to occupational exposure and are usually not clear about the types of exposure by which the carcinogenic hazard would be significant.
3 Most importantly, the listing of carcinogens is not subject to regulatory or judicial review. Even if an expert body errs or delays delisting a chemical, these independent organizations cannot be forced to correct or even address the error.

Although the HCS exempts food from MSDS and label requirements, premixes used in workplaces are not. Although many toxicologists have doubted the carcinogenicity of saccharin for almost two decades, it was only recently removed from the NTP report on carcinogens. Therefore, products containing saccharin had been considered hazardous products in the workplace under HCS. MSDSs and labels with cancer warnings were required on such materials as syrups for diet beverages that contained saccharin to which food service workers may be exposed. These questionable cancer warnings were required to be continued as long as the expert body listed saccharin.

The only recourse in preparing MSDSs in these situations is to present the refuting data as allowed by Appendix B. Besides saccharin, the number of "listed" carcinogens with significant mechanistic and pharmacokinetic data indicating low carcinogenic hazard is increasing. In the OSHA Inspection Procedures Compliance Guide (OSHA, 1986, 1990), some relief is given in that IARC 2B carcinogens, unlike 1A and 2A, need not have carcinogenic label warnings. Recent OSHA guidance has clarified the treatment of listed carcinogens under the new NTP criteria (Federal Register, 1996). The guidance indicates that those substances "known to be carcinogenic" or "reasonably anticipated to be carcinogenic" are to be considered carcinogenic under the OSHA HCS (OSHA, 1998).

OSHA guidance also clarifies how the certain "Exposure Circumstances" in the IARC monographs and NTP report on carcinogens should be handled under the

OSHA HCS. The example in the guidance refers to the consumption of alcoholic beverages listed by both IARC and NTP as known to be carcinogenic in humans. The guidance states that the listing certainly should result in a conclusion that ethanol is a hazardous substance, but because ingestion is not a route of exposure anticipated in the workplace, industrial products containing ethanol would not be required to have a carcinogen warning (OSHA, 1998).

Regarding data on chemicals not reviewed by expert bodies, the Appendix B (29 CFR 1910.1200) advice on animal data, human data, and statistical significance should be used as a guide. Traditionally, toxicologists have had trouble dealing with conflicting data. Appendix B does recognize the need for the use of scientific judgment. Although clarifying information has not been forthcoming from OSHA or the courts, most toxicologists are moving to a weight of the evidence approach (Ashby *et al.*, 1990).

### Europe

Carcinogens are those substances listed as R 45, R 49 and some of the substances listed as R 40 (R 40 also applies to some mutagens and certain effects observed on acute exposure) on Annex I. R 45 is used for category 1 and 2 carcinogens. R 49 is used for carcinogens that present a carcinogenic risk only by the inhalation route. R 40 is used for category 3 carcinogens.

The criteria for these categories is as follows:

1 Category 1: substances known to be carcinogenic to man.
2 Category 2: substances that should be regarded as carcinogenic to man; those with sufficient evidence to provide a strong presumption of carcinogenicity in humans.
3 Category 3: substances which cause concern of possible carcinogenic effects; those with some evidence in animals but insufficient to classify as category 2.

Paragraph 4.2.1 of the DSD requires manufacturers to provisionally classify the material if they have evidence that any of the above criteria are met. The manufacturer is required to submit the new information to the competent authority of the member state in which the substance is introduced into commerce so that the commission might update the classification of the substance.

### Canada

The WHMIS regulations incorporate substances considered A1a, A1b or A2 carcinogens by ACGIH (ACGIH, 2000) and Group 1 and 2 carcinogens as determined by the IARC (IARC, 2000).

### Japan

Carcinogenicty is a prescribed hazard under the MITI/MHW guidelines. Several carcinogens are listed in Groups 1 and 2 of the Ordinance on Prevention of Hazards due to Specified Substances (MHW/MITI, 1993; OECD, 1998).

## Teratology and reproductive toxicity

Fortunately, not many chemicals have been indicated to have the potential to cause reproductive harm in the workplace. Due to the low number of occupational incidents, known reproductive hazards are apparently well controlled in the workplace.

When necessary, there are a variety of test protocols and strategies available to the toxicologist. In vitro screening systems are more available for use as a priority-setting tool (Collins, 1987).

Evaluation of reproductive and teratology data is one of the more difficult tasks of a toxicologist. Actually, the diversity of conclusions toxicologists can make from the same set of data has been controversial. The influence of maternal toxicity and the danger of false-positive conclusions have been eloquently discussed (Johnson, 1987; OECD, 1998). See Table 11.5 for the proper reference to HCS, DSD and WHMIS definitions.

### *United States*

Teratogens and reproductive toxins are loosely defined in Appendix A of the HCS (Table 11.2). Toxicologists must use professional judgment when reviewing scientific evidence on these effects to determine the potential for hazard under workplace conditions (OECD, 1998). Further guidance is available in the ANSI labeling standard or substances classified by other jurisdictions such as the EU. Sowinski *et al.* (1987) offer criteria for classification.

### *Europe*

The DSD indicates the use of R 60, R 61, R 62 and R 63 for certain reproductive effects (see Table 11.5). R 61 or R 63 can also be indicated for teratogenic effects. Substances are classified as harmful (Xn) or toxic (T) in Annex I depending on the strength of the evidence. Unlike criteria for delayed effects on target organs, the European Commission has not established dose levels in animal studies at which a reproductive or teratogenic effect would be considered to be sufficiently serious to warrant classification. As with other hazards, there is heavy reliance on Annex I listing for these effects as opposed to industry scientists weighing the evidence under the US system.

### *Canada*

Reproductive and teratogenic effects are classified as D2A by WHMIS regulations. The definitions apply to adequate evidence of occurrence of death or other serious malformations or dysfunction of the embryo or fetus or loss of fertility or other serious reproductive adverse effects from workplace experience or appropriate animal studies. Detailed criteria for judging animal data are not given.

### *Japan*

Effects on human reproductivity, growth of the fetus and abnormalities are considered prescribed hazards under the MHW/MITI guidelines, but again without detailed criteria (MHW/MITI, 1993).

## Genotoxicity

Genotoxicity and mutagenicity data are useful information to predict more serious hazards of chemicals. The HCS, DSD and WHMIS regulations differ widely on how this data can trigger a hazard classification.

### *United States*

Chromosomal damage and mutations are listed only as an example of reproductive effects under HCS. OSHA does not address the issue of genotoxicity as a predictive test for carcinogenicity. OSHA guidance indicates that *in vitro* studies, such as an Ames bacterial mutagenicity test, are useful pieces of information but not definitive findings of hazard (OSHA, 1998).

Genotoxicity tests have been the subject of many reviews (Mason *et al.*, 1991). The regulations and guidance materials do not discuss potency of genotoxic effects or differences between results of *in vivo* or *in vitro* tests. Small but statistically significant effects from *in vitro* tests are problematic, especially in the absence of *in vivo* toxicology or pharmacokinetic data. Scientific judgment needs to guide the toxicologist here. One possible source of sensible judgment can be found in European and Canadian regulations or in Sowinski *et al.* (1987). These references cite the importance of *in vivo* data. Certainly, if a product with a high workplace exposure were found to be a "hot" genotoxicant *in vitro*, warnings about potential cancer as well as birth defects may be prudent in the absence of *in vivo* data.

### *Europe*

The criteria in the DSD refer to *in vivo* data. Three categories of classification are possible. T/R 46 is used for substances in categories 1 and 2 and Xn/R 40 is used for category 3 substances.

### *Canada*

WHMIS criteria differentiate between germ cell and somatic cell effects observed in *in vivo* tests. Germ cell mutagens are classified as D2A. Substances that cause somatic cell mutations or chromosome aberrations are classified as D2B.

### *Japan*

The MITI/MHW guidelines describe mutagenicity as a prescribed hazard. The Japanese guideline is unique among worldwide definitions as it states that strong mutations observed *in vitro* are covered. The Ministry of Labor maintains lists of new and existing substances that are defined as mutagens (MHW/MITI, 1993).

## Evaluation of untested mixtures

The major hazard communication regulations provide fairly specific guidance about hazard classifications of untested mixtures. Many of these requirements are very conservative, but some sections are overly simplistic and seem to overlook possible complexities of hazards arising due to chemical or biological interactions. The rules on untested mixtures are essential, however, so that all data on components are considered when evaluating untested mixtures. The various percentage cut-offs or thresholds specified by the HCS, DSD and WHMIS requirements are summarized in Table 11.11. Due to the complexity of these regulations, however, new practitioners in the field are encouraged to read the details of the regulations on untested mixtures.

### United States

The HCS conservatively specifies that untested mixtures are assigned the same hazards of any components contained at greater than 1 per cent, with the exception of carcinogens, where the threshold of 0.1 per cent is assigned. The rule does not specifically mention synergy or antagonism between components, or even the possibility of additive hazards contributed by components that are similar in chemical structure. However due to the performance nature of the HCS, any known interactions would be expected to be considered in the hazard evaluation of a mixture.

Although this rule is not overly conservative in all cases and may not be sufficiently conservative in a few cases such as strong sensitizers, it is certainly overly conservative in a number of examples:

1 *Corrosives*. The HCS defines irritation and corrosion as separate hazards. However, corrosion should be considered the most serious hazard on the continuum of nonirritant–mild irritant–severely irritant–corrosive. Dilution studies show that most corrosives are merely irritants when tested at concentrations of 5, 10, or even as high as 50 per cent in a nonirritating vehicle.
2 *Toxic*. Similarly, the 1 per cent rule for acute toxins is overly conservative, depending on the $LD_{50}$ and $LC_{50}$. Certainly, the $LD_{50}$ of a 10 per cent aqueous solution of a toxic chemical that has an $LD_{50}$ of 100 mg/kg would be 1,000 mg/kg and, therefore, would not be considered toxic by scientific principles.

Of course, the more complex the mixture, especially mixtures of two or more irritating or toxic components, the less certain one can predict the hazards of mixtures and therefore the conservatism of the regulations is more justified.

Model studies and threshold studies may be useful to predict irritation and sensitization hazards more accurately than by using the 1 per cent rule (Steinberg *et al.*, 1982). In addition, as data are acquired on similar mixtures, untested mixtures can be looked on as mixtures of tested mixtures rather than mixtures of single components.

Due to the lack of rigorous definition of a chemical substance in the OSHA HCS, creative approaches to addressing the untested mixture rule may be appropriate applications of scientific judgment. For instance, consider the composition of mixture 1, which is a mixture of equal parts of four ingredients A, B, C, and D. If component A is an eye irritant, mixture 1 needs to be considered an irritant if mixture 1 is not tested for eye irritation. However, if mixture 1 is tested in rabbits and found not to be an eye irritant, then the data on component A are superseded by data on the mixture. If a new product is developed, mixture 2, which is equal parts of A, B, C, and D, as well as new component E, which is not irritating, mixture 2 would need to be considered an eye irritant when viewing it as a mixture of the single components A, B, C, D, and E, because it contains 20 per cent of the irritant A.

However, mixture 2 can also be looked upon as a mixture of 80 per cent nonirritating mixture 1 and 20 per cent nonirritating component E. Viewing the composition of a mixture at different conceptual levels can lead to more accurate hazard classification information that is both scientifically valid and legally defensible.

Additional data used in creative interpretation of the hazards of mixtures should be well documented. Reasonable scientific judgment should be accepted by OSHA inspectors and federal judges. Data on similar product mixtures, model mixtures, and threshold testing of hazardous components may be useful to document creative hazard evaluation of mixtures.

A word of caution: the OSHA guidance specifies that the level of testing on mixtures should be the same as the level of testing on the components (OECD, 1998). This works better for acute hazards than chronic hazards. If only rabbit testing for irritation was conducted on components, for example, then rabbit testing alone on the mixtures is sufficient. However, if the irritating properties of components have been confirmed on humans, then rabbit data alone may not be sufficient to establish lack of irritant properties of a mixture.

In cases of well-studied components, especially for chronic hazards, the hazard is almost impossible to refute by testing mixtures. For instance, the liver toxicity of xylene has been described by oral, dermal, and inhalation studies in a variety of test animal species and exposure periods. Refuting the liver toxicity for a mixture that contains xylene in a single dermal study in rats would not be considered good scientific judgment.

An important "performance" exception is spelled out in the HCS. The 0.1 and 1 per cent rules do not apply if data are available to indicate that exposure under workplace conditions of any chemical contained in the mixture, regardless of concentration, will exceed the PEL, TLV, or Short-Term Exposure Limit (STEL).

### *Europe*

The DPD lists rather complex criteria for untested mixtures. Table 11.11 attempts to summarize the most important criteria, but the practitioner is encouraged to refer directly to the text of the DPD in order to understand the nuances and complexity of these requirements. The DPD is unique in that it considers the effects of acutely toxic materials, corrosives and irritants to be additive. The legend to Table 11.11 gives the appropriate citations in the DPD for this information. Several mathematical formulas are given to determine if components of various toxic or irritant potency are present at sufficient concentration to trigger a hazard classification for the preparation or mixture. Although the DPD does allow testing in some cases on the mixture or preparation itself, the regulations also specify in great detail limited circumstances under which data on a mixture can be applied to a similar mixture.

### *Canada*

Although the WHMIS hazard definitions more closely resemble the EU DSD, Canadian regulators chose to mimic the simpler OSHA HCS percentage thresholds for hazards of components of mixtures. Canada specifies either 0.1 or 1 per cent thresholds

*Table 11.11* Percentage thresholds for health hazards of untested mixtures specified in the OSHA HCS; the EU DPD, Annex I and Canadian WHMIS requirements[a]

| *Hazard* | *United States* | *Europe*[b] | | *Canada* | |
|---|---|---|---|---|---|
| | *Percentage threshold* | *R phrase* | *Percentage threshold* | *Hazard class* | *Percentage threshold* |
| *Health effects* | | | | | |
| Acute toxicity (lethality) | | | | | |
| Very toxic | 1.0 | T + /26, 27, 28 | T + > 7; T: 1–7; Xn: 0.1–1 | D1A | Formula[c] |
| Toxic | 1.0 | T/23, 24, 25 | T > 25; Xn: 3–25 | D1B | Formula |
| Harmful | N/A | Xn/20, 21, 22 | Xn > 25 | N/A | N/A |
| Acute toxicity (symptoms) | 1.0 | T + /39 | T + > 10 T: 1–10; Xn: 0.1–1.0 | N/A | N/A |
| | | T/39 | T > 10; Xn: 1–10 | | |
| | | Xn/39 | Xn > 10 | | |
| | | 40 | 10 | | |
| | | 65 | Not yet assigned | | |
| | | 67 | Not yet assigned | | |
| Skin corrosion | 1.0 | 34 | 10; R 36/38: 5—10 | E | 1.0 |
| | | 35 | 10; R 34 > 5; R 38 > 1 | | |
| Skin irritation | 1.0 | 38, 66 | 20 | D2B | 1.0 |
| Eye irritation | 1.0 | 36 | 20 | D2B | 1.0 |
| | | 41 | R 41 > 10; R 36: 5–10 | | |
| Skin sensitization | 1.0 | 43 | 1.0 | D2B | 1.0 |
| Respiratory irritation | 1.0 | 37 | 20 | N/A | N/A |
| Respiratory sensitization | 1.0 | 42 | 1.0 | D2A | 0.1 |
| Delayed effects | 1.0 | 33 | 1.0 | D2A | 1.0 |
| | | T/48 | T > 10; Xn: 1–10 | D2B | 1.0 |
| | | Xn/48 | 10 | | |
| Carcinogenicity | 0.1 | 40 | 1.0 | D2A | 0.1 |
| | | 45, 49 | 0.1 | | |
| Teratogenicity | 1.0 | 61, 63 | T > 0.5; Xn > 5 | D2A | 0.1 |

Table 11.11 (*continued*)

| *Hazard* | *United States* | *Europe*[b] | | *Canada* | |
|---|---|---|---|---|---|
| | *Percentage threshold* | *R phrase* | *Percentage threshold* | *Hazard class* | *Percentage threshold* |
| Reproductive effects | 1.0 | 60, 61, 62, 63 | T > 0.5; Xn > 5 | D2A | 0.1 |
| Mutagenicity | 1.0 | 40 | 10 | D2A | 0.1 |
| | | 46 | 0.1 | D2B | 1.0 |

[a] Example of how to read this table: for substances which are considered by the EU to be very toxic (T + ) with R phrase 26, 27 and/or 28, any mixture which contains more than 7 per cent of the substance would be considered very toxic (T + ), any mixture which contains between 1 and 7 per cent would be considered toxic (T), and any mixture which contains 0.1–1.0 per cent would be considered harmful (Xn). Appropriate R phrases would also be required.

[b] In the EU the hazards of acute toxicity, corrosion and irritation are considered by the Preparations Directive to be additive. (1) Where more than one acutely toxic material is contained in a mixture consult Article 3, Paragraph 5 (a), (b) and (c). (2) Where more than one corrosive or irritating substance is contained in a mixture consult Article 3, Paragraph 5(d), (e), (f), (g), (h) and (i).

[c] In WHMIS regulations the formula for calculating the $LD_{50}$ or $LC_{50}$ of a mixture is given in Section 45.

for hazards of components (see Table 11.11). One creative exception is that for acute lethality, a formula is provided in Section 45 of the regulations to estimate the $LD_{50}$ or $LC_{50}$ from data on components.

### *Japan*

No general guidance is given on untested mixtures under the MHW/MITI guidelines. Lists of substances under the Ordinance on Prevention of Hazards due to Specified Substances specify a percentage composition where a product is considered hazardous (MHW/MITI, 1993; OECD, 1998).

## The toxicologist as a communicator

MSDSs have a wide audience comprised of various levels of expertise and education. The main audience is the ordinary worker who handles the chemicals. Industrial hygienists and other toxicologists will also read MSDSs. Therefore, the toxicologist on the MSDS writing team will need to consider the entire audience spectrum. Ideally, the toxicologist needs to communicate simply for the worker and completely for the health and safety professional. Consideration should be given to using summary bullets with simple language, then more detailed explanations in following paragraphs or later pages of the MSDS. There is much literature on communication techniques for hazardous chemicals. A good source for more information is a report on the OSHA web site carried out by the University of Maryland (Environment Health and Education Center, 1997).

## International aspects

Based on the differences of hazard definitions and percentage thresholds for hazardous components of mixtures, MSDS writing teams which work for multinational companies certainly have a challenge in providing MSDSs which meet all national requirements while achieving a level of consistency in warnings to all customers. Legal consultation is advisable before establishing compliance programs. All options are problematic. A worst-case MSDS using the most conservative hazard definition may cause customer perception problems. Separate MSDSs for each jurisdiction using each local requirement would cause problems with multinational customers who would receive different MSDSs on the same product. This may also cause product liability problems in the US.

The need for a GHS was endorsed by the 1992 United Nations Conference on Environment and Development (UNCED). The harmonization process is being coordinated by the OECD with the cooperation of representatives of regulatory agencies from OECD member countries as well as the United Nations' Committee on the Transport of Dangerous Goods (UNCETDG) and the International Labor Organization (ILO). The OECD process is attempting to tackle the task of harmonizing health, environmental and physical hazard definitions. The criteria that have been agreed to at the time of this writing for acute toxicity are listed in Table 11.12 and for all other hazard classifications reviewed so far in Table 11.13 (OECD, 1998). Detailed white papers on each of these hazards are available on the OECD web site. In many cases the OECD work groups were able to agree on compromise criteria for acute hazards. A notable advance made for many hazards was that tiers of severity were suggested. The OECD report and white papers are an excellent source of more detailed information on hazard communication issues.

As this chapter was going to press, final drafts were submitted to discussion participants on a two class system for organ toxicants and percentage thresholds of hazardous components in untested mixtures. Before a harmonized system can be implemented worldwide and end the confusion of different hazard communication requirements, existing legislation will need to be amended adopting the OECD definitions. This will certainly be a daunting task.

## Communication of environmental hazards

While the OSHA HCS and WHMIS requirements do not include environmental data, the EU MSDS directive requires environmental toxicity and environmental fate data. Additionally, the EU DSD provides the authority to classify materials based on environmental hazards. Transport regulations that also classify materials based on environmental hazards include: IMDG (international sea transport), ADR/RID (European land transport), DOT (US land transport) and TDG (Canadian land transport).

The DSD addresses two types of environmental information, environmental toxicity and environmental fate. Use of the sixteen-section format ANSI MSDS standard

*Table 11.12* Schematic presentation of the integrated hazard classification system for human health and the environment (reprinted from OECD, 1998)[a]

| *Endpoint* | *Hazard classes and criteria* | | | | |
|---|---|---|---|---|---|
| *Acute toxicity* | *Class 1* | *Class 2* | *Class 3* | *Class 4* | *Class 5* |
| Oral (mg/kg) | 5 | 50 | 300 | 2000 | 5000 (or equivalent doses for other routes) |
| Dermal (mg/kg) | 50 | 200 | 1000 | 2000 | Criteria: indication of significant effect in human; any mortality at Class 4; significant clinical signs at Class 4; indications from other studies |
| *Inhalation*[b] | | | | | |
| Gas (ppm) | 100 | 500 | 2500 | 5000 | |
| Vapor (mg/l)[c,d] | 0.5 | 2.0 | 10 | 20 | |
| Dust/mist (mg/l 4 h)[e] | 0.05 | 0.5 | 1.0 | 5 | |

[a] For the convenience and comparison of the various endpoints, the scheme and criteria for classifying each toxic end-point are presented in the following diagram. The criteria have been drastically abridged and the end-point chapters must be consulted for the specific details to avoid misunderstanding.

[b] Inhalation cut-off values are based on 4-h testing exposures. Conversion of existing inhalation toxicity data that has been generated according to 1-h exposures should be divided by a factor of 2 for gases and vapors and 4 for dusts and mists.

[c] Saturated vapor concentration may be used as an additional element to provide for specific health and safety.

[d] For some chemicals the test atmosphere will not just be a vapor but will consist of a mixture of liquid and vapor phases. For other chemicals the test atmosphere may consist of a vapor that is near the gaseous phase. In these latter cases, classification should be based on ppm as follows: Class 1 (100 ppm), Class 2 (500 ppm), Class 3 (2,500 ppm), Class 4 (5,000 ppm).

[e] The values for dusts and mists should be reviewed to adapt to any future changes to OECD test guidelines with respect to technical limitation in generating, maintaining and measuring dust and mist concentrations in respirable form.

*Table 11.13* OECD criteria suggested for harmonization of hazard definitions on irritation, sensitization, germ cell mutagenicity, carcinogenicity, reproductive toxicity, and aquatic toxicity (OECD, 1998).

| *Endpoint* | *Class* | *Criteria* |
|---|---|---|
| Dermal irritation/corrosion | 1A | Destruction of tissue observed within 1 h after 3 min exposure |
| | 1B | Destruction of tissue observed within 14 days after a 1-h exposure |
| | 1C | Destruction of tissue observed within 14 days after a 4-h exposure |
| | 2 | Reversible adverse effects. Mean Draize score for erythema or edema 2.3–4.0 |
| | 3 | Reversible adverse effects. Mean Draize score for erythema or edema 1.5–2.3 |
| Eye irritation/corrosion | 1 | Irreversible damage to cornea, iris, conjunctiva within 21 days of exposure in at least one animal. Mean Draize score in two of three animals. Corneal opacity $>$ 3, iritis $>$ 1.5 |
| | 2A | Mean Draize score in two of three animals. Corneal opacity 1–3, iritis 1–1.5, Rrdness $>$ 2, chemosis $>$ 2 which are reversible in 21 days |
| | 2B | Mean Draize score in two of three animals. Corneal opacity 1–3, iritis 1–1.5, redness $>$ 2, chemosis $>$ 2 which are reversible in 7 days |
| Respiratory sensitization | 1 | Evidence of specific respiratory hypersensitivity in humans or positive results from animal tests |
| Dermal sensitization | 1 | Evidence of specific respiratory hypersensitivity in humans or positive results from animal tests |
| Germ cell mutagenicity | 1A | Positive evidence from epidemiological studies |
| | 1B | Positive results in:<br>*In vivo* heritable germ cell tests in mammals<br>Human germ cell tests<br>*In vivo* somatic mutagenicity tests combined with some evidence of germ cell mutagenicity |
| | 2 | May induce heritable mutations in human germ cells<br>Positive evidence from tests in mammals and somatic cell tests<br>In vivo somatic genotoxicty supported by *in vitro* mutagenicity |
| Carcinogenicity | 1A | Known human carcinogen based on human evidence |
| | 1B | Presumed human carcinogen based on demonstrated animal carcinogenicity |
| | 2 | Suspected carcinogen or limited evidence of human or animal carcinogenicity |
| Reproductive toxicity | 1A | Known human reproductive or developmental toxicant |
| | 1B | Presumed human reproductive or developmental toxicant |
| | 2 | Suspected human reproductive toxicant |
| | Additional | Effects on or via lactation |
| Acute aquatic toxicity | 1 | $LC_{50} < 1$ mg/l |
| | 2 | $LC_{50}$ 1–10 mg/l |
| | 3 | $LC_{50}$ 10–100 mg/l |
| Chronic aquatic toxicity | 1 | Acute toxicity $<$ 1 mg/l and (lack of rapid biodegradability or log $K_{ow} > 4$ unless BCF $<$ 500) |

Table 11.13 (*continued*)

| *Endpoint* | *Class* | *Criteria* |
|---|---|---|
| | 2 | Acute toxicity 1–10 mg/l and (lack of rapid degradability log $K_{ow} > 4$ unless BCF < 500) unless chronic toxicity > 1 mg/l |
| | 3 | Acute toxicity 10–100 mg/l and (lack of rapid degradability log $K_{ow} > 4$ unless BCF < 500) unless chronic toxicity > 1 mg/l |
| | 4 | Acute toxicity > 100 mg/l and (lack of rapid degradability log $K_{ow} > 4$ unless BCF < 500) unless chronic toxicity > 1 mg/l |

(ANSI, 1993) provides for global requirements, displaying environmental toxicity and fate data in Section 12, environmental hazard classifications in Section 3 and transport classifications in Section 14.

Environmental toxicity is adverse biochemical or physiological effects produced when a test material is exposed to living organisms (Hooper and Hoel, 1996). Toxicity testing is divided into short-term acute and long-term chronic testing, but does not include physical effects. Since one test substance may produce different effects in different organisms and plants, it may be necessary to test a variety of organisms to review the anticipated ecotoxicity of a test material. The DSD recommends testing on three types of organisms: freshwater fish, freshwater invertebrates (water fleas) and freshwater algae, each representing a different level in the environmental food chain.

Ecotoxicity test results are usually expressed as the lethal concentration ($LC_{XX}$), effect concentration ($EC_{XX}$) or inhibition concentration ($IC_{XX}$) required to produce a lethal or sublethal effect in XX per cent of the test organisms. $LC_{50}$ represents the concentration producing death in 50 per cent of the test organisms. "Effect" includes sublethal effects like immobility or lack of growth. "Inhibition" includes sublethal effects in populations of organisms that may be difficult to individually differentiate. It should be noted that aquatic testing guidelines are designed for the evaluation of water soluble, non-volatile pure chemicals. Since many chemicals do not meet these ideal physical and chemical properties, guidelines have been developed to assist decisions in sample preparation and test results interpretation (ASTM, 1997; OECD, 2000). The DSD classifies all materials with an $EC_{50}$, $IC_{50}$ or $LC_{50}$ in any organism of less than or equal to 1 ppm as R 50, very toxic to aquatic organisms. All materials with any of these toxicity values in any organism in the range of 1–10 ppm are classified as R 51, toxic to aquatic organisms, and all materials with any of these toxicity values in any organism in the range of 10–100 ppm as R 52, harmful to aquatic organisms.

Environmental fate is the breakdown, transformation, accumulation and movement of a test material in the environment. Environmental fate data generally focus on biodegradation and bioaccumulation, but other characteristics should not be overlooked in a more complete evaluation of a chemical's presence in the environment.

Biodegradation is the ability of biological microorganisms in a test environment to breakdown or transform a test material under controlled conditions and in a defined time span, often 28 days. Biodegradation tests often use carbon dioxide production or oxygen consumption to account for carbon transfer from the material through a complete breakdown to water, carbon dioxide and mineral salts. This carbon transfer may be diverted for use in the production of more microorganisms. Some biodegradation tests also use a rate function to evaluate biodegradability levels requiring 60 per cent biodegradation within 10 days of reaching a 10 per cent biodegradation level. Poorly soluble materials may be expected to degrade over a reasonable time span, but it is unlikely that rapid degradation criteria will be achieved in 10 days or less.

Biodegradation test results are expressed as a percentage of the decrease in Dissolved Organic Carbon (DOC), the increase in carbon dioxide ($CO_2$) or utilization of oxygen ($O_2$) in a closed system. Methods using DOC are not suitable for poorly soluble chemicals. Biodegradation test results are also dependent on the microorganisms and their ability or option to replicate in the test system. The production of more microorganisms can alter the biodegradation potential when measuring DOC, $CO_2$ or $O_2$. The DSD classifies all materials that biodegrade less than 60 per cent under these controlled conditions as R 53, may cause long-term adverse effects in the aquatic environment.

Bioaccumulation is the increased concentration of a chemical within a living organism or plant, in excess of the concentration found in the environment. Long-term bioaccumulation tests can be conducted with organisms, measuring the concentration of a chemical in the tissues of the organisms. These tests either require radiotracer methods or a sophisticated analytical method to detect the chemical at low concentrations. One screening study is the octanol/water partition coefficient test where the chemical concentrates in either the octanol phase or the water phase. If the chemical significantly accumulates in the octanol phase, the chemical is considered potentially bioaccumulative. Biological limitations for molecular size and weight also impact the potential for a chemical to bioaccumulate and must be taken into account.

Bioaccumulation test results are expressed as BCF values for bioaccumulation testing with organisms and log values for octanol/water coefficient testing. BCF values greater than or equal to 100 are considered bioaccumulative and log $P$ (or log $K_{ow}$) values from 3 to 7 are considered potentially bioaccumulative. BCF values take precedence, if available, over log $P$ values. The DSD also classifies materials that bioaccumulate or potentially bioaccumulate as R 53, may cause long-term adverse effects in the aquatic environment.

## Evaluation of environmental effects of mixtures

Most regulatory classification schemes for environmental hazards of mixtures rely on data on chemical components. Data from mixture testing is often useful information for customer requirements, but some regulatory requirements do not permit the use of mixture data. Aquatic toxicity mixture data is useful in comparisons of mixture data results with component-based toxicity evaluations. Component-based evaluations often under-predict or over-predict the mixture toxicity, depending on the water solubility of each component, the expected toxicity, and significant effects of one component on another. Mixture data is a practical interpretation of adverse effects on the aquatic environment.

Under the DPD, aquatic toxicity mixture data may be used for classification of environmental hazards. Component-based toxicity evaluations are also subject to classification in the absence of mixture data, and are based on the total presence of various toxic levels of components. Criteria for mixtures are listed in Table 11.14. For transport

*Table 11.14* Criteria for environmental hazard classifications of mixtures under the EU DPD

| *Mixture classification R phrase* | *R phrase* | *Criteria for component data* | *Criteria based on component data* |
|---|---|---|---|
| R 50 | Very toxic to aquatic organisms | $LC_{50} < 1$ ppm | 25% R 50 components |
| R 51[a] | Toxic to aquatic organisms | $LC_{50} < 10$ ppm | 25% R 50 + R 51 components |
| R 51[a] | Toxic to aquatic organisms | $LC_{50} < 10$ ppm | 2.5% R 50 components |
| R 52[a] | Harmful to aquatic organisms | $LC_{50} < 100$ ppm | 25% R 50 + R 51 + R 52 components |
| R 52[a] | Harmful to aquatic organisms | $LC_{50} < 100$ ppm | 2.5% R 50 + R 51 components |
| R 52[a] | Harmful to aquatic organisms | $LC_{50} < 100$ ppm | 0.25% R 50 components |
| R 53 | May cause long-term adverse effects in the aquatic environment | N/A | 25% R 53 components |

[a] R 51 and R 52 are to be used only in combination with R 53 when both criteria are met.

classifications, component-based toxicity evaluations are encouraged. Toxicity mixture data is only permitted under the transport regulations in select cases with review by the competent authorities.

Biodegradation mixture data is limited in its usefulness, since biodegradation is best viewed as a function of a pure chemical. A test following the haphazard breakdown of the components of a mixture may not reveal the complete details for predicting the environmental fate of the individual components of the mixture. Biodegradation mixture data is accepted in some venues in the US and Canada, but is not accepted under the EU DPD. Component-based biodegradation evaluations are subject to classification under the DPD, based on the accumulated presence of greater than or equal to 25 per cent of non-biodegradable components. Biodegradation data is not currently subject to classification under transport regulations with the exception of international bulk transport in marine vessels (MARPOL 73/78).

Bioaccumulation mixture data is inappropriate, since bioaccumulation is strictly a function of a pure chemical. Component-based bioaccumulation evaluations are subject to classification under the DPD, based on the accumulated presence of greater than or equal to 25 per cent bioaccumulative or potentially bioaccumulative components. Component-based bioaccumulation evaluations are also used by various transport regulations.

## Global harmonization and environmental data

Global harmonization of existing criteria for the classification and regulation of the environmental hazards of chemicals and mixtures of chemicals is being coordinated by the OECD and UN Committee of Experts on the Transport of Dangerous Goods, with comments and suggestions from a variety of interested industry groups. The environmental criteria under development would require the same information for environmental toxicity and fate currently subject to classification under various hazard communication or transport regulations, but the criteria for classification would be slightly different from any current requirements. The suggested harmonized criteria are listed in Table 11.15. Acute ecotoxicity by itself would be regulated, as well as combinations of acute toxicity, nonbiodegradation and bioaccumulation. Chronic data would be used to clarify classifications based on these combinations. Poorly soluble materials would be subject to further interpretation and clarification when the additional guidance documents are available (OECD, 1998).

## Labeling for transportation

Transportation specialists in conjunction with toxicologists, hazard communication specialists, and chemical regulatory specialists, are responsible for complying with classifying, labeling, marking, placarding, and manifesting requirements mandated by the DOT Hazardous Material Regulations. The DOT, created in 1967, was granted the responsibility for regulating the shipment of hazardous materials in the US. The responsibility was intended to oversee classification, labels, placards, packaging, handling, stowage and emergency procedures. Significant transportation accidents in the early 1970s prompted the consolidation of regulations and the development of the Hazardous Materials Transportation Act (HMTA) in 1975. This act granted the DOT the authority to designate and regulate hazardous materials. It also established registration requirements for shippers, civil and criminal penalties, and federal preemption of state and local regulations.

The regulations require a hazardous material to be classified in one of nine hazard classes. Additionally, materials are assigned four-digit numerical identities known as UN numbers. Specific chemicals often have unique UN numbers, while chemical mixtures often have generic UN numbers. Fire departments and other emergency responders keep references and important information on these UN numbers and hazard classes. Most chemical products are classified in generic hazard classes based on physical, chemical or toxicological properties.

Besides requiring a material identity on the label, the regulations also require specific pictorial symbols for primary and subsidiary hazard classifications (like flammable, toxic and corrosive). Most hazard classes are divided into three packing groups based on the severity of the physical, chemical or toxicological characteristics. Hazard classes

*Table 11.15* Harmonized criteria proposed for environmental hazards by the OECD work group

| *Endpoint* | *Class* | *Criteria* |
|---|---|---|
| Acute aquatic toxicity | 1 | $LC_{50} < 1$ ppm |
| | 2 | $LC_{50} < 10$ ppm |
| | 3 | $LC_{50} < 100$ ppm |
| Chronic aquatic toxicity | 1 | $LC_{50} < 1$ ppm and (lack of rapid biodegradation or BCF $> 500$[a])[b] |
| | 2 | $LC_{50} < 1$ ppm and (lack of rapid biodegradation or BCF $> 500$[a])[b] |
| | 3 | $LC_{50} < 10$ ppm and (lack of rapid biodegradation or BCF $> 500$[a])[b] |
| | 4 | Poorly soluble with no toxicity $\leq$ level of water solubility and (lack of rapid biodegradation or BCF $> 500$[a])[b] |

[a] Or log $P > 4$ (when no BCF available).

[b] Unless chronic toxicity $> 1$ ppm.

are listed in Table 11.16 and hazard classes using acute toxicological characteristics are listed in Tables 17 and 18. Specific information on each hazard class may be found in Title 49 CFR (49 CFR 170–178).

Most companies use integrated computer and printing systems to generate a single product label that meets OSHA HCS and DOT HMTA requirements. Regulations of other organizations that need to be considered include: ICAO/IATA (international air transport) (ICAO, 1999; IATA, 1999), IMDG (international sea transport) (IMO, 1998), ADR/RID (European land transport) (Economic Commission for Europe, 1998), TDG (Canadian land transport) (Transport Canada, 1998), ADG (Australian land transport) (Australian Code for the Transport of Dangerous Goods by Road and Rail, 1998) and Official Mexican Standards, called Normas or NOMS (Mexican nonbulk land transport) (Official Mexican Standards, 1995). Programs exist in the US (CHEMTREC) and Canada (CANUTEC) to assist emergency responders in dealing with accidents occurring during the transportation of chemical shipments.

## The toxicology of exposure limits

The risk assessment processes used to set workplace exposure limits use the same basic principles for setting safe exposure levels to food additives, drugs, or pesticides described in previous chapters.

Traditional processes, still largely in effect today, will use a no-effect level in animals and, with the use of an appropriate safety factor, establish a safe exposure level. Many workplace exposure limits are set based on human experience of discomfort, usually irritation. It has been estimated that 70 per cent of all exposure limits are based on irritation (Bos *et al.*, 1992).

Monitoring for chemicals in the workplace that have exposure limits is the work of industrial hygienists. For more details on the subject, the reader is referred to textbooks on industrial hygiene. The international regulatory scene for exposure limits is well covered by Kortsha (1991). The history of industrial hygiene (Clayton, 1991) and legislative trends (Baier, 1991) are well reviewed elsewhere.

*Table 11.16* United Nations transportation hazard classes[a]

| *Class number* | *Description of class* |
|---|---|
| 1 | Explosives |
| 2 | Gases |
| 3 | Flammable liquids |
| 4 | Flammable solids, spontaneously combustible substances and water-reactive substances |
| 5 | Oxidizing substances and organic peroxides |
| 6 | Toxic and infectious substances |
| 7 | Radioactive substances |
| 8 | Corrosive substances |
| 9 | Miscellaneous dangerous substances and articles |

[a] Extracted from 49 CFR 173 (label specifications can be found in Sections 172.411–172.446). Applicable for ICAO/IATA (international air), IMDG (international sea), ADR/RID (European domestic), DOT (US domestic), TDG (Canadian domestic), ADG (Australian domestic) and Mexican (NOM domestic) regulations.

*Table 11.17* Transportation Class 6.1 (toxic substances). Definition and packing group criteria[a]

| *Packaging group* | *Oral toxicity $LD_{50}$ (mg/kg)* | *Dermal toxicity $LD_{50}$ (mg/kg)* | *Inhalation toxicity by dusts and mists $LC_{50}$ (1 h) (mg/l)* | *Inhalation toxicity by vapors* |
|---|---|---|---|---|
| I | $< 5$ | $\leq 40$ | $\leq 0.5$ | $V \geq 10\ LC_{50}$ and $LC_{50} \leq 1000$ ml/m$^3$ |
| II | $> 5– \leq 50$ | $> 40– \leq 200$ | $> 0.5– \leq 2$ | $V \geq LC_{50}$ and $LC_{50} \leq 3000$ ml/m$^3$, and criteria for packing Group I is not met |
| III solids | $> 5– \leq 200$ | $> 200– \leq 1000$ | $> 2– \leq 10$ | $V \geq 1/5\ LC_{50}$ and $LC_{50} \leq 5000$ ml/m$^3$, and criteria for packing Groups I or II are not met |
| III liquids | $> 5– \leq 500$ | $> 200– \leq 1000$ | $> 2– \leq 10$ | |

[a] From 49 CFR 173.132 and 173.133. V = the saturation vapor concentration in ml/m$^3$ at 20°C and standard atmospheric pressure. Applicable for ICAO/IATA, IMDG, ADR/RID, DOT and TDG regulations.

*Table 11.18* Transportation Class 8 (corrosive substances). Definition and packing group criteria[a]

| |
|---|
| *Packing Group I: substances presenting great danger* |
| Materials that cause full thickness destruction of intact skin tissue within an observable period of up to 60 min starting after the exposure time of 3 min or less |
| *Packing Group II: substances presenting medium danger* |
| Materials that cause full thickness destruction of intact skin tissue within an observable period of up to 14 days starting after the exposure time of more than 3 min but not more than 60 min |
| *Packing Group III: substances presenting minor damage* |
| Materials that cause full thickness destruction of intact skin tissue within an observable period of up to 14 days starting after the exposure time of more than 60 min but not more than 4 h |
| Materials that do not cause full thickness destruction of intact skin tissue but exhibit a corrosion rate on steel or aluminum surfaces exceeding 6.25 mm/year at a test temperature of 55°C |

[a] Applicable for ICAO/IATA, IMDG, ADR/RID, DOT and TDG regulations.

## United States

OSHA promulgates PELs under OSH Act authority. The original air contaminant limits issued by OSHA in 1971 were largely adopted from recommendations issued by ACGIH and ANSI. Many were set at levels using small margins of safety from levels known to cause occupational discomfort. Some are set using larger safety factors from no effect levels in studies in laboratory animals. OSHA attempted to comprehensively update PELs in 1989. The Federal Register document on this action is a good source of the rationale used to set exposure limits (Federal Register, 1989). A court decision on a legal case brought by labor unions said that OSHA's review was not sufficient, and overturned the new PELs in 1992. OSHA is now planning to issue a set of criteria which will address the issues in a court case so that they can amend current PELs to make them more protective on an as needed basis (Federal Register, 1999b). It is important to note that OSHA must consider practical considerations before setting exposure limits under provisions of the OSHA Act. Use of more sophisticated mathematical risk assessment procedures has become more popular. The EPA, under TSCA, has occasionally found the need to set NCELs. The NCELs set by the EPA are usually made lower than the PELs for similar chemicals. It will be interesting to see if OSHA adopts a more conservative approach in the future.

The exposure recommendations considered the most up to date are published annually by the Threshold Limit Values for Chemical Substances Committee of ACGIH that has been in existence since 1941. This activity of ACGIH is recognized throughout the world and the exposure recommendations of ACGIH form the underpinnings of workplace exposure regulations in many countries. An annual publication announces changes being considered for the following year (ACGIH, 1999). ACGIH defines a number of terms in establishing safe exposure levels. A TLV is defined as the airborne concentration of substances and conditions under which it is believed that nearly all workers may be repeatedly exposed day after day without adverse health effects. TLVs are established by ACGIH based on available information from industrial experiences and experimental human and animal studies. TLVs are expressed as Time Weighted Average concentration (TWA) for a conventional 8-h workday and 40-h workweek. ACGIH also uses the concept of STEL which is the TWA concentration a worker can be exposed without adverse effects for 15 min. A ceiling (TLV-C) level is defined as the airborne concentration that should not be exceeded during any part of the work exposure. ACGIH will use a "skin" notation for chemicals where significant dermal absorption can occur. The general methodology for establishing TLVs is discussed in ACGIH's annual publication and rationale for decisions on individual chemicals is documented in a compendium (ACGIH, 1993).

The annual publication also lists Biological Exposure Indices (BEIs) for certain chemicals. BEIs are reference values for certain measurements in biological specimens from workers, such as urine and exhaled air, to determine their level of exposure to these chemicals. BEIs can be based on the relationship between the intensity of exposure and biological levels of the determinant or biological levels and health effects. The data used to set BEIs come from controlled or field studies with humans. Due to pharmacokinetic differences between species, animal studies are not useful to set BEIs.

### *Europe*

Workplace exposure limits have yet to be addressed by harmonizing legislation in the EU. National requirements are still in effect.

One of the major requirements is in the UK regulation entitled Control of Substances Hazardous to Health (COSHH, Health and Safety Commission, 1985). Besides listing exposure limits, COSHH includes many provisions similar to the OSHA HCS such as training of employees and written workplace hazard communication programs. COSHH also requires a detailed hazard and risk assessment of chemicals used in the workplace. The German Research Society also annually publishes a report entitled Maximum Concentrations (MAK) at the Workplace and Biological Tolerance Values (BAT) for Working Materials. MAK and BAT values are analogous to ACGIH TLVs and BEIs. National standards are also in effect in several other European countries including Finland, France, Netherlands and Sweden (Kortsha, 1991).

### *Canada*

Federal legislation on workplace safety in Canada has traditionally lagged behind the activity of requirements in the highly industrialized province of Ontario. The WHMIS regulatory system addressed many industrial hygiene issues at the federal level in 1987. The Canada Dangerous Substances Regulations adopt the current exposure limits published by ACGIH.

### *Australia*

Worksafe Australia has the authority to establish standards for workplace safety under NOHSC. Standards for several hundred chemicals have been published and a fair amount of related research and industrial surveys are ongoing (NOHSC, 1994e).

## Conclusions and future considerations

Hazard communication regulations have been profoundly successful in that they have fundamentally changed the way industry operationally deals with chemicals. Work forces are trained to read MSDSs before working with a new chemical. Workplace and transport labels with hazard symbols offer immediate recognition of the acute health hazards to workers, carriers, customers and the general public. All this was accomplished via a variety of performance standards in many corners of the world that allow flexibility in addressing complex physical and health hazard information. Even though MSDSs and labels are not required all over the world, almost all chemical products are labeled and accompanied by an MSDS no matter what the destination might be due to systems in place with manufacturers to satisfy the requirements of customers in the regulated countries.

This is not to say that improvements in hazard communication regulations and compliance programs are not desirable. However, except for the changes in the DSD and DPD in the EU and in the international effort for the regulation of the transport of dangerous goods, existing regulations have been fairly dormant over the last 5 years. OSHA seems to have abandoned plans for various improvements since requesting comments in a Federal Register (1990) notice that solicited ideas for improvement of MSDS and label requirements (Long, 1992). Some of the important issues to be addressed were:

1 MSDS format
2 Electronic distribution of MSDSs
3 The danger of over-warning

OSHA seems to be relying more on consensus processes to achieve advances in hazard communication. The ANSI standards on labeling and MSDSs are periodically updated with OSHA input. OSHA has also strongly supported the OECD process for hazard classification harmonization. Consistent worldwide requirements would take much confusion out of international commerce of chemicals. The OECD process has expended much effort in a laudable start, however, tough issues such as rules for untested mixtures and scientific judgment await resolution. Adoption by national legislative bodies and regulatory authorities is another formidable obstacle.

Scientific advancements will continue to provide a challenge for authorities that oversee hazard communication. These regulations need to be improved in the scientific aspects of toxicology testing and interpretation. However, this cannot be done unless responsible scientists provide the regulatory agencies with new ideas backed up with data. A number of important topics that need to be addressed in the future include:

1 Computer prediction via SAR models for hazards.
2 Risk-based rather than hazard-based communication.
3 Dealing with conflicting data and increased use of weight-of-evidence approach.
4 Communication of environmental hazards.
5 Improved approaches for mixtures.
6 Substitution of *in vitro* test methods for animal testing.

It is hoped that continued improvement contributes in no small way to a reduction in occupational injuries and illnesses.

## Websites of interest

| | |
|---|---|
| *Government agencies* | |
| Department of Justice (Canada) | www.canada.justice.gc.ca |
| DOT (Hazmat) | www.hazmat.dot.gov |
| ECE Transport (EU) | www.unece.org/trans |
| EPA (US) | www.epa.gov |
| Health Canada | www.hc.gc.ca |
| Government Printing Office (US) | www.gpo.gov |
| Japan (MITI) | www.miti.go.jp |
| OSHA (US) | www.osha.gov |
| Transport Canada | www.tc.gc.ca |
| WorkSafe Australia | www.worksafe.gov.au |
| *Organizations and expert bodies* | |
| ACGIH | www.acgih.org |
| CANUTEC | www.tc.gc.ca/canutec |
| CHEMTREC | www.cmahq.com |
| IARC | www.iarc.org.fr |
| IATA | www.iata.org |
| IMO | www.imo.org |
| NTP | www.ntp-server.niehs.gov |
| OECD | www.oecd.org |

## Acknowledgements

We would like to thank our colleagues at Lubrizol for their technical advice and comments, particularly John Blickensderfer, Tracy Harris, Marlin McKinley, Bill Starr and Mick Wragg. Special thanks go to Linda Toth for typing this manuscript and to Jennifer Hanna for many hours of editorial assistance. And finally, this chapter is dedicated to the memory of our good friend and colleague Dave Taylor who was an inspiration to us all.

## References

ACGIH. American Conference of Governmental Institutional Hygienists. *Documentation of the Threshold Limit Values and the Biological Exposure Indices*. 6th ed., 1993.

ACGIH. American Conference of Governmental Industrial Hygienists. *TLVs and BEIs*, 2000.

ANSI. American National Standards Institute. *American National Standard for the Precautionary Labeling of Hazardous Industrial Chemicals*, 1976. ANSI Z 129.1-1976.

ANSI, American National Standards Institute. *American National Standard for Hazardous Industrial Chemicals – Material Safety Data Sheets – Preparation*, 1993, ANSI Z400.2–1993.

ANSI. American National Standards Institute. *American National Standard for Hazardous Industrial Chemicals – Material Safety Data Sheets – Preparation*, 1998. ANSI Z400.1-1998.

ANSI. American National Standards Institute. *American National Standard for Hazardous Industrial Chemicals – Precautionary Labeling*, 2000. ANSI Z129.1-2000.

Ashby J, Doerrer NG, Flamm FG, Harris JE, Hughes DH, Johannsen FR, Lewis SC, Krivanek ND, McCarthy JF, Moolenaar RJ, Raabe GK, Reynolds RC, Smith JM, Stevens JT, Teta MJ, Wilson JD. A scheme for classifying carcinogens. *Regul Toxicol Pharmacol* 1990;12 270–295.

ASTM. *Standard Practice for Aquatic Toxicity Testing of Lubricants: Sample Preparation and Results Interpretation*, 1997. ASTM D6081-97.

Australian Code for the Transport of Dangerous Goods by Road and Rail, 6th ed., 1998.

Baier EJ. Legislation and legislative trends. In: Clayton GD, Clayton FE, editors. *Patty's Industrial Hygiene*. New York: John Wiley and Sons, Inc., 1991.

Ballard AM. Nonanimal test for chemical corrosivity accepted for federal regulatory purposes. *Chem Reg* Rep 2000;24:590.

Bos PMJ, Zuart A, Reunzel PGJ, Bragt PC. Evaluation of the sensory irritation test for the assessment of occupational health risk. *Crit Rev Toxicol* 1992;21:423–447.

Buehler EV. Delayed contact hypersensitivity in the guinea pig. *Arch Dermatol* 1965;91:171–177.

Campbell DI, George EL, Hall LL, Stara JF. Dermal irritancy studies with palladium, platinum, lead and manganese compounds. *Arch Environ Health* 1975;30:168–170.

Clayton GD. Industrial hygiene: retrospect and prospect. In: Clayton GD, Clayton FE, editors. *Patty's Industrial Hygiene*. New York: John Wiley and Sons, Inc., 1991.

Code of Federal Regulations. Title 29, Section 1910.1200, 1999a.

Code of Federal Regulations. Title 16, Section 1500.42, 1999b.

Code of Federal Regulations. Title 49, Parts 170–178, 1999c.

Collins TFX.. Developmental toxicity protocols utilized in the safety evaluation of chemicals. In: Mehlman MA, editor. *Advances in Modern Environmental Toxicology, Volume X, Safety Evaluation:-Toxicology Methods, Concepts and Risk Assessment.* Princeton, NJ: Princeton Scientific Publishing Company, 1987.

Controlled products regulations under the Hazardous Products Act for Occupational Exposures (WHMIS). SOR/88-66. *Canadian Gazette Part II.* 1988;122(2).

DHEW. *A Recommended Standard: an Identification System for Occupational Hazardous Materials.* DHEW (NIOSH) Publication No. 75-126, 1974.

DOL. *Report of the Standards Advisory Committee on Hazardous Materials Labeling to the Assistant Secretary of Labor of Occupational Safety and Health*, 1975.

Draize JH, Woodward G, Calvery HO. Methods for the study or irritation and toxicity of substances applied topically to the skin and mucous membranes. *J Pharmacol Exp Ther* , 1944;82:377–390.

Economic Commission for Europe. Inland Transport Committee. *European Agreement Concerning the International Carriage of Dangerous Goods by Road (ADR) and Protocol of Signature*, 1998.

ECVAM. European Centre for the Validation of Alternative Methods. *Statement on the Validity of the Local Lymph Node Assay for Skin Sensitisation Testing*, 2000.

Environment Health and Education Center. *Hazard Communication: A Review of the Science Underpinning the Art of Communication for Health and Safety*. Maryland: The University of Maryland Medical School, 1997.

EPA. *Summary of the Second EPA Workshop for Dermal Carcinogenicity Testing by the Dermal (Cutaneous) Route*. Research Triangle Park, NC: EPA, 1988, 18–19 May.

European Commission. Council Directive 67/548/EC on the approximation of the laws, regulations and administrative provisions relating to the classification, packaging and labeling of dangerous substances, 1967.

European Commission. Council Directive 88/379/EC on the approximation of the laws, regulations and administrative provisions relating to the classification, packaging and labeling of dangerous preparations, 1988.

European Commission. Council Directive 91/155/EC defining and laying down the detailed arrangements for the system of specific arrangements relating to dangerous preparations in implementation of Article 10 of the EU, 1991.

European Commission. Commission Directive 98/98/EC. Twenty-fifth adaptation for technical progress to Council Directive 67/548/EC, 1998.

European Commission. Council Directive 99/45/EC on the approximation of the laws, regulations and administrative provisions relating to the classification, packaging and labeling of dangerous preparations, 1999.

European Commission. Commission Directive 2000/32/EC. Twenty-sixth adaptation for technical progress to Council Directive 67/548/EC, 2000a.

European Commission. Commission Directive 2000/33/EC. Twenty-seventh adaptation for technical progress to Council Directive 67/548/EC, 2000b.

Federal Register. Air contaminants, final rule. *Federal Register* 1989;54:2332.

Federal Register. Hazard communication: request for comments and information. *Federal Register* 1990;55:20580.

Federal Register. National Toxicology Program (NTP) revises criteria and process for listing substances in the biennial report on carcinogens. *Federal Register* 1996;61:50499.

Federal Register. Notice of availability of report: the local lymph node assay. *Federal Register* 1999a;64:14006.

Federal Register. Permissible exposure limits (PELs) for air contaminants. *Federal Register* 1999b; 63:61284–61286.

Gad SC, Chengelis CP. Dermal sensitization. In: Gad SC, Chengelis CP, editors. *Acute Toxicology Testing Perspectives and Horizons*. Caldwell, NJ: The Telford Press, 1988.

Government Institutes. *The OSHA Compliance Course*. Rockville, MD: Government Institutes, 1991. pp. 20850.

Health and Safety Commission. *Control of Substances Hazardous to Health.* (COSHH) Consultative Document, 1985.

Health Canada. *Guidelines for the Disclosure of Toxicological Information on a Material Safety Data Sheet*, 1997.

Heidemann, C.A., Simkins, L.K., 1991. Hazard communication and worker right-to-know programs. In: Clayton, G.D., Clayton, F.E. eds. *Patty's Industrial Hygiene*, New York: John Wiley and Sons, Inc.

Hooper, D.L., Hoel, D.L, 1996. Lubricants, The Environment and ASTM D02, SAE *Technical Paper Series 961727.*

IARC, 2000. Overall Evaluations of Carcinogenicity to Humans.

IATA, 1999. Dangerous Goods Regulations, 41st Edition.

ICAO, 1999. Technical Instructions for the Safe Transport of Dangerous Goods by Air.

International Maritime Organization, 1998. International Maritime Dangerous Goods Code. 29th Amendment.

Johnson, E.M., 1987. Recognition of teratogens in occupational setting. In: Mehlman MA, ed. *Advances in modern environmental toxicology, volume X, Safety evaluation: toxicology methods, concepts and risk assessment*, Princeton Scientific Publishing company.

Karol MH, Stadler J, Underhill D, Alarie Y. Monitoring delayed-onset pulmonary hypersensitivity in guinea pigs. *Toxicol Appl Pharacol* 1981;61:277–285.

Kay JH, Calandra JC. Interpretation of eye irritation tests. *J Soc Cosmet Chem* 1962;13:281.

Kimber I. The local lymph node assay. In: Marzulli FN, Maibach HI, editors. *Dermatotoxicology*, 5th ed. New York: Taylor and Francis, 1996.

Klaassen CD. Principals of toxicology. In: Klaassen CD, Amdur MO, Doull J, editors. *Casarett and Doull's Toxicology: the Basic Science of Poisons,* 3rd ed. New York: J. MacMillian Publishing Company, 1986.

Kortsha GX. Industrial hygiene abroad. In: Clayton GD, Clayton FE, editors. *Patty's Industrial Hygiene*. New York: John Wiley and Sons, Inc., 1991.

Long JR. Standard for MSDSs in the offing. *Chem Eng News 1992;70(20): 7–11.*

Magnusson B, Kligman AM. The identification of contact allergens by animal assay, the guinea pig maximization test method. *J Invest Dermatol* , 1969;52:268–276.

Mason JM, Langebach R, Shelby MD, Zeiger E, Tennant RW. Ability of short-term tests to predict carcinogenesis in rodents, *Ann Rev Pharmacol Toxicol* , 1991;30:149–168.

McCreesh AH, Steinberg M. Skin irritation testing in animals. In: Marzulli FN, Maibach HI, editors. *Dermatotoxicology*, 2nd ed. Washington, DC: Hemisphere Publishing Corporation, 1983.

Mintz BW. *OSHA History, Law and Policy*. Washington, DC: The Bureau of National Affairs, Inc., 1984.

MHW/MITI. *The Guidelines for Providing Information Pertaining to the Safety of Chemical Substances*. Ministry of Health and Welfare. Ministry of International Trade and Industry. March 26, 1993.

NTP. Ninth Report on Carcinogens. Research Triangle Park, NC: National Institute of Environmental Health Sciences, 2000.

NOHSC. National model regulations for the control of workplace hazardous substances. *NOHSC* 1994a:1005.

NOHSC. National code of practice for the preparation of material safety data sheets. *NOHSC* 1994b:2011.

NOHSC. Approved criteria for classifying hazardous substances. *NOHSC* 1994c:1008.

NOHSC. List of designated hazardous substances. NOHSC 1994d:10005.

NOHSC. National code of practice for the control of workplace hazardous substances. *NOHSC* 1994e:2007.

Official Mexican Standards. Normas or NOMS. Diario Official De La Federacion, 1995.

OECD. *Guidelines for the Testing of Chemicals*. Paris Cedex: OECD Publications, 1993.

OECD. *Harmonized Integrated Hazard Classification System for Human Health and Environmental Effects of Chemical Substances*. Paris Cedex: OECD Publications, 1998.

OECD. *Guidance Document on Aquatic Toxicity Testing of Difficult Substances and Mixtures*. Paris Cedex: OECD Publications, 2000.

OSHA. US Department of Labor. Inspection procedures for the hazard communication standard. *OSHA Instruction* 1986;CPL 2-2:38A.

OSHA. US Department of Labor. Inspection procedures for the hazard communication standard. *OSHA Instruction* 1998;CPL 2-2:38D.

Pepellco WE. Effect of exposure route on potency of carcinogens. *Regul Toxicol Pharmacol* 1991;13:3–17.

Phillips L, Steinberg M, Maibach HI, Akers WA. A comparison of rabbit and human skin response to certain irritants. *Toxicol Appl Pharmacol* 1972:21;369–382.

Robinson MK, Stotts J, Danneman PJ, Nusair TL, Bay PHS. A risk assessment process for allergic contact sensitization. *Fd Chem Toxicol* 1989;27(7):479–489.

Rothstein MA. *West's Handbook Series, Occupational Safety and Health Law*, 3rd ed. St. Paul: West, 1990.

Secretary of Labor vs. American Cyanamid. 1992. 2/7/92:OSHRC; 1497.

Sowinski EJ, Broecker B, Faccini JT, Beekhuizen S, Weil CS, Gelbke HP, Fleig H, Matheson DW, Granville GC, Rozenboom DM, Carney IF, Pell S, Munn A, Aubrun JC, Reape MJ. Criteria for identifying and classifying carcinogens, mutagens, and teratogens. *Regul Toxicol Pharmacol* 1987;7:1–20.

Steinberg M, Wolfe GW, McCreesh AH. Correlating the inflammatory response between animal and human skin. *J Toxicol-Cut Ocul Toxicol* 1982;1(1):33–48.

Switzerland, 1969, 21 March. Federal law on trade in toxic substances (toxicity law) and order relating to toxic substances (19 September 1983) and order concerning prohibited toxic substances (23 December 1971).

Transport Canada. *Transport of Dangerous Goods Regulations*, 1998.

Chapter 12

# Federal air and water regulations

## SDWA, CAA, HAPS and ozone regulation

*Shayne C. Gad*

There is a very broad and diverse fabric of legislation and regulation and regulatory practice which concerns the protection of people and wildlife from potential toxic hazards in the environment. Other chapters in this text have presented those cases which are focused directly on specific manufacturers and products. There is, however, also a body of regulations which focuses on the principal general means of environmental exposure – air and water. These laws generally depend on other laws or unspecified means to generate the toxicity data that they legislate on. In fact, more of them specifically require or provide guidelines for testing or data generation.

Table 12.1 summarizes the principle US laws and this chapter will seek to examine these laws and also some general environmental regulatory practices.

### Clean Air Act

- Title: Clean Air Act (CAA, 1976)
- Agency: EPA
- Year passed: 1970; amended 1974, 1977, 1978, 1980, 1981, 1982, 1983, 1990, and 1997
- Groups regulated: state and local governments

The Clean Air Act (CAA)is a US law administered by the US Environmental Protection Agency (EPA), which regulates the emission of certain toxic chemicals and particulate matter. Although the principal enforcement provisions are the responsibility of state and local governments, overall administrative responsibility rests with the EPA. This act requires criteria documents for air pollutants and sets both national air quality standards and standards for sources that create air pollutants, such as motor vehicles, power plants, etc. Important actions taken under this law include standards for a phased-out elimination of lead in gasoline and the setting of sulfuric acid air emission guidelines for existing industrial plants. Regulated pollutants include: sulfur dioxide, carbon monoxide, nitrogen oxides, hydrocarbons, ozone, photochemical oxidants, and particulate materials. Similar legislation exists in the UK under the Clean Air Act of 1956, supplemented in 1968.

*Table 12.1* Federal laws related to air and water exposure to toxic substances

| *Legislation* | *Agency* | *Area of concern* |
|---|---|---|
| Clean Air Act (1970, amended 1974, 1977, 1990) | EPA | Air pollutants |
| Hazardous Materials Transportation Act (1972) | DOT | Transport of hazardous materials |
| Clean Water Act (formerly Federal Water Pollution Control Act; 1972, amended 1977, 1978, 1987) | EPA | Water pollutants |
| Marine Protection, Research and Sanctuaries Act (1972) | EPA | Ocean dumping |
| Dangerous Cargo Act (1952) | DOT, USCG | Water shipment of toxic materials |
| Safe Drinking Water Act (1974, amended 1977, 1986) | EPA | Drinking water, contaminants |
| Federal Mine Safety and Health Act (1977) | DOL, NIOSH | Toxic substances in coal and other mines |
| Comprehensive Environmental Response, Compensation, and Liability Act (1981) | EPA | Hazardous substances, pollutants, and contaminants |
| Oil Pollution Act (1990) | DOT | Oil pollution |
| Pollution Prevention Act (1990) | EPA | Toxics use reduction |

## Clean air acts

UK legislation, passed in 1956 and 1968, prohibits the emission of dark smoke from any chimney or trade premises. The Act also enabled local authorities to declare smoke control areas, in which the emission of any smoke is an offense.

## Synopsis of law

Section 112 of the CAA provides a list of 189 hazardous air pollutants, to which the EPA may add or delete pollutants. The EPA must establish national emissions standards for sources that emit any listed pollutant. The original 1970 version of Section 112 required the standards to provide "an ample margin of safety to protect the public health from such hazardous air pollutants". The implication of this language, that standards were to be set without regard to the costs of emissions control, generated interim debate from the beginning and contributed to the EPA's glacial pace of implementation.

The EPA finally attempted to escape the strict interpretation of the CAA when it issued a standard for vinyl chloride in 1986 (Reed, 1986). In this instance the agency claimed it could consider costs and declined to adopt a standard that would ensure safety. A court set aside this standard because the EPA had improperly considered costs, but it did make clear that for determining what emissions level was safe, even for a carcinogen, the EPA need not eliminate exposure.

Additionally, the decision said the EPA could consider costs in deciding what, if any additional margin of protection to prescribe (Natural Resources Defense Council, Inc. vs. United States Environmental Protection Agency, 1987).

The 1990 amendments to the CCA responded to the difficulties presented by the strict approach of Section 112. The amendments replaced the health-based standard with a two-tiered system of regulation. The EPA must first issue standards that are technology-based, designed to require the "maximum degree of emission reduction achievable" [Maximum Achievable Control Technology (MACT)] [CAA Section 112(d)(2)]. If the MACT controls are insufficient to protect human health with an "ample margin of safety" the EPA must issue residual risk standards [CAA Section 112(f)]. The 1990 amendments essentially define an ample margin of safety for carcinogens by requiring the EPA to establish residual risk standards for any pollutant that poses a lifetime excess cancer risk of greater than one in one million.

The 1997 amendments to the CAA tightened the controls on ozone levels (to 0.08 ppm) and added strict limits on airborne particulates, particularly fine particulates. New particulate matter standards regulate particles 2.5 μm or smaller in diameter with an annual limit of 15 $\mu g/m^3$.

A commonly cited example of a law that required health-based, or risk-based, standards for pollution control is Section 112 of the 1970 CAA, which required the EPA to set emission standards for hazardous air pollutants [under the National Emissions Standards for Hazardous Air Pollutants (NESHAPS) program] that would "protect public health" with an "ample" margin of safety. Implementation of this standard of safety for carcinogenic air pollutants proved to be so troublesome for the agency that between 1970 and 1990 NESHAPS were set for only seven air pollutants. The difficulty in setting the risk-based standards was that the statute provided no indication of what an "ample margin of safety" was or how such a concept might be applied to carcinogens, given that the agency considered carcinogens to act by a no-threshold mechanism. The 1990 amendments to the CAA replaced the NESHAPS health-based standards with technology-based standards for controlling hazardous air pollutants using specified emissions control technology. The statute specifies that after installation of the control technology, health-based standards must be set to further control emissions where unacceptable risks remain.

The CAA Amendments of 1990 specifically require that primary air quality standards completely protect the public health and that the standards incorporate sufficient safety margins. There is an implicit assumption in the CAA Amendments that a "no-effects" level exists for every pollutant and for each adverse health effect. Despite the enhanced susceptibility of high-risk subpopulations to the toxic effects of pollutants, Finklea and co-workers stated that high-risk groups were not, to any great extent, considered in any quantitative or analytic sense in the derivation of the original National Ambient Air Quality Standards (NAAQS). The rational of the EPA was that "adequate protection for the larger susceptible population segments and the margins of safety included would also ensure protection for the large number of relatively small susceptible segments of the population for which we have little or no quantitative exposure information". Thus, high-risk segments of the population were often not specifically considered in the standard derivation process because there was not enough evidence to offer a precise assessment of risk, and because it was thought that they made up only a negligible percentage of the population. In contrast, recent analyses by the EPA, as with NAAQS for lead, utilize the concept of high-risk population in a more quantitative way by estimating the fraction of the susceptible subpopulation (children) that would be protected at different air levels of lead.

The sections of the US Code for the amendments are:

- CAA Amendments of 1990
- Title I – provisions for attainment and maintenance of national ambient air quality standards
- Title II – provisions relating to mobile sources
- Title III – Hazardous air pollutants
- Title IV – Acid deposition control
- Title V – permits
- Title VI – Stratospheric ozone protection
- Title VII – provisions relating to enforcement
- Title VIII – miscellaneous provisions

In trying to assess the role of high-risk groups in the derivation of environmental health standards, it is useful to consider the extent to which the EPA has utilized the concept of high-risk groups within the standard setting process. Perhaps the most common approach utilized by the EPA and other regulatory agencies has been the implementation of safety (or uncertainty) factors. While this approach implicitly recognizes that certain people are more sensitive to pollutants than others, it is inherently imprecise. The precise difference in sensitivity between a statistically "normal" individual and groups at increased risk will vary for the different causes of the high-risk condition and for different pollutants.

The EPA has utilized the uncertainty factor approach in attempting to deal with protection of the high-risk individuals, as illustrated by the national drinking water standards for noncarcinogenic chlorinated hydrocarbon insecticides and herbicides (29). These substances were tested in two animal species – rat and dog. Chronic toxicity testing provided an estimate of the lowest level of pollutant (on a milligram of dose per kilogram of body weight) that the animal could ingest with either minimal or no toxic effects. In the absence of data to indicate a basis for an alternative choice, the species that was the most sensitive to the substance was chosen to derive the standard, implying that humans are as sensitive as the most sensitive animal species. In the absence of supporting human exposure data (as with most cases, with the exception of methoxychlor), an uncertainty factor of 500 was applied to the minimally toxic dose in the most sensitive animal species (i.e., the minimally toxic dose was divided by 500). It should be noted that this methodology differs from that of the RFD reference dose approach described earlier, in which case a factor of 100 would have been applied. This number was taken to be the total amount of insecticide (or herbicide) to which a human could be exposed each day, over an unspecified period of time, without suffering any adverse health effects.

Next, amounts of the chlorinated hydrocarbons normally consumed via the diet were derived from market basket surveys. If this amount were substantially less than 1/500 of a known toxic dose in the most sensitive animal species tested (i.e., the assumed acceptable level of intake for humans), then a drinking water limit was established that would permit 20 per cent of that safe limit to be consumed via water.

Several questions occur when evaluating such a methodological scheme. For example, on what basis can we assume that the most sensitive humans have the same degree of responsiveness and the most sensitive animals? Why was 500 chosen as a safety factor? What assurances exist that it would provide sufficient protection for the general population as well as high-risk groups for these chemicals? Who, in fact, are the groups considered at increased risk?

The main problem with such an approach is its lack of specificity in identifying susceptible subpopulations, the extent of their susceptibility, and, most important, what fraction is protected by different standard levels. It should be realized, however, that when only limited data are available imprecise safety factors are the only realistic options available. Still, this approach will result in uncertain levels of protection. Alternative approaches must be developed to reduce the magnitude of that uncertainty.

The EPA has specifically evaluated the increased sensitivity of specific high-risk groups with respect to several toxic substances including carbon monoxide, lead, nitrates, nitrogen dioxide, ozone, and sulfur dioxide (Table 12.2). Following is a detailed description of the EPA's consideration of high-risk groups in the derivation of drinking water standards for nitrates and cadmium. The role of high-risk groups in deriving standards for carcinogens is also discussed.

There are two categories of regulations for toxic substances in the air: (1) the ambient air regulations, which refer to the general atmospheric concentrations; and (2) emissions from industrial sources and automobiles.

### *Ambient air regulations*

The EPA has developed standards for toxic air pollutants under the term NAAQS. There are two types of standards: primary and secondary. The former is designed to protect the public health; the latter is meant to protect public welfare, such as the effects of air pollution on vegetation, materials and visibility. NAAQS have been established for the six sources of air pollutants presented in Table 12.2.

*Table 12.2* National Ambient Air Quality Standard (EPA, 2000)

| *Air pollutant* | *Primary standard* | *Secondary standard* |
|---|---|---|
| *Particulate matter (< 10 μm)* | | |
| Annual mean (arithmetic) | 50 μg/m$^3$ | 50 μg/m$^3$ |
| 24-h average | 150 μg/m$^3$ | 150 μg/m$^3$ |
| *Sulfur dioxide* | | |
| Annual mean (arithmetic) | 0.03 ppm (80 μg/m$^3$) | |
| 24-h average[a] | 0.14 ppm (365 μg/m$^3$) | |
| 3-h average[a] | | 0.5 ppm (130 μg/m$^3$) |
| *Carbon monoxide* | | |
| 8-h average[a] | 9 ppm (10 mg/m$^3$) | No standard |
| 1-h average[a] | 35 ppm (40 mg/m$^3$) | No standard |
| *Nitrogen dioxide* | | |
| Annual mean (arithmetic) | 0.053 ppm (100 μg/m$^3$) | 0.053 ppm (100 μg/m$^3$) |
| *Ozone* | | |
| Maximum daily 1-h average | 0.12 ppm (235 μg/m$^3$) | 0.12 ppm (235 μg/m$^3$) |
| *Lead* | | |
| Maximum quarterly average | 1.5 μg/m$^3$ | 1.5 μg/m$^3$ |

[a] Not to be exceeded more than once per year.

### *Toxic pollutants emissions from mobile sources*

The CAA Amendments of 1990 set the emission standards for vehicles in two tiers. The Tier I standards set the limit for $NO_x$ tailpipe emissions at 0.6 g/mile per automobiles and must meet this standard for 100,000 miles or 10 years. Light-duty trucks must meet this standard for 75,000 miles or 7 years. The standard for hydrocarbons is set at 0.4 g/mile. Both standards have been enacted beginning in 1994. Tier II tailpipe emission standards for nitrogen oxides and hydrocarbons will be set at 0.2 and 0.125 g/mile, respectively, beginning in the year 2003. The Tier I standard for carbon monoxide will start at 10 g/mile. The vehicle must meet the limit for 50,000 miles or 5 years.

## *Emissions from industrial sources*

Hazardous air pollutants from a stationary source involve a stationary source located within a contiguous area that emits or has the potential to emit 10 tons/year or more of any hazardous air pollutant or 25 tons or more per year of any combination of hazardous air pollutants. The EPA may establish a lesser quantity for such a source on the basis of the potency of the air pollutant, persistence, potential for bioaccumulation, and other relevant factors.

Listed below are individual classes of compounds that are air toxicants which the EPA is required to regulate under the new CAA:

*Compound/CAS registry number:*

Acetaldehyde [75-07-0]
Acetamide [60-35-5]
Acetonitrile [75-0-8]
Acetophenone [98-86-2]
2-Acetylaminofluorene [53-96-3]
Acrolein [107-02-8]
Acrylamide [979-06-1]
Acrylic acid [79-10-7]
Acrylonitrile [[107-13-1]
Allyl chloride [107-05-1]
4-Aminobiphenyl [92-67-1]
Aniline [62-53-3]
*o*-Anisidine [90-04-0]
Asbestos [1332-21-4]
Benzene (including benzene from gasoline) [71-43-2]
Benzidine [92-87-5]
Benzotrichloride [98-07-7]
Benzyl chloride [100-44-7]
Biphenyl [92-52-4]
Bis(chloromethyl)ether [542-88-1]
Bis(2-ethylhexl)phthalate (DEHP) [117-81-7]
Bromoform [75-25-2]
1,3-Butadiene [106-99-0]
Calcium cyanamide [156-62-7]
Caprolactam [105-60-2]
Captan [133-06-2]
Carbaryl [63-25-2]
Carbon disulfide [75-15-0]
Carbon tetrachloride [56-23-5]
Carbonyl sulfide [463-58-1]
Catechol [120-80-9]
Chloramben [133-90-4]
Chlordane [57-74-9]
Chlorine [7782-50-5]
Chloroacetic acid [79-11-8]
2-Chloroacetophenone [532-27-4]
Chlorobenzene [108-90-7]
Chlorobenzilate [510-15-6]
Chloroform [67-66-3]
Chloromethyl methyl ether [107-30-2]
Chloroprene [126-99-8]
Cresols/cresylic acid (isomers and mixture) [1319-77-3]
*m*-Cresol [108-39-4]
*o*-Cresol [95-48-7]
*p*-Cresol [106-44-5]
Cumene [98-82-8]
2,4-D, salts and esters [94-75-7]
DDE [3547-04-4]
Diazomethane [334-88-3]
Dibenzofurans [132-64-9]
1,2-Dibromo-3-chloropropane [96-12-8]
Dibutylphthalate [84-74-2]

*(continued)*

*Compound/CAS registry number:*

1,4-Dichlorobenzene [106-46-7]
3,3-Dichlorobenzidene [91-94-1]
Dichloroethyl ether [bis(2-chloroethyl)ether] [111-44-4]
1,3-Dichloropropene [542-75-6]
Dichloryos [62-73-7]
Diethanolamine [111-42-4]
*N,N*-Diethyl aniline (*N,N*-dimethylaniline) [121-69-7]
Diethyl sulfate [64-67-5]
3,3-Dimethoxybenzidine [119-90-4]
Dimethyl aminoazobenzene [60-11-7]
3,3′-Dimethyl benzidine [119-93-7]
Dimethyl carbamoyl chloride [79-44-7]
Dimethyl formamide [68-12-2]
1,1-Dimethyl hydrazine [57-14-7]
Dimethyl phthalate [131-11-3]
Dimethyl sulfate [77-78-1]
4,6-Dinitro-*o*-cresol, and salts [53-45-21]
2,4-Dinitrophenol [51-28-5]
2,4-Dinitrotoluene [121-14-2]
1,4-Dioxane (1,4-diethyleneoxide) [123-91-1]
1,2-Diphenylhydrazine [122-66-7]
Epichlorohydrin (1-chloro-2,3-epoxypropane) [106-89-8]
1,2-Epoxybutane [106-88-7]
Ethyl acrylate [140-88-5]
Ethyl benzene [100-41-4]
Ethyl carbamate (urethane) [51-79-6]
Ethyl chloride (chloroethane) [75-00-3]
Ethylene dibromide (dibromoethane) [106-93-4]
Ethylene dichloride (1,2-dichloroethane) [107-06-2]
Ethylene glycol [107-21-1]
Ethylene imine (aziridine) [151-56-4]
Ethylene oxide [75-21-8]
Ethylene thiourea [96-45-7]
Ethylidene dichloride (1,1-dichloroethane) [75-34-4]
Formaldehyde [50-00-0]
Heptachlor [76-44-8]
Hexachlorobenzene [118-74-1]
Hexachlorobutadiene [87-68-3]
Hexachlorocyclopentadiene [77-47-4]
Hexachloroethane [67-72-1]
Hexamethylene-1,6-diisocyanate [822-06-0]
Hexamethylphosphoramide [680-31-9]
Hexane [110-54-3]
Hydrazine [302-01-2]
Hydrochloric acid [7647-01-0]
Hydrogen fluoride (hydrofluoric acid) [7664-39-3]
Hydrogen sulfide [7783-06-4]
Hydroquinone [123-31-9]
Isophorone [78-59-1]
Lindane (all isomers) [58-89-9]
Maleic anhydride [108-31-6]
Methanol [67-56-1]
Methoxychlor [72-43-5]
Methyl bromide (bromomethane) [74-83-9]
Methyl chloride (chloromethane) [74-87-3]
Methyl chloroform (1,1,1-trichloroethane) [71-55-6]
Methyl ethyl ketone (2-butanone) [78-93-3]
Methyl hydrazine [60-34-4]
Methyl iodide (iodomethane) [74-88-4]
Methyl isobutyl ketone (hexane) [108-10-1]
Methyl isocyanate [624-83-9]
Methyl methacrylate [80-62-6]
Methyl tert-butyl ether [1634-04-4]
4,4′-Methylene bis(2-chloroaniline) [101-14-4]
Methylene chloride (dichloromethane) [75-09-2]
Methylene diphenyl diisocyanate (MDI) [101-68-8]
4,4′-Methylenedianiline [101-77-9]
Naphthalene [91-20-3]
Nitrobenzene [98-95-3]
4-Nitrobiphenyl [92-93-3]
4-Nitrophenol [100-02-7]
2-Nitropropane [79-46-9]
*N*-Nitroso-*N*-methylurea [684-93-5]
*N*-Nitrosodimethylamine [62-75-9]
*N*-Nitrosomorpholine [59-89-2]
Parathion [56-38-2]
Pentachloronitrobenzene (quintobenzene) [82-68-8]
Pentachlorophenol [87-86-5]
Phenol [108-95-2]
*p*-Phenylenediamine [106-50-3]
Phsogene [75-44-5]
Phosphine [7803-51-2]
Phosphorus [7723-14-0]
Phthalic anhydride [85-44-9]
Polychlorinated biphenyls (aroclors) [1336-36-3]
1,3-Propane sultone [1120-71-4]
β-Propiolactone [57-57-8]
Propionaldehyde [123-38-6]
Propoxur (Baygone) [114-26-1]

*(continued)*

| *Compound/CAS registry number:* |
|---|
| Propylene dichloride (1,2-dichloropropane) [78-87-5] |
| Propylene oxide [75-56-9] |
| 1,2-Propylenimine (2-methyl aziridine) [75-55-8] |
| Quinoline [91-22-5] |
| Quinone [106-51-4] |
| Styrene [100-42-5] |
| Styrene oxide [96-09-3] |
| 2,3,7,8-Tetrachlorodibenzo-*p*-dioxin [1746-01-6] |
| 1,1,2,2-Tetrachloroethane [79-34-5] |
| Tetrachloroethylene (perchloroethylene) [127-18-4] |
| Titanium tetrachloride [7550-45-0] |
| Toluene [108-88-3] |
| 2,4-Toluene diamine [95-80-7] |
| 2,4-Toluene diisocyanate [584-84-9] |
| *o*-Toluidine [95-53-4] |
| Toxaphene (chlorinated camphene) [8001-35-2] |
| 1,2,4-Trichlorobenzene [120-82-1] |
| 1,1,2-Trichloroethane [79-005] |
| Trichloroethylene [79-01-6] |
| 2,4,5-Trichlorophenol [95-95-4] |
| 2,4,6-Trichlorophenol [88-06-2] |
| Triethylamine [121-44-8] |
| Trifluralin [1582-09-8] |
| 2,2,4-Trimethylpentane [540-84-1] |
| Vinyl acetate [108-05-4] |
| Vinyl bromide [593-60-2] |
| Vinyl chloride [75-01-4] |
| Vinylidene chloride (1,1-dichloroethylene) [75-35-4] |
| Xylenes (isomers and misture) [1330-20-7] |
| *m*-Xylene [108-38-3] |
| *o*-Xylene [95-47-6) |
| *p*-Xylene [106-42-3] |
| Antimony compounds |
| Arsenic compounds |
| Beryllium compounds |
| Cadmium compounds |
| Chromium compounds |
| Cobalt compounds |
| Coke oven emissions |
| Cyanide compounds |
| Glycol ethers |
| Lead compounds |
| Manganese compounds |
| Mercury compounds |
| Fine mineral fibers |
| Nickel compounds |
| Polycyclic organic matter |
| Radionuclides (including radon) |
| Selenium compounds |

## HAPS

Hazardous Air Pollutant (HAP) programs reduce air toxics emissions by regulating industry and implementing emission reduction programs for vehicles and other sources. State or local permits for major industrial sources of air pollution limit HAPs emissions. In many cases, these permits include federal requirements for control equipment to reduce emissions.

HAPs, also called toxic air pollutants or air toxics, are those pollutants known or suspected to cause serious health effects, including cancer. HAPs can also be damaging to the environment. They are different from pollutants with ambient standards. HAPs may exist as particulate matter, as gases or as gases absorbed into particles. They include metal fumes, smoke, other particles, and vapors from fuels, coatings and other sources.

HAPs come from natural sources such as forest fires and volcanoes, and from human sources, both stationary and mobile. Two types of stationary sources generate routine (as opposed to accidental) HAP emissions: point sources, such as aluminum plants or wood cabinet makers; and area sources, such as wood stoves or auto body shops. Mobile sources are major contributors to HAP emissions. They can be classified as on- or off-road: cars and trucks or lawnmowers.

Some HAPs have a cumulative effect on the body. They may also have environmental effects, including contamination of water, soil, or food. HAPs can be gases, particulates or aerosols (fine solid particles and liquid droplets combined). They include a wide variety of chemicals such as Polychlorinated Biphenyls (PCBs), benzene, dioxins, furans, formaldehyde, trichloroethylene, and trace metals such as lead, arsenic and mercury. The Canadian Environmental Protection Act has a priority list of 44 such substances that are potentially toxic and need further assessment.

### *Emission reduction measures for HAPs*

Only limited data is available on HAPs. Emission reduction measures that target volatile organic compounds and particulate matter will reduce HAP levels.

### *Global warming*

Global warming is the result of an accumulation of "greenhouse gases" such as carbon dioxide, methane, nitrous oxide and ChloroFluoroCarbons (CFCs) in the atmosphere. These gases normally act like an insulating blanket around the earth.

With the rapid rise in population, with increased agricultural and industrial activity, and increased consumption of coal, oil and gasoline, the levels of these gases in the atmosphere are rising. Higher concentrations of greenhouse gases trap heat in the lower atmosphere and on the earth's surface. As a result, over the next 50–100 years, the earth's temperature could increase by as much as 3°C. This would change weather patterns and seasonal temperatures. Autumn and winter would be warmer; summers would be drier. Sea levels across the globe might rise 1 m, as polar ice caps melt.

A rise in sea level would have significant implications for low lying coastal areas. Global warming would adversely affect our health, fisheries, the forest ecosystem and our agricultural industry.

## Clean Water Act

- Title: CWA
- Agency: EPA
- Years passed: 1972; amended 1977-1983, 1987, 1988, 1990, 1992; reauthorized in 1997; originally the Federal Water Control Act
- Groups regulated: industry

The CWA is a US legislation; a 1977 amendment of the earlier Federal Water Pollution Control Act (FWPCA, 1972). It is administered by the EPA and, in addition to providing for funding for municipal sewage treatment plants, it authorizes the regulation of emissions from municipal and industrial sources. The objective of this act is to restore and maintain the chemical, physical, and biological integrity of US waters. The provisions of the act are as follows:

1 To eliminate the discharge of pollutants into navigable waters.
2 To achieve an interim goal of water quality for the protection and propagation of fish, shellfish, and wildlife.

3 To prohibit the discharge of toxic pollutants in toxic amounts.
4 To develop and implement waste treatment processes for adequate control of sources of pollutants.
5 To provide federal financial assistance to construct publicly owned waste treatment works.
6 To develop the technology necessary to eliminate the discharge of pollutants in navigable waters and the oceans.

Some important actions taken under this statute include setting standards for emissions of organics from smelter operations and publishing priority lists of toxic pollutants. This act also allows the US Government to recover clean-up and other costs as damages from the polluting agency, company or individual.

### *Synopsis of law*

The EPA has had responsibility for regulating toxic water pollutants since 1972. As originally enacted, Section 307 of the FWPCA required the EPA to publish within 90 days and periodically add to a list of toxic pollutants for which effluent standards (discharge limits) would then be established.

Section 307 (a)(4) of the FWPCA originally specified that in establishing standards for any listed pollutant the EPA was to provide an "ample margin of safety", a difficult criterion to meet for most toxic pollutants and arguably impossible for any known to be carcinogenic (Bruser *et al.*, 1981). The law also mandated both a rapid timetable and a complex procedure for standard setting. The EPA's slow implementation of these instructions precipitated a series of lawsuits. The agency eventually reached a court-sanctioned settlement that fundamentally altered federal policy toward toxic pollutants of the nation's waterways (NRDC, 1984). The settlement allowed the EPA to act under other provisions of the act that allow consideration of economic costs and technological feasibility in setting limits. Congress incorporated the terms of this settlement in 1977 amendments to the statute.

In 1987, Congress again amended the FWPCA to toughen standards for toxic pollutants. Under the 1977 amendments, the EPA had developed health-based "water quality criteria" for 126 compounds it had identified as toxic. These criteria described desirable maximum contamination levels which, because the EPA's discharge limits were technology based, generally were substantially lower than the levels actually achieved (Heineck, 1989). The 1987 amendments gave these advisory criteria real bite by requiring that states incorporate them in their own mandatory standards for water quality and impose on operations discharging into below-standard waterways additional effluent limits. The following are listed as toxic pollutants:

*Inorganics*
Antimony and compounds
Arsenic and compounds
Asbestos
Beryllium and compounds
Cadmium and compounds
Copper and compounds
Cyanides
Lead and compounds
Mercury and compounds
Nickel and compounds
Selenium and compounds
Silver and compounds
Thallium and compounds
Zinc and compounds

(*continued*)

*Organics*

Acenaphthene
Acrolein
Acrylonitrile
Aldrin/dieldrin
Benzene
Benzidine
Carbon tetrachloride
Chlordane
Chlorinated benzenes (other than dichlorobenzenes)
Chlorinated ethanes (including 1,2-dichloroethane, 1,1,1-trichloroethane, and hexachloroethane)
Chlorinated naphthalene
Chloroalkyl ethers (chloromethyl, chloroethyl, and mixed ethers)
Chlorophenols (2-chlorophenol, 2,4-dichlorophenol, pentachlorophenol)
Chlorophenylphenyl ether
DDT and metabolites
Dichlorobenzenes (1,2-, 1,3-, and 1,4-dichlorobenzenes)
Dichlorobenzidine
Dichloropropene
2,4-Dimethylphenol
Dinitroltoluene
Diphenylhydrazine
Endosulfan and metabolites
Endrin and metabolites
Ethylbenzene
Fluoranthene
Haloethers [chlorophenylphenyl ether, bromophenylphenyl ether, bis(dichloroisopropyl)ether, bis(chloroethoxy)methane, and polychlorinated diphenyl ethers]
Halomethanes (methylene chloride, methyl chloride, methyl bromide, bromoform, dichlorobromomethane, trichlorofluoromethane, and dichlorodifluoromethane)
Heptachlor and metabolites
Hexachlorobutadiene
Hexachlorocyclohexane (all isomers)
Hexachlorocyclopentadiene
Isophorone
Naphthalene
Nitrobenzene
Nitrophenols (2,4-dinitrophenol and dinitrocresol)
Nitrosamines
PCBs
Pentachlorophenol
Phenol
Phthalate esters
Polynuclear aromatic hydrocarbons (benzanthracenes, benzopyrenes, benzofluoranthene, chrysene, dibenzanthracenes, and indenopyrenes)
2,3,7,8-Tetrachlorodibenzo-*p*-dioxin (TCDD)
Tetrachloroethylene
Toluene
Toxaphene
Trichloroethylene
Vinyl chloride

## Safe Drinking Water Act

- Title: SWDA
- Agency: EPA
- Year passed: 1974; amended 1986
- Groups regulated: water suppliers

The SWDA is a US act setting standards for drinking water to protect the public health. It authorizes research relating to causes, diagnosis, treatment, control and prevention of human diseases and other impairments resulting, directly or indirectly, from contaminants in drinking water (SDWA, 1974).

### *Synopsis of law*

The 1974 SDWA was enacted to ensure that public water supply systems "meet minimum national standards for the protection of public health". Under the SDWA, the EPA is required to regulate any contaminants "which may have an adverse effect on human health" (Douglas, 1976). The EPA was to establish national primary drinking water regulations for public water systems. For each contaminant of concern, the agency was to prescribe a Maximum Contaminant Level (MCL) or a treatment technique for its control.

The 1974 Act prescribed a two-stage process. The EPA first was required to promulgate interim national primary drinking water regulations, uniform minimum standards that would "protect health to the extent feasible (taking costs into consideration)". These interim regulations were later supplanted by regulations formulated on the basis of a series of reports by the National Academy of Sciences (NAS). The charge to the NAS Safe Drinking Water Committee was to recommend the MCLs necessary to protect humans from any known or anticipated adverse health effects. In turn, the EPA was to specify MCLs as close as feasible to the levels recommended by NAS. By 1986, the EPA had established MCLs for 23 contaminants but treatment techniques for none (Gray, 1986).

In that year, Congress amended the SDWA to cover more contaminants, to apply more pressure to states and localities to clean up their drinking water supplies, and to strengthen the EPA's enforcement role. The EPA was required to adopt regulations for a total of 83 contaminants within 3 years (including all but one of those originally regulated). It was directed to prescribe regulations for two treatment techniques for public water systems: filtration and disinfection. In translating recommended MCLs, now Maximum Contamination Level Goals (MCLGs), into feasible and enforceable regulations, the EPA was directed to assume installation of the best available technology.

Over the past decade, continuing criticism of the SDWA has prompted yet another attempt at reform. Congress failed to pass amendments at the end of the 1994 session, but further revisions to the SDWA are expected in the near future. Congress will likely abandon the current requirement that the EPA issue standards for 25 additional contaminants every 3 years in order to permit the EPA to choose contaminants for regulation based on scientific criteria. There is also support for a cost–benefit approach to standard setting that would release the current feasibility requirement for standards. The financial burden on smaller municipal water systems for achieving compliance is a major political issue.

In accordance with the 1986 Amendments to the SDWA, the EPA has imposed regulations and guidelines for the MCLs and MCLGs of certain chemicals in the drinking water. MCL is defined as the maximum permissible level of a contaminant in water which is delivered to any user of a public water system. MCLG is a nonenforceable concentration of a drinking water contaminant that is protective of adverse human health effects and allows an adequate margin of safety. These contaminants include metals, anions, volatile organics, synthetic organic chemicals, pesticides, radionuclides, and coliforms and other bacteria that can produce adverse health effects on humans.

Table 12.3 lists the MCL and MCLG values of toxic substances, including regulated inorganic, organic, radionuclides, and aggregate properties. Drinking water standards for radionuclides and organics are listed in Tables 12.4 and 12.5, respectively. The MCLG values represent the desired concentration levels of contaminants at which no known or anticipated adverse health effects occur, thus allowing an adequate margin of

*Table 12.3* Drinking water standards for inorganic chemicals (EPA, 1978)

| *Inorganic contaminants* | *MCL (mg/l)* | *MCLG (mg/l)* |
|---|---|---|
| Antimony | 0.006 | 0.006 |
| Arsenic | 0.05 | – |
| Asbestos (fibers/l > 10 μm length) | 7 MFL | 7 MFL |
| Barium | 2 | 2 |
| Beryllium | 0.004 | 0.004 |
| Bromate | 0 | 0 |
| Cadmium | 0.005 | 0.005 |
| Chloramine[a] | 4 | 4 |
| Chlorine | 4 | 4 |
| Chlorine dioxide | 0.8 | 0.3 |
| Chlorite | 1 | 0.08 |
| Chromium, total | 0.1 | 0.1 |
| Copper (at tap) | Treatment technique; 1.3 (action level) | 1.3 |
| Cyanide | 0.2 | 0.2 |
| Fluoride[b] | 4 | 4 |
| Hypochlorite (regulated as chlorine) | – | 4 |
| Hypochlorous acid (regulated as chlorine) | – | 4 |
| Lead (at tap) | Treatment technique; 0.015 (action level) | 0 |
| Mercury (inorganic) | 0.002 | 0.002 |
| Nickel | 0.1 | 0.1 |
| Nitrate (as N) | 10 | 10 |
| Nitrite (as N) | 1 | 1 |
| Nitrate and nitrite (both as N) | 10 | 10 |
| Selenium | 0.05 | 0.05 |
| Sulfate | 500 | 500 |
| Thallium | 0.002 | 0.0005 |

[a] Measured as free chlorine.
[b] Under review.

safety. MCL and MCLG values should be as close as feasible and should be achieved by using the best available treatment techniques.

While the MCLG for the microbial agents *Giardia lamblia, Legionella,* total coliform bacteria, viruses, and standard plate count has been set at zero, the MCL is based on the treatment technique. It is critical that no bacteria, viruses or other microbial agents be found in the drinking water supply.

*Table 12.4* Drinking water standards for radionuclides

| *Radionuclides* | *MCL* | *MCLG* |
|---|---|---|
| β particles and photo activity | 4 mrem | 0[a] |
| Gross α-particle activity | 15 *p*Ci/l | 0[a] |
| Radium-226 and -228 (combined) | 5 *p*Ci/l | 0[a] |
| Radon | 300 *p*Ci/l | 0 |
| Uranium | 20 μg/l | 0 |

[a] Proposed in 1991, no final decision has been taken.

*Table 12.5* Drinking water standards for organics

| *Organics* | *MCL (mg/l)* | *MCLG (mg/l)* |
|---|---|---|
| Acifluofen | – | 0 |
| Acrylamide | Treatment technique | 0 |
| Acrylonitrile | – | 0 |
| Adipate (diethylhexyl) | 0.4 | 0.4 |
| Alachlor | 0.002 | 0 |
| Aldicarb | 0.007 | 0.007 |
| Aldicarb sulfone | 0.007 | 0.007 |
| Aldicarb sulfoxide | 0.007 | 0.007 |
| Atrazine | 0.003 | 0.003 |
| Bentazon | – | 0.02 |
| Benzene | 0.005 | 0 |
| Benzo(a)pyrene (PAH) | 0.0002 | 0 |
| Bromodichloromeethane (THM) | 0.08 | 0 |
| Bromoform (THM) | 0.08 | 0 |
| Carbofuran | 0.04 | 0.04 |
| Carbon tetrachloride | 0.005 | 0 |
| Chloral hydrate | 0.06[a] | 0.04 |
| Chlordane | 0.001 | 0 |
| Clhorodibromomethane (THM) | 0.06 | 0.08 |
| Chloroform (THM) | 0.08 | 0 |
| Cyanazine | – | 0.1001 |
| 2,4-D | 0.07 | 0.07 |
| Dalapn | 0.2 | 0.2 |
| Di(2-ethylhexyl)adipate | 0.4 | 0.4 |
| Dibromochloropropane (DBCP) | 0.0002 | 0 |
| Dichloroacetic acid | 0.06 | 0 |
| *o*-Dichlorobenzene | 0.6 | 0.6 |
| *m*- Dichlorobenzene | 0.6 | 0.6 |
| *p*- Dichlorobenzene | 0.075 | 0.075 |
| 1,2-Dichloroethane | 0.005 | 0 |
| 1,1-Dichloroethylene | 0.007 | 0.007 |
| *cis*-1,2-Dichloroethylene | 0.07 | 0.07 |
| *trans*-1,2- Dichloroethylene | 0.1 | 0.1 |
| Dichloromethane | 0.005 | 0 |
| 1,2-Dichloropropane | 0.005 | 0 |
| 1,3-Dichloropropane | – | 0 |
| Di(2-ethylhexyl)phthalate (PAE) | 0.006 | 0 |
| Dinoseb | 0.007 | 0.007 |
| Diquat | 0.02 | 0.02 |
| Endothall | 0.1 | 0.1 |
| Endrin | 0.002 | 0.002 |
| Epichlorohydrin | Treatment technique | 0 |
| Ethylbenzene | 0.7 | 0.7 |
| Ethylene dibromide (EDB) | 0.00005 | 0 |
| Glyphosphate | 0.7 | 0.7 |
| Heptachlor | 0.0004 | 0 |
| Heptachlor epoxide | 0.0002 | 0 |
| Hexachlorobenzene | 0.001 | 0 |
| Hexachlorobutadiene | 0.001 | – |
| Hexachlorocyclopentadiene | 0.05 | 0.05 |
| Lindane | 0.0002 | 0.0002 |
| Methoxychlor | 0.04 | 0.04 |

Table 12.5 (*continued*)

| *Organics* | *MCL (mg/l)* | *MCLG (mg/l)* |
|---|---|---|
| Monochlorobenzene | 0.1 | 0.1 |
| Oxamyl(Vydate) | 0.2 | 0.2 |
| PCBs | 0.0005 | 0 |
| Pentachlorophenol | 0.001 | 0 |
| Picloram | 0.5 | 0.5 |
| Simazine | 0.004 | 0.004 |
| Styrene | 0.1 | 0.1 |
| 2,3,7,8-TCDD(Dioxin) | $3 \times 10^{-8}$ | 0 |
| Tetrachloroethylene | 0.005 | 0 |
| Toluene | 1 | 1 |
| Toxophene | 0.003 | 0 |
| 2,4,5-TP | 0.05 | 0.05 |
| Trichloroacetic acid | 0.06[a] | 0.3 |
| 1,2,4-Trichlorobenzene | 0.07 | 0.07 |
| 1,1,1-Trichloroethane | 0.2 | 0.2 |
| 1,1,2-Trichloroethane | 0.005 | 0 |
| Vinyl chloride | 0.002 | 0 |
| Xylenes | 10 | 10 |

[a] Total for all haloacetic acids cannot exceed 0.06 mg/l level. THM, trihalomethane; PAH, polynuclear aromatic hydrocarbons; PAE, phthalate esters.

## SMCLs

Secondary drinking water standards are unenforceable federal guidelines with regard to taste, odor, color, and other nonaesthetic effects of drinking water. Federal law does not require water systems to comply with them. States may, however, adopt their own enforceable regulations. Secondary Maximum Contaminant Levels (SMCLs) are presented in Table 12.6.

Concentration levels of carcinogenic chemicals in drinking water supplies must not exceed specific values as defined under the "Health Advisories". Such "no-exceed"

*Table 12.6* Secondary drinking water standards

| *Chemicals/aggregate properties* | *SCML (mg/l)* |
|---|---|
| Aluminum | 0.05–0.2 |
| Chloride | 250 |
| Color | Fifteen color units |
| Copper | 1.0 |
| Corrosivity | Noncorrosive |
| Fluoride (under review) | 2.0 |
| Foaming agents | 0.5 |
| Iron | 0.3 |
| Manganese | 0.05 |
| Odor | 3 threshold odor number |
| PH | 6.5–8.5 SU |
| Silver | 0.1 |
| Sulfate | 250 |
| Total Dissolved Solids (TDS) | 500 |
| Zinc | 5 |

concentrations of cancer-causing pollutants in the water have been defined for adult (70 g) and child (10 g) for 1 day, 10 days, and long-term consumption, respectively. Any further discussion on the subject is beyond the scope of this book. For further information regarding drinking water regulations and "health advisories" readers may call the Safe Drinking Water Hotline at +1-800-426-4791 or EPA's Office of Water at +1-202-260-1332.

## Precautionary principle

Across all of these regulations, as well as many more presented in this book, risks and risks–benefit analyses are performed to provide a means of setting allowable exposure limits. An approach which developed in Europe and is now advocated for use in the US is the precautionary principle (Bodansky, 1994).

As Sonja Boehmer Christiansen has pointed out, the precautionary principle evolved out of the German socio-legal tradition, created in the heyday of democratic socialism in the 1930s, centering on the concept of good household management. This was regarded as a constructive partnership between the individual, the economy and the state to manage change so as to improve the lot of both society and the natural world upon which it depended for survival. This invested the precautionary principle with a managerial or programmable quality, a purposeful role in guiding future political and regulatory action.

As Boehmer Christiansen argues, the German concept of vorsorgeprinzip means much more than the rough English translation of foresight planning. It absorbs notions of risk prevention, cost effectiveness but in a looser economic framework, ethical responsibilities towards maintaining the integrity of natural systems, and the fallibility of human understanding. The right of nature means, in part, giving it room to accommodate human interference, so precaution presumes that mistakes can be made. For the Germans, therefore, precaution is an interventionist measure, a justification of state involvement in the day to day lives of its land and its citizenry in the name of good government. Social planning in the economy, in technology, in morality and in social initiatives all can be justified by a loose and open ended interpretation of precaution. As we shall see, it is precisely the unravellability that makes precaution both feared and welcomed (Raffensperger and Fickner, 1999).

Throughout the late 1970s and early 1980s these notions of care and wise practice have been extended to six basic concepts now enshrined in the precautionary principle.

- *Preventative anticipation*: a willingness to take action in advance of scientific proof of evidence of the need for the proposed action on the grounds that further delay will prove ultimately most costly to society and nature, and, in the longer term, selfish and unfair to future generations.
- *Safeguarding* of ecological space or environmental room for maneuver as a recognition that margins of tolerance should not even be approached, let alone breached. This is sometimes known as widening the assimilative capacity of natural systems by deliberately holding back from possible but undesirable resource use.
- *Proportionality* of response or cost-effectiveness of margins of error to show that the selected degree of restraint is not unduly costly. This introduces a bias to conventional cost benefit analysis to include a weighting function of ignorance,

and for the likely greater dangers for future generations if life support capacities are undermined when such risks could consciously be avoided.

- *Duty* of care, or onus of proof on those who propose change: this raises profound questions over the degree of freedom to take calculated risks, thereby to innovate, and to compensate for possible losses by building in ameliorative measures. Formal duties of environmental care, coupled to an extension of strict liability for any damage, no matter how unanticipated, could throttle invention, imagination and growth. Alternatively, when creatively deployed such strictures could encourage imagination and creativity in technology, economic valuation, technological advance and unusual forms of ameliorative compensation. Hence the concept of proportionality can be regarded either as a deadweight or a touchstone for the visionary (Costanza and Perrings, 1990).
- *Promoting* the cause of intrinsic natural rights: the legal notion of ecological harm is being widened to include the need to allow natural processes to function in such a manner as to maintain the essential support for all life on earth. The application of ecological buffers in future management gives a practical emphasis to the thorny ethical concept of intrinsic natural rights.
- *Paying* for past ecological debt: precaution is essentially forward looking but there are those who recognize that in the application of care, burden sharing, ecologically buffered cost effectiveness and shifting the burden of proof, there ought to be a penalty for not being cautious or caring in the past. This suggests that those who have created a large ecological burden already should be more "precautious" than those whose ecological footprints have to date been lighter. In a sense this is precaution put into reverse: compensating for past errors of judgment based on ignorance or an unwillingness to shoulder an unclearly stated sense of responsibility for the future. This element of the principle is still embryonic in law and practice, but the notion of "common but differentiated responsibility" enshrined in the UN Framework Convention on Climate Change, and the concept of conducting precaution "according to capabilities" as laid down in Principle 15 of the Rio Declaration reflect to some extent these ideas.

By no means all of these interpretations are formally approved in international law and common practice. At present the line is to act prudently when there is sufficient scientific evidence and where action can be justified on reasonable judgments of cost effectiveness and where inaction could lead to potential irreversibility or demonstrate harm to the defenders and future generations. In substance, the application is usually derived for chemicals whose effects are potentially toxic, persistent or bioaccumulative (i.e., concentrating in the food chain from one predator to another), or where certain combinations or concentrations of chemicals could alter the physical and chemical state of soil or water. In this sense the notion in international affairs is mostly one of prevention, and justification of some action rather than to claim scientific uncertainty as a reason for delay.

According to the "precautionary principle," governments should take precautions to protect public health and the environment, even in the absence of clear evidence of harm and notwithstanding the costs of such action. Supporters of this principle assume that the consequences of scientific false positives, which ultimately cost society money, are greater than the consequences of false negatives, which cost society lives. Therefore, scientific uncertainty should become a basis for environmental protection regulation. Professor Cross contends that the principle is based on the fallacy that actions have no consequences beyond those intended and argues that regulators should also consider new and collateral risks created by regulations to determine whether the regulations are in the best interest of public health and the environment.

Environmental regulation entails physical risks of its own, including perverse side effects that undermine the intended benefits of the regulation. Alternatives to regulated activities can produce unanticipated physical risks. Similarly, regulating a substance may result in the loss of substantial public health or environmental benefits that the substance provided. Risks of remediation are a third set of physical risks of environmental regulation that Professor Cross identifies. Often efforts to clean up environmental hazards are counterproductive, doing more harm than good. The precautionary principle fails to take these risks into account.

There are also indirect risks of the precautionary principle, such as the health risks of political resource misallocation (Holm and Harris, 1999). Regulators using the precautionary principle address public health problems without any comparative risk analysis and focus on the problem they face first, rather than the problem with the greatest magnitude of environmental risks. Another indirect risk of the precautionary principle is the health risk produced by the costs of regulation. Unnecessary and expensive regulation harms the national economic welfare by decreasing productivity, lowering wages, and raising prices. This depletes resources available for expenditures on public health. Therefore, the public suffers an increase in morbidity and mortality. Because the precautionary principle does not address such risks, its application could cause more health harm than good.

Professor Cross favors a comparative risk analysis, combined with some of the tenets of the precautionary principle, that first identifies and evaluates all problems and then distributes scarce resources to remedy the greatest public health and environmental problems. The precautionary principle should not be abandoned, nor should it be misapplied to prevent or burden all government action. Cross also discusses the Risk Trade-Off Analysis (RTA) of John Graham and Jonathan Wiener. The RTA has the potential of unduly preoccupying public health decision-making with the risks of regulation. Professor Cross concludes that the goal should be better regulation, not zero regulation. To that end, a system that more evenly balances the risks of a product or an activity with the risks of regulation is preferable to the rigid precautionary principle.

## References


Bodansky D. Precautionary principles in environmental law. In: O'Riordan T, Cameron J, editors. *Interpreting the Precautionary Principle*. London: Earthscan Publications, 1994. pp. 203–228.

Bruser J, Harris R, Page T. Waterborne carcinogens: an economist's view. *The Scientific Basis of Health and Safety Regulation*. Washington, DC: Brookings Institution, 1981.

Clean Air Act (1976), 42 U.S.C. § 7401 *et seq.*

Costanza R, Perrings C. A flexible assurance bonding system for improved environmental management. *Ecol Econ* 1990;2:57–75.

Douglas I. Safe Drinking Water Act of 1975 – history and critique. *Environ Affairs* 1976;5:501.

EPA Proposed interim primary drinking water regulations and Environmental Protection Agency. *Fed Reg* 1978;43(130):29135–29137 (to be codified at 40 C.F.R. § 141).

EPA. National Ambient Air Quality Standards, 2000. www.epa.gov/air/criteria/html.

Federal Water Pollution Control Act Amendments of 1972, 33 U.S.C. § 307.

Gray K. The Safe Drinking Water Act Amendments of 1986: now a tougher act to follow. *Environ L Rep* 1986;16:10338.

Heineck D. New clean water act toxics control initiatives. *Nat Resources Environ* 1989;1:10.

Holm S, Harris J. Precautionary principle stifles discovery. *Nature* 1999;400:398.

Natural Resources Defense Council, Inc. vs. United States Environmental Protection Agency. 1984;595 F. Suppl 1255 (S.D.N.Y).

Natural Resources Defense Council, Inc. vs. United States Environmental Protection Agency. 1987;824 F.2d 1146 (D.C.Cir).

Reed PD. The trial of hazardous air pollution regulation. *Environ L Reg* 1986;16:10066–10072.

Raffensperger C, Fickner J. *Protecting Public Health and the Environment: Implementing the Precautionary Principle*. Los Angeles, CA: Island Press, 1999.

Safe Drinking Water Act (1974), 42 U.S.C. §§ 300f–300j-9.

Chapter 13

# Understanding the Safe Drinking Water and Toxic Enforcement Act of 1986

## (California's Proposition 65)

*Stephen B. Harris and Judith L. Garfield*

### Introduction

In November of 1986, California voters overwhelmingly ratified Proposition 65, the Safe Drinking Water and Toxic Enforcement Act of 1986 (California Code of Regulations, 1986). The law's objective is to protect California's drinking water sources and human health by regulating how chemicals known to the state to cause cancer, birth defects, or other reproductive toxicity are dealt with. Proposition 65 requires the Governor to formally register and inform the public and/or employees of private business – at least annually – with a list of such chemicals. The Governor selected California's Environmental Protection Agency and the Office of Environmental Health Hazard Assessment (OEHHA) to carry out the proposition.

On February 27, 1987, the Governor published the first list of chemicals; it included only twenty-six identified carcinogens and three reproductive toxicants. As of December 2000, 724 chemicals have been listed; 465 for cancer and 259 for reproductive toxicity (OEHHA, 2001). The list consists of a constellation of chemicals including dyes, solvents, pesticides, drugs, food additives, and by-products of certain processes. These chemicals may be naturally occurring or synthetic, and some are found in common household products while others are specialty chemicals used in industrial applications. Chemicals recognized as producing cancer are labeled "carcinogens", and those producing birth defects or other reproductive harm are labeled "reproductive toxicants".

While the proposition describes many ways to improve the safety of drinking water, it mainly focuses on two problems (California Code of Regulations, 1986), which are presented in this chapter: First, businesses need to warn people that they may be or may have been exposed to certain chemicals. Companies are forbidden to deliberately expose anyone to a listed chemical without first providing a "clear and reasonable warning" to the individual. The warning requirement takes effect 12 months after the chemical in question is officially listed.

Second, companies – for profit or nonprofit – are not allowed to intentionally permit a significant amount of a listed chemical to be released into any drinking water source, which includes surface water, ground water, or land. The law affects businesses with ten or more employees in California or businesses that sell products in California. However, governmental organizations and drinking water utilities are not bound to this law. The discharge ban takes effect 20 months after the offending chemical is listed.

## How chemicals are listed

The State of California's OEHHA relies upon published scientific literature when considering whether a chemical is potentially harmful. As described in Proposition 65, there are two ways for the government to place a chemical on the list: First, a chemical must be identified as causing cancer or reproductive toxicity by a state or federal agency or a formally organized group or program or, second, a chemical must be clearly shown to cause cancer, birth defects, or reproductive harm, as determined by the state's "qualified experts" (two independent committees of scientists and health professionals appointed by the governor).

To authorize the list of carcinogens, the Science Advisory Board of OEHHA's Carcinogen Committee named the following agencies: United States Environmental Protection Agency (EPA), United States Food and Drug Administration (FDA), National Institutes of Occupational Safety and Health (NIOSH), National Toxicology Program (NTP), and the International Agency for Research on Cancer (IARC). To address the list of reproductive toxicants, the Science Advisory Board of OEHHA's Developmental and Reproductive Toxicant Identification Committee (DART) uses the same organizations, except that IARC is only allowed to evaluate transplacental carcinogenicity.

Both committees have the power to remove any of their respective agencies if they believe the agency no longer has the expertise to identify toxic chemicals. However, chemicals previously listed will remain on the roster.

## The identification process

OEHHA initially determines what chemicals an agency has identified. Then the chemicals are either placed on a list or written up in a report or document that shows similar findings. As part of the process (which may be held publicly if required), an advisory committee reviews the written material. Additionally, the material may be made available to the public for review and comment before the chemical is officially deemed a carcinogen or reproductive toxicant. The information is next published either in the *Federal Register* or elsewhere. Finally, an authority of the agency must agree that the chemical is carcinogenic, and sign either the list, report, or document. The agency may also adopt the information as final for its own regulatory purposes.

The procedure (OEHHA, 1997) used by an agency to formally identify chemicals that cause cancer or reproductive toxicity is summarized below. If data have been determined to be scientifically valid (according to generally accepted principles) for a chemical but an agency has not evaluated the data, the chemical does not satisfy the criteria – even though scientifically valid data exists. Chemicals that have been formally identified must meet the following criteria:

- There is enough evidence from human studies to show a direct relationship between the chemical and cancer or reproductive toxicity.
- There is enough evidence from animal studies to show that the chemical causes cancer or reproductive toxicity. For carcinogens and reproductive toxicants, this includes exposing various animal species and strains to the chemical using different routes and dose levels, and exposure conditions (frequency and duration). Then for carcinogens the data shows increased tumor frequency, early age of onset of

tumors, or formation of tumors at a specific site. Or, the evidence may include a single study that results in high tumor frequency for a tumor type. For reproductive toxicity, this also includes maternal toxicity data.

Once there is enough evidence to warrant listing a chemical, the following steps need to be taken. At least 60 days prior to adding a chemical to the list, OEHHA publishes an announcement in the California Regulatory Notice Register identifying the agency (e.g., EPA) prepared to list the chemical and the name of chemical about to be listed. The Carcinogen or the DART Committee (whichever is most appropriate) is then given copies of the notice along with at least 30 days to review the proposed action.

Within those 30 days, everyone who is concerned about the status of the chemical, including any member of the proper committee, must submit to OEHHA in writing – along with supporting documentation – any objections they have to adding the chemical to the official list. If a committee member's objections show there is not enough evidence to support the criteria that was previously discussed, OEHHA then reviews the objections. If OEHHA finds that the evidence is flawed, it refers the chemical back to the appropriate committee so they can decide whether they think the chemical has in fact been clearly shown to cause cancer or reproductive toxicity. If it turns out they agree that the evidence is faulty, OEHHA may reconsider placing the chemical on the list.

Then again, the lead agency may take a second look at a chemical on its own or by request from a concerned third party, including any member of the appropriate committee. OEHHA may also ask the committee for advice about chemicals that are being reassessed so as to better decide whether the chemical should stay on the list. Until OEHHA accepts the recommendation, the chemical remains on the list.

## Warning requirements

Proposition 65 requires that businesses provide a clear and reasonable warning whenever people are exposed to a chemical listed as a carcinogen or reproductive toxicant. While businesses are not required to change in any way to reduce exposures, they must inform those who have been exposed. But what constitutes an exposure? Broadly defined, it is when a substance is inhaled, ingested, or absorbed through the skin from any or all of the following ways – air, water, food, consumer products, and any other environmental exposure as well as occupational or workplace exposures. Warnings are required under three different circumstances: chemical exposure in consumer products or services, the workplace, and the environment.

To warn the public about unhealthy consumer products, businesses may place a label on the product or the container it is packaged in, identify the product at the retail outlet (e.g., shelf labels, signs, or menus), and/or identify the product using signs (e.g., public advertising). To warn employees about occupational exposures, employers must provide employees access to a worker safety program that fully complies with the information, training, and labeling requirements of the federal or California hazard communication standards for the chemical in question. Warnings may also be presented on signs in the workplace or on labels affixed to a product containing a listed chemical. These alternative warnings need only reach employees who have been exposed at their workplace. All other exposures that do not meet the definition of consumer or occupational exposures are considered environmental exposures.

For potential environmental exposures, preplanned warnings must be designed to inform people before they are exposed and, obviously, must make it clear that the chemical is known to cause cancer, birth defects, or other reproductive harm. For example, a California mini-market may comply with the law by positioning a sign in a highly visible location that states the following: "Chemicals known to the State of California to cause cancer, birth defects, or other reproductive harm may be present in foods or beverages sold or served here".

## Discharge prohibition

Proposition 65 adheres to the fundamental philosophy that industry is responsible for proving that the chemicals it uses are safe. In this way, government agencies do not have to prove that the chemicals cause harm and, consequently, should be regulated. Hence, while companies may have to change the way they do business to comply with the discharge rule, they are only responsible for their own chemical emissions, not those that result from the actions of others. A company cannot be held liable for chemicals that are found in drinking water supplies or water that does not meet primary drinking water standards.

There are only a few exceptions to the above restrictions. If a business can show that the chemical exposure or discharge it caused poses "no significant risk" of cancer, warnings about discharges into drinking water are not required. This risk level is defined as a cancer risk of not more than one excess case of cancer in 100,000 individuals exposed over a 70-year lifetime. If a "No Observable Effect Level" (NOEL) on reproduction is seen (assuming 1,000 times the exposure level in question), the business is exempt as well. In this framework, the NOEL is the highest dose level that has not been linked with observable reproductive toxicity in humans and animals.

Additionally, discharges that do not cause a significant amount of the chemical to enter any drinking water source are also excused, as long as the chemical that is being discharged complies with all other laws and applicable regulations. To help communities decide whether to excuse exposures or discharges for which they are responsible, OEHHA instituted guidelines for what it considers safe daily intake levels for particular chemicals that represent the "no significant risk" level and the one one-thousandth NOEL.

## Quantitative risk

A quantitative risk assessment is conducted on a listed chemical to determine the level of daily exposure that poses no significant risk. Predetermined scientific standards are used to make a valid assessment. For animal bioassays, the following stipulations must be considered: for animal bioassays, the following questions must be answered: is the study protocol carefully designed, to what extent is animal exposure similar to human exposure, is there is a chronological exposure pattern, how long should the study last, how pure is the test material, what is the size of the control and exposed groups, what is the route of exposure, and how often do tumors or birth defects occur?

The epidemiological data that results from human studies must be of high enough quality to determine whether the study was appropriate (considering the above factors), how the exposed and reference (control) groups were selected, how reliable was the measurement of exposure, and how thoroughly the study was followed up. Biases and confounding factors need to be identified and quantified. The risk assessment is based on the results of the most sensitive study for all exposure routes.

Assuming there is no carcinogenic threshold dose (i.e., doses below which there is no carcinogenic response) determined in the cancer bioassays, no-threshold models are used to develop a quantitative risk estimate. Furthermore, in the absence of mechanistic data from the cancer bioassays to support a threshold mechanism, it has been assumed by the USEPA and most regulatory agencies that any dose of a carcinogen in animal studies is associated with some increased risk in humans.

After identifying the most suitable cancer bioassay(s) for developing a cancer risk assessment, the high doses to which animals were exposed must be transformed into equivalent doses in humans (low doses). Numerous mathematical models have been designed to help convert the data to estimate human cancer risk. These models vary in the way they assume cancer arises or who is prone. Thus, these models can produce a variety of risk estimates. OEHHA uses a linearized multistage model (based on the biological theory of a particular result, such as DNA damage, that is responsible for the carcinogenic response caused by the chemical) (Crump, 1979).

If when individual tumors first appear, survival is reduced and it is uncertain what reason or reasons led to tumor development, time-to-tumor models may be used. This type of model is based on the assumption that exposure to a carcinogen may have an effect on the amount of time it takes before a tumor arises, and this process may be dose dependent (World Health Organization, 1978). Human cancer potency is derived from data on either human or animal cancers. Potency is expressed as the slope of the dose-response curve for tumor induction.

When good-quality physiologic, pharmacokinetic, and metabolic data can be collected, they may be used to determine risk of interspecies, interdose, and interroute extrapolations. When cancer risk is applied to the general population, it is based on a human body weight of 70 kg, and when cancer risk applies to a certain subpopulation, the following assumptions are made:

| | |
|---|---|
| Man (18 + years of age) | 70 kg |
| Woman (18 + years of age) | 58 kg |
| Woman with conceptus | 58 kg |
| Adolescent (11–18 years of age) | 40 kg |
| Child (2–10 years of age) | 20 kg |
| Infant (0–2 years of age) | 10 kg |

As noted previously, the risk level represents no significant risk and is calculated to result in one excess case of cancer in an exposed population of 100,000. This assumes a lifetime exposure at a particular level, except where public health considerations support a different level. These include chemicals in food produced by cooking, which are necessary to make the food palatable or to avoid microbiological contamination; chlorine disinfection that complies with all state and federal safety standards necessary to conform with sanitation requirements; and clean-up and resulting discharge that are ordered and supervised by an appropriate governmental agency or court of law.

## No significant risk levels

OEHHA may determine what it considers reasonable exposure levels that pose no significant risk. These levels are based upon a risk assessment (using previously determined guidelines) that is either carried out by the lead agency or reviewed by OEHHA. For example, the EPA identifies chemicals as causing developmental or reproductive toxicity in implementing its Toxic Release Inventory (TRI) program (i.e., Section 313 of the Emergency Planning and Community Right-to-Know Act of 1986 (EPCRA)). Therefore, in 1994, the EPA added a number of chemicals to the TRI list and published its additions to the list in the *Federal Register* 1994; 59:1788–1859 and 1994; 59:61432–61485). OEHHA reviewed the list and determined that for several TRI chemicals, the regulatory criteria were met. The chemicals were then placed on the Proposition 65 list of chemicals known to cause reproductive toxicity.

Subsequently, OEHHA continued the process of considering whether the remaining chemicals should be added to the list. One such chemical, dimethyl chlorothiophosphate did not meet the scientific criteria, so it was not listed (http://www.oehha.ca.-gov/prop65 CRNR_notices/admin_listing/process_procedures/notice8.html).

## Carcinogen exposure levels

"Exposure level" is defined by the chemical concentration necessary to be absorbed, inhaled, or ingested by an individual before it is considered unhealthy. This also includes to what extent people are responsible for their exposure. For example, if a person chooses not to wear gloves when using dangerous chemicals and then become exposed, that person can be held accountable. This does not include someone who is exposed to a listed chemical from any other source or product.

"Lifetime exposure" means the rate at which a person can be estimated to be exposed to a given chemical over a 70-year lifetime. Under these circumstances, the level of exposure to a listed carcinogen is determined by multiplying the level in question (concentration) times the individual's reasonably anticipated rate of exposure to the chemical over a lifetime of 70 years.

Currently, for an exposure expected to affect the general population in any geographic area, the following assumptions are made: a person ingests 2 l of drinking water per day, inhales 20 $m^3$ of air per day, and has a life span of 70 years. For an exposure that may affect a certain subpopulation in any geographic area, specific data (if available) relating to that subpopulation are used to determine the level. For an exposure that is plausibly expected to affect the embryo or fetus, the gestation period is considered to be 9 months.

| Subpopulation | Water (l/day) | Air ($m^{3/}$day) |
|---|---|---|
| Man (18 + years of age) | 2 | 20 |
| Woman (18 + years of age) | 2 | 20 |
| Woman with conceptus | 2 | 20 |
| Adolescent (10–18 years of age) | 2 | 20 |
| Child (2–10 years of age) | 2 | 15 |
| Infant (0–2 years of age) | 1 | 4 |

For workplace exposures, a worker inhales 10 $m^3$ of workplace air/8 h/day, 40 h/week, and 50 weeks/year over an assumed 40-year period. The exposed individual from the general population who occasionally enters the workplace inhales 1.25 $m^3$ of air for 1 h/month for a 70-year lifetime. Where data is available for exposure to consumer products, lifetime exposure is calculated using the average rate of intake or exposure for average users of the consumer product – not on a per capita basis for the general population.

## Reproductive toxicant exposure levels

Although similar to carcinogens, the rate at which someone is exposed to reproductive toxicants is based on the pattern and duration of exposure that directly effects reproduction. The assumptions described above for carcinogens are also used for reproductive toxicants (the "no significant level" is that level of exposure determined in animal bioassays which, when multiplied by 1,000, will not produce birth defects or other reproductive damage in humans) or until more specific and scientifically valid data become available. Where exposure of the mother to a listed reproductive toxicant effects either the embryo or fetus, the exposure level is based on the reasonably anticipated rate of exposure for the mother during the 9-month period.

## Enforcement

Violators of Proposition 65 can be slapped with significant penalties if they do not prevent a listed chemical from being discharged or do not provide a required warning that a listed chemical is present. Violators may be subject to sanctions and civil penalties of up to $2,500 per day for each infringement, in addition to any other penalties established by law. The State Attorney General, district attorneys, and city attorneys that represent more than 750,000 people may bring legal actions to cover civil penalties. Additionally, city prosecutors may file for legal action if they win the consent of the district attorney or if officials fail to prosecute.

Besides lawsuits by public prosecutors, private citizens may also enforce Proposition 65's provisions through private lawsuits. Citizens can take legal action if they notify the appropriate public official of the violation with a 60-day notice, and no public official is at that time already prosecuting the alleged perpetrator. Furthermore, citizens do not need to show actual personal injury from the violation to bring the suit. Governmental employees are also required to disclose violations should they learn that a hazardous waste is being illegally released or is threatened to be illegally released. Employees who fail to comply face criminal penalties.

## Conclusion

Proposition 65 has provided an effective means for reducing certain chemical exposures that may not have been adequately controlled under existing federal or state laws. For instance, enforcement action has resulted in reducing the amount of lead in ceramic tableware. Certain chemicals released into the air, including ethylene oxide, hexavalent chromium, and chloroform from facilities in California, have been significantly reduced. Other chemicals on the list are no longer used as ingredients in some commonly used products: As an example, trichloroethylene is no longer used in most correction fluids, toluene has been removed from many nail care products, and foil caps on wine bottles no longer contain lead. Finally, Proposition 65 may be most well known for its resulting widespread warning campaign to educate the public on the dangers of drinking alcohol during pregnancy and its potential dangers to the unborn child.

While government officials have worked hard to design and carry out this proposition, the responsibility for safeguarding our water supply lies with the work of knowledgeable professionals. Presently, there is a serious shortage of people entering the toxicology field. We need to find ways to attract high school and college students to the field to promote toxicology not only as an excellent career choice but one with many growth opportunities. Additionally, working professionals need to continually update their skills and knowledge to do their best work. This could be accomplished by having members of governmental agencies and other organizations invite toxicologists in other sectors to give lectures and present workshops. With these improvements, the agencies will have a larger pool of qualified applicants to choose from, and those already in the sector will have more tools to make the best decisions about safe drinking water issues.

## References

California Code of Regulations. Title 22. Social Security. Division 2. Department of Social Services-Department of Health Services. Part 2. Health and Welfare Agency-Department of Health Services Regulations. Subdivision 1. Health and Welfare Agency. Chapter 3. Safe Drinking Water and Toxic Enforcement Act, 1986.

Crump, K.S. (1979) Dose response problems in carcinogenesis. *Biometrics* 35, 157–167.

Office of Environmental Health Hazard Assessment. *Procedure for Prioritizing Candidate Chemicals for Consideration Under Proposition 65 by the ''States Qualified Experts''*. Sacramento, CA: OEHHA, California Environmental Protection Agency, May 1997.

Office of Environmental Health Hazard Assessment. Personal communication (2001).

World Health Organization (1978). *Environmental Health Criteria, No. 6: Principles and Methods for Evaluating the Toxicity of Chemicals*. World Health Organization, Geneva, vol. 1, pp. 25–61.

Chapter 14

# Oversight regulations

*Shayne C. Gad*

Toxicological safety assessment studies performed for regulatory agencies must comply not only with guidelines and regulations concerning scientific aspects and the information needs of regulatory bodies, but must also comply with a host of other regulations concerning data integrity, animal welfare, worker safety, and waste disposal. This chapter will address the various regulations imposed by governmental agencies that surround the conduct of toxicological safety assessment studies. Included in this chapter is a discussion of Good Laboratory Practices (GLPs) regulations, applicable both within the US and in other countries, and US laws concerning laboratory use of animals; occupational exposure to hazardous chemicals in laboratories; and environmental compliance for hazardous, biomedical, and radioactive waste disposal.

## GLPs

In 1976, the US Food and Drug Administration (FDA) first proposed GLPs in an attempt to ensure the integrity of data generated during nonclinical laboratory studies (FDA, 1978). The regulations came in the wake of a realization that certain basic laboratory practices were not routinely followed by all laboratories performing safety assessment studies. In many cases, personnel involved in the conduct of a study were no longer employed by the laboratory, and the data collected was insufficient for determining what had happened during the study. The essence of the GLPs require that a well-defined, organizational structure exists and that a paper trail be established, through which the course of events for a study can be reconstructed. Personnel not directly involved in the conduct of a study must monitor each study to assure the testing facility management and regulatory agencies that the facilities, equipment, personnel, methods, practices, records, and controls do, in fact, conform to the requirements of the GLPs.

The FDA GLPs were revised and the final rule was adopted on September 4, 1987 (FDA, 1987a,b). On August 17, 1989, the US Environmental Protection Agency (EPA) issued two final rules for its Good Laboratory Practice Standards (EPA, 1976a,b,1989a–d). One rule addressed studies conducted to meet pesticide registration requirements under the Federal Insecticide, Fungicide and Rodenticide Act (FIFRA, EPA, 1979, 1983), and the other rule addressed studies required for industrial chemicals under the Toxic Substances Control Act (TSCA).

In other countries, similar regulations were adopted. The Organization for Economic Cooperation and Development (OECD, 1981, 1982, EEC 1987, 1988) and the Japanese regulatory agencies: the Ministry of Health and Welfare (MHW, 1987); the Ministry of International Trade and Industry (MITI, 1984); and the Ministry of Agriculture, Forestry and Fisheries (MAFF, 1984) adopted the use of GLPs for their regulatory studies during the early 1980s. The UK also has promulgated similar regulations (UK GLP Monitoring Authority, 1990, 1991, 1992, 1995, 1997). While the GLPs for the various agencies differ slightly, they are essentially the same.

As with most legal documents, the GLPs are open to a fair degree of interpretation. Sponsors (the study initiator or supporter) maintain or contract quality assurance personnel to evaluate a laboratory's compliance with the GLPs. Laboratories must have personnel who are not involved in study conduct to review study procedures and data to ensure GLP compliance. Thus, an entire profession has evolved around the interpretation and application of the GLPs. Many circumstances do not fit neatly into the letter of the law. The views and interpretations presented in this discussion of the GLPs represent those of the authors and are based upon practical application, scientific validity, and the underlying essence of the GLPs to ensure that study data are collected in such a way that the integrity of the data is preserved and that the course of event for a study can be reconstructed.

The beginning of the paper trail for studies performed in accordance with the GLPs is the study protocol. All studies must have a protocol. The protocol indicates the scope of the work to be performed and the parties involved. The protocol must be approved by the study sponsor and signed and dated by the study directory. Any change to the protocol must include the reason for the change and be signed and dated by the study director.

The study director is the individual charged with the overall responsibility for technical conduct of the study and with ensuring that all GLPs are followed. The study directory must have the training, education, and experience necessary to serve as the single point of control for the study. The study director has the ultimate responsibility for the implementation, documentation, and interpretation of the study. All events that could compromise the outcome of a study must be acknowledged by the study director.

The study director is designated by the management of the laboratory performing the study. The role of the study director becomes increasingly complicated when several laboratories are involved in study conduct. How does the study director oversee functions at other laboratories? How can the study director ensure GLP compliance at other laboratories? Often a representative of the sponsor coordinates the activities of the various laboratories. Should this individual be the study directory? In this case, which laboratory's management designates the study director?

This dilemma is probably best rectified by recognition of a main testing facility, that being the facility managing the administration or application of the material being tested. The study director, appointed by the main testing facility's management, must be aware of the aspects of the GLPs that may be compromised by conducting studies at more than one facility. These GLP issues should first be addressed in the protocol by specifying the laboratories involved, the scope of work they are to perform, record retention and reporting requirements for each of the laboratories, and quality assurance responsibilities for each of the laboratories. In the GLP compliance statement, which will be discussed later, the study director should indicate the role that each laboratory assumed in meeting the requirements of the GLPs.

In addition to appointing a study director, the testing facility management must ensure that a sufficient number of trained personnel and adequate resources are available to complete a study in a timely manner. The management establishes and approves the practices and procedures necessary to conduct the studies in a manner that ensures the quality and integrity of the data. The practices and procedures must be formalized in written documents called *Standard Operating Procedures* (SOPs).

The GLPs indicate some specific procedures that require SOPs, such as observations of the test systems (test organisms), collection and identification of specimens, and maintenance and calibration of equipment. SOPs must be readily available in the study areas so that personnel can easily consult them, if necessary, while performing a procedure. Revisions to SOPs must be documented in such a way that identification of the version of an SOP in use at any given time is possible.

The GLPs are not specific when it comes to the contents of SOPs, except for the maintenance and calibration of equipment. Equipment SOPs must indicate the frequency and procedures for testing, calibrating, or standardizing equipment as well as the personnel responsible for these functions. The GLPs require that detailed records be kept for all equipment inspections, maintenance, testing, calibration, and standardizing operations. Records of both routine and nonroutine maintenance must be kept for all pieces of equipment used in the generation, measurement, or assessment of data and for facility environmental control equipment. Routine maintenance records must indicate whether or not the maintenance procedures followed SOPs. Nonroutine maintenance records must include information regarding the nature of the problem and how the problem was discovered (Gad and Taulbee, 1996).

For certain pieces of equipment, the advantages of keeping detailed maintenance records is clear. The records can be used to support, after the fact, that the piece of equipment was functioning properly at its time of use. For example, with a piece of equipment such as a balance, a log of calibrations or verifications using standard weights can provide reasonable assurance that weights recorded during a study were accurate. On the other hand, is a record necessary for the routine cleaning of a microscope or the nonroutine replacement of a microscope's ocular? Would these maintenance records add in any way to the validity of histopathological findings? Clearly, if a malfunction in a piece of equipment would not compromise study integrity, then an SOP should not be needed.

Another critical aspect of the GLPs is the integrity of the chemicals used in a study. Chemicals can be broadly classified as standard reagents or study specific compounds. The GLPs require that reagents and solutions be labeled as to their identity, titer or concentration, storage requirements, and expiration date. Additional requirements are imposed on study-specific compounds.

Study-specific compounds, referred to as *substances* or *articles*, depending on the version of the GLPs, can be placed into one of two or three categories, again depending on the version of the GLPs. The FDA GLPs recognize two categories: *test articles* and *control articles*. The EPA GLPs (both FIFRA and TSCA) recognize three categories: *test substances*, *control substances*, and *reference substances*. The OECD GLPs recognize two categories: *test substances* and *reference substances*, but include *control substance*in parentheses after *reference substance*, implying that they are the same. The three Japanese GLPs refer only to *test and control articles or substances.*None of this would be much of a problem, except that test, control, and reference substances or articles must be characterized prior to use in a study. Therefore, one must know whether a material is considered a test, control, or reference substance or article (Taulbee and DeWoskin, 1993).

The *test article or substance* is the compound, mixture, device, etc., under study, which is regulated by, for which a permit is sought, etc. Clear enough. *Control substance*, as defined in the FIFRA GLPs means "any chemical substance or mixture, or any other material other than a test substance, feed, or water, that is administered to the test system of the purpose of establishing a basis for comparison with the test substance with a carrier" or a *carrier*, defined in the FIFRA GLPs as "any material, including but not limited to feed, water, soil, nutrient media, with which the test substance is combined for administration to a test system", what, then, is a control substance if one compares the response of organisms receiving the test substance with organisms receiving only the vehicle or carrier? Frequently, corn oil is used as a carrier for oral administration, and the control group receives the corn oil without the test substance. Should the corn oil also be classified as a control substance and be characterized? (Interestingly, various parts of guidelines use the term *vehicle*, and therefore it is not synonymous with the GLP term *carrier*.)

A key part of the definition of *control substance* is "for the purpose of establishing a basis for comparison". A vehicle or carrier is not administered for the purpose of establishing a comparison, but is necessary for the administration of the test substance. The purpose of administering a vehicle or carrier to a control group is to ensure that any effects seen in groups receiving the test substance are not attributable to the vehicle or carrier. Classically, materials that are administered to establish a basis for comparison have been called *positive controls*. Positive control substances are used in testing because they are known to elicit an effect that may be suspected with the test substance. Only substances clearly administered as a positive control should be considered "control substances" and be subjected to characterization requirements.

To further sink into the mire of substance definition, *reference substance* is defined in the FIFRA GLPs as "any chemical substance or mixture, or analytical standard, or material other than a test substance, feed, or water, that is administered to or used in analyzing the test system in the course of a study for the purposes of establishing a basis for comparison with the test substance for known chemical or biological measurements". So, the only difference between a control substance and a reference substance is the inclusion of analytical standards.

As stated earlier, test, control, and reference substances need to be characterized. Characterization typically includes determining the identity, purity, and composition. Determining purity requires an analytical standard for comparison. The analytical standard also needs to be characterized, which includes purity determination, which requires an analytical standard which...

There are other requirements for test, control, and reference substances. The stability of the test, control, or reference substance must be determined, but may be determined during the course of the study. If, however, a substance is determined to be unstable, then the integrity of the study may be compromised. Stability under conditions of storage at the test site must also be known. Storage stability is often known prior to the study in vague terms, such as ambient temperature storage or frozen storage. Again, if a substance is not stable under the exact storage conditions at the testing facility, a study may be compromised.

Solubility, if pertinent to study conduct, is another parameter that must be determined prior to beginning the experimental portion of a study. Solubility can be affected by many factors, such as temperature; ideal solubility and the solubility obtainable during a study may differ considerably. For an aquatic toxicology study, where solubility is a crucial parameter, mixing times are usually scant compared to the amount of time a chemist may allow for solubility determinations at a bench. Does determination of ideal solubility add to study integrity?

Labeling requirements for test, control, or reference substances include name, Chemical Abstract Service (CAS) number or code, batch number, expiration date, and storage conditions. If the study has an experimental duration of more than 4 weeks, then a reserve sample must be collected and retained.

The GLPs require that the receipt and distribution of each batch of test, control, or reference substance is documented. The accountability of test substance may be critical in reconstructing study events. If results suggest that an improper dose or concentration level was used, records showing the amounts used for formulations may be able to support the presupposition.

Facilities must be adequate to ensure that test, control, or reference substances are stored properly and are not contaminated. Standard operating procedures that address the receipt, identification, storage, handling, mixing, and method of sampling for test, control, or reference substances are required. Handling practices must be sufficient to prevent cross-contamination among various substances.

Mixtures of a test, control, or reference substance with a carrier that are used for administration to a test system must be analyzed for uniformity or homogeneity, concentration, stability, and, if relevant, solubility. These analyses ensure that the specified amount of a test, control or reference substance is being administered to the test system. Normally acceptable parameters for these analyses, such as all homogeneity analyses being within 10 per cent of the target concentration, should be specified in an SOP, but must also be left open to the discretion of the study director. Of course, these analyses do not ensure that a test system actually receives the proper amount of a test, control, or reference substance, only that the proper amount is in the mixture with the carrier.

A testing laboratory must have adequate facilities for the housing and care of the test systems. Test systems are any animal, plant, microorganism, or chemical or physical matrix to which the test, control or reference substance is applied or administered for study. Separation of test systems must be sufficient for the isolation of individual projects. Separation can be accomplished through the use of different rooms or protective barriers. Additionally, there must be sufficient space for the storage of supplies for the care and maintenance of the test systems.

All studies are concluded by the issuance of a final report. Many requirements for the final report are listed in the GLPs. The final report must be signed and dated by the study director. The final report, all raw data and supporting documentation, protocols and specimens must be retained, with the exception of various samples that may not afford further evaluation. Archiving facilities must be adequate to prevent deterioration and allow for orderly storage and expedient retrieval. Materials must be retained for as long as a sponsor holds a permit for research or marketing to which the study is applicable, at least 5 years following submission in support of a research or marketing permit, or 2 years if the study is not submitted in support of an application, whichever is longer.

All applications of toxicological safety assessment studies in support of a research or marketing permit must include a compliance statement that is signed by the sponsor, study director, and in the case of FIFRA submissions, the applicant. The compliance statement must indicate that all requirements of the GLPs were followed, must state any aspects of the GLPs that were not followed, or must state that the person was not a study sponsor, did not conduct the study, and has no knowledge as to whether the GLPs were followed.

Submission of a false compliance statement is a criminal offense. Therefore, the various ambiguities in the GLPs and their interpretation have legal implications. Study directors, laboratory management, and study sponsors should be aware of these gray areas and if in doubt, should express such in the compliance statement (FDA, 1993).

Legislation (Directive 91/507/EEC) has been put in place with the aim of ensuring that safety tests are conducted in conformity with GLP as laid down by Directives 87/18/EEC (EEC, 1987) and 88/320/EEC (EEC, 1988). In the context of this requirement, the following are considered as safety tests which must conform to GLP: single-and-repeated-dose toxicity; reproductive (including embryo/foetal and perinatal) toxicity; tests for mutagenic and carcinogenic potential; local tolerance studies; toxicokinetic studies which provide systemic exposure data for the aforementioned studies, and pharmacodynamics studies designed to investigate potential adverse side effects. Furthermore, studies which provide general or specific data for safety assessment, e.g., validation of virus removal or inactivation for biological/biotechnological products, should also be carried out in accordance with GLP.

The principles of GLP should ensure that the results produced by these tests are comparable and of high quality. Tests do not have to be repeated within the EC because there are no differences in laboratory practice. This reduces animal experimentation in accordance with Directive 86/609/EC (EEC, 1986), regarding the protection of animals used for experimental and other scientific purposes. Directive 88/320/EC established the necessity for a harmonized system for study audit and laboratory inspection to ensure that GLP is being applied. Member states are required to designate authorities responsible for monitoring GLP compliance. Guidance for compliance monitoring procedures and the conduct of laboratory inspections and study audits is given in the Annex to this directive. All major GLP systems have also been added to or amended to cover computerized data collection and electronic signatures (DIA, 1988; FDA, 1988, 1993, 1994, 1997; EPA, 1995).

## Occupational health and safety issues

Most workplaces have various health and safety issues that must be addressed, from forklift safety in warehouse operations to hand and wrist safety in office settings. The laboratory, particularly the laboratory performing toxicological safety assessment studies, is confronted with many of the common safety issues, plus some rather unique concerns.

For many workplaces where hazardous chemicals are used, safety regulations set by the US Occupational Safety and Health Administration (OSHA) must be followed. For many years, the OSHA standard "Hazard Communication" also known as *Hazcom* or *Right-to-Know*, required all employers where hazardous chemicals are used to develop safety procedures and inform workers of the dangers.

Laboratories had difficulty complying with many of the Hazard Communication requirements simply because of the sheer number of chemicals typically found in the laboratory. On the other hand, the quantities of hazardous chemicals typically found in laboratories are vastly smaller than those found in industrial settings. OSHA recognized these differences and, in 1990, established a new standard, "Occupational Exposure to Hazardous Chemicals in Laboratories".

The major impact of the laboratory safety standard was to require each laboratory to develop a Chemical Hygiene Plan (CHP). The CHP is a written program that specifies procedures, equipment, and work practices necessary to protect employees from the hazards of chemical exposure. The laboratory standard supersedes the Hazard Communication standard and any chemical-specific standards except for the requirements to limit exposures to specified levels, prohibit eye and skin contact, and perform exposure monitoring and medical surveillance.

Most laboratories have a limited scope of operation. The chemicals that will be used in the laboratory are known, as are the dangers associated with the chemicals. The laboratory involved in performing toxicological safety assessments, however, is continually faced with new test materials, many of which do not yet have any toxicological data. In this case, the CHP must be developed with the impact of the unknown in mind. The importance of establishing good, overall hygiene procedures becomes paramount.

Another health and safety concern for laboratories performing toxicological safety studies is the "Bloodborne Pathogens" standard which OSHA instituted in 1991. This OSHA standard pertains to the dangers associated with pathogens transmitted through human blood. While toxicological safety studies do not involve the use of human blood, many laboratories performing safety assessments may also analyze samples derived from clinical studies. The handling of human plasma samples would necessitate compliance with the requirements of the Bloodborne Pathogens standard, including the development of an Exposure Control Plan and employee training. Many aspects of the Bloodborne Pathogens standard overlap with requirements concerning biomedical or infectious waste, which will be discussed later.

Some toxicological safety assessment studies, especially those pertaining to the metabolism or fate of a chemical, involve the use of radioactively labeled test material. Typically, test materials tagged with $^{14}C$ or $^{3}H$ are used, but other radioactive isotopes may be more appropriate in certain circumstances. Laboratories must obtain a license that permits the use of radioactive isotopes. Byproduct licenses are usually issued by the US Nuclear Regulatory Commission (NRC), although various minimal uses may be regulated at the state level.

Laboratories using radioactive isotopes must develop and implement radiation safety programs. While OSHA has some requirements concerning radiation safety, the majority of requirements are governed by the NRC. In the case of some of the minimal-use licenses, where a state has an agreement with the NRC, a state agency such as the Department of Health may oversee radiation safety requirements.

The NRC recently revised its requirements; the effective date for the new requirements is January 1, 1994. The current emphasis for radiation safety is for the development of a radiation protection program commensurate with the activities of the licensee. The programs should take into account safety aspects from the time of receipt through disposal.

Radiation exposure is virtually nonexistent with $^{14}C$ and $^{3}H$ unless the material enters the body. Therefore, making sure that work surfaces are not contaminated through the use of wipetests is the best defense against exposure. Biomonitoring through the radioanalysis of urine samples allows the determination of whether work practices are adequate for worker safety.

## Environmental compliance

Generally, a toxicology laboratory produces three types of hazardous waste: hazardous chemical waste, radioactive waste, and biomedical or infectious waste. Any waste fitting two or more of these categories is considered a "mixed" waste. Classifying and appropriately segregating hazardous chemical waste is an essential aspect of the hazardous chemical waste disposal process. Legitimate waste disposal companies are unwilling to dispose of unknown chemicals. Hazardous waste disposal may be regulated by federal, state, or local laws. With decreasing jurisdiction, requirements may become more and more stringent, not less stringent.

Hazardous chemical waste is typically defined by properties such as flammability or toxicity. Many reagents used by a toxicology laboratory clearly require disposal as hazardous chemical waste, such as organic solvents. Most hazardous waste laws recognize a "conditionally exempt small quantity generator" status which exempts companies producing less than 100 kg of hazardous waste per month from the many regulations except proper disposition of the waste. Generators producing between 100 and 1,000 kg of hazardous waste per month would nominally need a permit as required under the Resource Conservation and Recovery Act (RCRA) if the waste is to be stored for more than a few months.

Disposal requirements for test materials or items contaminated with test material are not well defined. Often, test material waste is produced before the degree of toxicity has been established. Test materials are not typically one of the select group of "listed" toxic materials that require disposal as a hazardous waste, because the "listed" materials have already been tested. Waste may also be classified as toxic if it meets certain toxicity requirements. The procedure for determining whether or not a waste meets the toxicity requirements has changed within the last few years, but typically compares the degree of extractable or leachable toxicity to that of a known quantity of a "listed" waste. Since the concentrations used in toxicological testing are usually low, the overall amount of compound being disposed of is small, and test material disposal would not typically meet the toxicity requirements.

Disposal of radioactive waste has become an extremely controversial and sensitive issue over the last few years. Radioactive waste sites in the US have been filled or closed, and new sites have not yet been opened. Facilities that generate low-level radioactive waste are being contaminated with less than 0.05 $\mu$Ci/g of $^{14}C$ or $^{3}H$ may, under NRC guidelines, be incinerated. Incineration, however, is also regulated by the EPA, and its regulations must be adhered to also. NRC guidelines also allow for liquid scintillation media with less than 0.05 $\mu$Ci/g of $^{14}C$ or $^{3}H$ to be disposed of without regard to its radioactivity. With the advent of "environmentally safe" liquid scintillation cocktails, these media no longer necessitate disposal as hazardous chemical waste.

NRC guidelines allow for liquid radioactive waste to be disposed of in sanitary sewer systems. Again, EPA, state, or local regulations may supersede NRC guidelines. Sanitary sewage disposal limits are set, both in terms of concentration and total amount, but fairly significant amounts of radioactivity may be disposed of via the sanitary sewer. Solid radioactive waste production can be minimized by using solid, nonabsorbent materials and decontaminating them in a liquid matrix, which can then be disposed of via the sanitary sewer.

Using isotopes with short half-lives and allowing material to decay in storage is another alternative for minimizing waste production. This is not always possible because of the structure of the chemical or the intended use of the radiolabeled material. The low energy decay of $^{14}C$ and $^{3}H$ make them relatively safe isotopes to work with, and use of other isotopes may greatly increase radiation hazards for employees.

Another controversial topic in waste disposal concerns biomedical waste. The EPA instituted a 1-year program to track medical waste, after which it decided to forgo imposing any federal regulations. Most states, however, have adopted regulations concerning the disposal and treatment of biomedical waste. Typical treatment techniques for biomedical waste are incineration or sterilization using an autoclave. Treatment facilities are typically being regulated more strongly than generators of biomedical waste.

The term *infectious waste* is frequently used as a synonym for biomedical waste. For the toxicology laboratory performing safety assessments, this is an unfortunate choice of terms. Infectious agents are not used in testing. In fact, facilities go through great pains to ensure that no infectious agents are present. Still, the problem of perception exists. If syringes used for oral gavage are disposed of with the normal solid waste stream, how does someone at a landfill know it was not used on an AIDS patient? How are vacutainers of rat blood differentiated from that of human blood? These issues of perception force the toxicology laboratory to handle its biomedical waste in accordance with regulations designed for the control of human infectious waste. There are now international standards (ISO 14000) governing these environmental concerns (Eascio *et al.*, 1996).

## Laboratory animal care and welfare

The remainder of this chapter considers the broad and general category of animal care and welfare. The concept of adequate animal care and the mandates either implied or specified by law will be discussed.

Quality science and research depends upon quality animal care. In fact, state-of-the-art research demands first class animal care. The commonly accepted ideas of animal welfare were in vogue long before the passage of the first version of the Animal Welfare Act 1966 (AWA). Federal regulations concerning the safety of human drugs have had several safety objectives (see Chapter 2), which have had a great impact on animal care. These safeguards are essential to assure the scientific staff that animal care meets certain standards. In no area of science is the need or the quest for knowledge more acute than in the area of animal welfare/animal rights. Immersed in this quest are the biological scientist, research administrators, regulatory officials, veterinarians, and the lay public in general.

There is no question that animal welfare should be and perhaps already is the foremost consideration in the justification for using animals as experimental subjects (Acred *et al.*, 1994). There is long-standing and growing opposition to the use of animals, primarily because of the gray area of distinction between animal welfare and animal rights (French, 1975; Hampson, 1979; Freudinger, 1985). The question of animal rights suggests that you can elevate the so-called animal rights (if in fact there are such rights) to a position of equal moral value with human rights (Russell and Burch, 1959). However, it is not possible to equate animal rights with human rights without diminishing human rights, especially the human right to better health. Most biologically oriented scientists and veterinarians understand the need to provide the best available environment, nutrition, housing and general care to all experimental subjects used in basic research, safety testing, and educational practices. There is no question, nor can there be any wavering concern that animals used in these endeavors must be provided with the best care available. If these practices are followed, the question of animal welfare is addressed. Without the use of animal models or animal subjects, major accomplishments in transplantation; surgical techniques; chemotherapy; or in pathogenesis of infectious diseases and cancer, diagnostic methods, or reproductive failures could never have occurred.

## *The law*

Historically, regulations covering the care and use of animals for experimental research have been developed from two main sources: the scientists themselves and humane societies that were formed to protect companion, farm, and laboratory animals (Russell and Burch, 1959).

The National Institutes of Health (NIH), since its inception in 1887 (Hygienic Laboratory at the US Marine Hospital in Staten Island), has led the way in developing guidelines for the proper care and use of laboratory animals. In the 1920s the Director of NIH was personally responsible for decisions concerning the use of animals in any given experiment. During World War II, the Committee for Medical Research and the National Research Council of the National Academy of Science joined in the effort to reduce war-related injury and disease. This research required the use of animals, which also led to standards for animal care. By the war's end, the major animal care issues were transferred to NIH from the Wartime Office of Scientific Research. By 1958, NIH's Division of Research Grants began a peer-review system of selecting the most meritorious grant applications for funding. Included in the examination process was the issue of care and use of laboratory animals.

In 1963, the first edition of the *Guide for the Care and Use of Laboratory Animals* was issued by the Animal Care Panel, later renamed the American Association for Laboratory Animal Science (AAALAC International, 1997a). The fifth and most recent edition of the guide was published in 1985 and remains the primary reference for research animal care and use in the US.

In 1966 Congress passed the Pet Protection Act in the wake of public outcry over the alleged misuse of animals. This act was the first version of what is now the AWA (LAWA, 1996). The AWA was revised in 1970 and in 1976 and underwent a major revision in 1985. The act gave the US Department of Agriculture (USDA) responsibility for implementing the new law, which applied to dogs, cats, rabbits, monkeys, guinea pigs, and hamsters. Recent changes in the interpretation of the law have brought all animals used for research purposes under the umbrella of the law. The USDA has issued Parts 1, 2, and 3 of the final regulations implementing the 1985 amendments of the AWA (NIH, 1985), which cover definitions (Part 1), regulations (Part 2), and standards (Part 3). It is Part 3 that gives specific guidelines for the humane handling, care, treatment, and transportation of animals used in research and teaching programs.

The Animal and Plant Health Inspection Service (APHIS) (part of USDA) was given responsibility for facility registration, licensing of animal suppliers, and inspection of licensed facilities to assure compliance with the AWA. Inspectors for APHIS inspect licensed facilities on at least an annual basis (unannounced) to ascertain the status of the facilities' physical plants, training of animal care personnel, and the overall care and welfare of research animals covered under the AWA. Table 14.1 presents current housing standards for laboratory animals.

Prior to 1985, the Public Health Service (PHS) policy on humane care and use of laboratory animals governed animal use by Awardee Institution (PHHS, 1983). The amended policy [which is in agreement with nine Code of Federal Regulations (CFR)] lists nine areas of concern that must be addressed and adhered to by the responsible institutional official (NIH, 1985; APHIS, 1989). Briefly, these areas are:

1 Care and use of animals must be in accordance with the AWA.
2 Experiments involving animals must be designed with consideration to their relevance to human or animal health, advancement of knowledge, or the good of society.
3 Animals selected for a procedure must be of the appropriate species, quality, and minimum number required to provide valid results.
4 Use must avoid or minimize discomfort, stress, and pain to the animals.
5 Appropriate sedation, analgesia, or anesthesia must be provided to the animals.
6 Animals that suffer severe or chronic pain must be euthanized as soon as possible.
7 Housing conditions must be appropriate for the species.
8 Investigators must be appropriately qualified and experienced for conducting procedures on living animals.
9 Exceptions to the principles of animal care and use should not rest with investigator(s), but with the Institutional Animal Care and Use Committee (IACUC).

Since 1989, great strides have been made to set and apply standards for animal care and welfare across all research facilities, regardless of funding sources. These standards are more "performance-based" than the previous "engineering-based" application of the guidelines, i.e., exact cage size, air flow, and painted surfaces within animal rooms. Also, suitable means of euthanasia have been characterized (see Table 14.2).

## *IACUC*

As described briefly above, the AWA (USDA, 1995) requires the animal research facility to form an IACUC for oversight of animal-related issues, following the standards set forth by the *Guide* (NIH, 1992). An active, well informed IACUC is one of the best guarantees that animal welfare policies are supported and adhered to. Although research facilities are sometimes criticized for self-policing (IACUC members are chosen by the management of the facility), there are numerous checks and outside [USDA, American Association for Accreditation of Laboratory Animal Care (AAALAC)] inspections that, along with practical considerations already noted, make most IACUCs effective. The required membership of an IACUC has been noted previously, and the total number of members depends on the size of the facility

*Table 14.1* Recommended housing conditions for selected laboratory animal species[a]

| *Species* | *Housing type*[b] | *Size* | *Minimum cage/pen floor area* | *Minimum cage/ pen height (cm)* | *Room temperature (°C)* |
|---|---|---|---|---|---|
| Mouse | Group | 10–15 g | 39–52 $cm^2$ | 13 | 18–26 |
| | | 15–25 g | 52–77 $cm^2$ | 13 | 18–26 |
| | | 25 g | 97 $cm^2$ | 13 | 18–26 |
| Rat | Group | 100 g | 110 $cm^2$ | 18 | 18–26 |
| | | 100–200 g | 110–148 $cm^2$ | 18 | 18–26 |
| | | 200–400 g | 148–258 $cm^2$ | 18 | 18–26 |
| | | 400–500 g | 258–387 $cm^2$ | 18 | 18–26 |
| | | 500 g | 452 $cm^2$ | 18 | 18–26 |
| Hamster | Group | 60–80 g | 65–84 $cm^2$ | 15 | 18–26 |
| | | 80–100 g | 84–103 $cm^2$ | 15 | 18–26 |
| | | 100 g | 123 $cm^2$ | 15 | 18–26 |
| Guinea pig | Group | 350 g | 387 $cm^2$ | 18 | 18–26 |
| | | 350 g | 651 $cm^2$ | 18 | 18–26 |
| Rabbit | Single | 2 kg | 0.14 $m^2$ | 36 | 16–22 |
| | | 2–4 kg | 0.14–0.27 $m^2$ | 36 | 16–22 |
| | | 4–5.4 kg | 0.27–0.36 $m^2$ | 36 | 16–22 |
| | | 5.4 kg | 0.45 $m^2$ | 36 | 16–22 |
| Rabbit (nursing)[c] | Single (plus litter) | 2–4 kg | 0.46 $m^2$ | 36 | 16–22 |
| | | 4–5.4 kg | 0.56 $m^2$ | 36 | 16–22 |
| Cat | Single | 4 kg | 0.27 $m^2$ | 61 | 18–29 |
| | | 4 kg | 0.36 $m^2$ | 61 | 18–29 |
| Dog | Single | 15 kg | 0.72 $m^2$ | Species–specific[d] | 18–29 |
| | | 15–30 kg | 1.1 $m^2$ | Species-specific[d] | 18–29 |
| | | 30 kg | 2.2 $m^2$ | Species-specific[d] | 18–29 |
| Monkey | Single | 1 kg | 0.14 $m^2$ | 51 | 18–29 |
| | | 1–10 kg | 0.14–0.39 $m^2$ | 51–76 | 18–29 |
| | | 10–25 kg | 0.39–0.72 $m^2$ | 76–91 | 18–29 |
| | | 25–30 kg | 0.72–0.90 $m^2$ | 91–117 | 18–29 |
| | | 30 kg | 1.4 $m^2$ | 117 | 18–29 |
| Ape | Single | 20 kg | 0.90 $m^2$ | 140 | 18–29 |
| | | 20–35 kg | 0.90–1.4 $m^2$ | 140–152 | 18–29 |
| | | 35 kg | 2.3 $m^2$ | 213 | 18–29 |

Table 14.1 (*continued*)

| *Species* | *Housing type*[b] | *Size* | *Minimum cage/pen floor area* | *Minimum cage/ pen height (cm)* | *Room temperature (°C)* |
|---|---|---|---|---|---|
| Poultry[e] | Single | < 0.25 kg | 0.02 $m^2$ | – | 16–27 |
| | | 0.25–1.5 kg | 0.02–0.09 $m^2$ | – | 16–27 |
| | | 1.5–3.0 kg | 0.09–0.18 $m^2$ | – | 16–27 |
| | | > 3 kg | 0.27 $m^2$ | – | 16–27 |
| Sheep/goat | Single | < 25 kg | 0.9 $m^2$ | – | 16–27 |
| | | 25– > 50 kg | 0.9–1.8 $m^2$ | – | 16–27 |
| Swine | Single | < 15 kg | 0.7 $m^2$ | – | 16–27 |
| | | 15–50 kg | 0.7–1.4 $m^2$ | – | 16–27 |
| | | 50–200 kg | 1.4–4.3 $m^2$ | – | 16–27 |
| | | > 200 kg | > 5.4 $m^2$ | – | 16–27 |
| Cattle | Single | < 75 kg | 2.2 $m^2$ | – | 16–27 |
| | | 75–350 kg | 2.2–6.5 $m^2$ | – | 16–27 |
| | | 350–650 kg | 6.5–11.2 $m^2$ | – | 16–27 |
| | | > 650 kg | > 13.0 $m^2$ | – | 16–27 |
| Horse | Single | Adult | 13.0 $m^2$ | – | 16–27 |

[a] Recommendations for primary enclosure from the *Guide* (NRC, 1996); larger animals may also require secondary enclosures for exercise, mating or other activities.
[b] Typical preference for housing; listed group-housed animals are frequently single housed according to study protocol.
[c] Space recommendation from AWA (USDA, 1995).
[d] Recommended cage height for dogs is 6 inches above the head during normal standing position.
[e] Based primarily on the chicken.
[f] Height recommendation not specified; allowance should be made for typical postures.

*Table 14.2* Common methods of euthanasia in the toxicology laboratory[a]

| *Method* | *Species* | *Major advantages* | *Major limitations* | *AVMA recommendation* |
|---|---|---|---|---|
| Barbiturate injection | Most | Rapid, safe, cheap | Requires training, restraint, drug control | Acceptable (preferred) method |
| Anaesthetic inhalation | Most | Rapid, multiple animals exposed | Initial irritation, hazardous to staff | Acceptable method for small species |
| Carbon dioxide inhalation | Several small laboratory animals | Rapid, safe, cheap, multiple animals | Some species stressed or very tolerant | Acceptable at high concentrations |
| Carbon monoxide inhalation | Most small | Rapid loss of consciousness | Hazardous to staff, difficult to detect | Acceptable method with appropriate generation |
| Microwave exposure | Mice, rats | Rapid, safe, brain enzymes fixed | Specialized training, equipment, costly | Acceptable method with appropriate equipment |
| Tricaine/bensocaine injection | Fish, amphibians | Rapid, safe | Costly | Acceptable method |
| Cervical dislocation | Birds, small rodents/ rabbits | Rapid, safe, cheap, no drug residue | Requires training, unpleasant for staff | Conditionally acceptable method when justified |
| Decapitation | Most small | Rapid, no drug residue | Requires training, some hazard, unpleasant | Conditionally acceptable method when justified |
| Gunshot | Large farm or wildlife species | Rapid, ease for certain species | Requires training, dangerous, unpleasant | Conditionally acceptable method when necessary |
| Electrocution | Foxes, sheep, swine, mink | Rapid, cheap, no drug residue | Requires special equipment, hazardous, unpleasant, severe contractions | Conditionally acceptable method in specialized instances |
| Pithing | Small amphibians | Rapid, no drug residue | Requires training, unpleasant | Conditionally acceptable method in specialized instances |
| Nitrogen/argon inhalation | Most small | Rapid, safe, readily available | Stressful in some species, must limit $O_2$ | Conditionally acceptable method is specialized instances |
| Exsanguination | Several | Safe, cheap | Very stressful | Unacceptable method without anesthesia |
| Rapid freezing | Several small | Safe, cheap | Very stressful | Unacceptable method without anesthesia |
| Air embolism injection | Several | Safe, cheap | Causes convulsions, other signs of distress | Unacceptable method without anesthesia |
| Drowning | Several | Safe, cheap | Very stressful, slow | Unacceptable method |

Table 14.2 (*continued*)

| *Method* | *Species* | *Major advantages* | *Major limitations* | *AVMA recommendation* |
|---|---|---|---|---|
| Strychnine dosing | Several | Possibly convenient | Causes convulsions, painful contractions | Unacceptable method |
| Chloroform injection | Several | Possibly convenient | Very hazardous to staff | Unacceptable method |
| Cyanide dosing | Several | Possibly convenient | Unpleasant, hazardous | Unacceptable method |
| Stunning (blow to head) | Several | Rapid, no drug residue | Unpleasant, sufficient force required | Unacceptable method without other lethal procedure |

[a] Summarized from the Report of the AVMA Panel on Euthanasia (AVMA, 1993).

and the extent of its research. Beyond the mandated composition of the groups, it is desirable to include members from diverse backgrounds among scientific and nonscientific disciplines. Outside experts may also serve as consultants. There may be need for subcommittees to carry out the many functions of the IACUC.

As noted in the *Guide*, the IACUC oversees and assesses the animal care program and the use of animals in research. At least every 6 months, it inspects the facility (with emphasis on all animal housing, testing and supply areas), reviews protocols and carefully records all findings. A written report (containing meeting minutes, documents reviewed, areas inspected and items discussed) signed by most or all members is issued to principle investigators and facility management detailing any deficiencies. Specifically, protocols and subsequent animal use and euthanasia are evaluated on the following bases (the *Guide*):

- Rationale of proposed animal use.
- Justification of the species used.
- Justification of the number of animals used.
- Availability of alternative methods (Bennett, 1994).
- Adequacy of staff training and experience.
- Atypical housing or husbandry requirements.
- Consideration of sedation, analgesia and anesthesia.
- Unnecessary duplication of testing.
- Use of repeated and/or major surgical techniques.
- Criteria for excessive pain or distress
- Method of euthanasia.

Standardized procedures in addition to specialized research projects must be considered by the IACUC. Standardized protocols should not require a full review each time they are employed, but any revisions affecting animal use must be considered by the IACUC. Specialized protocols should each be reviewed by the IACUC. There should be a review/approval form for each project, which should be maintained as part of study records. These records should also be maintained in facility files, available for independent review, such as a USDA inspection. The IACUC report must reflect adherence to the AWA, with minor and major deviations being noted. A major deficiency is one that threatens the heath, well being or safety of the animals.

Identification of animal welfare deficiencies need not only originate from formal inspections, however. Any staff member should be encouraged to present concerns to the IACUC. Technical or nontechnical personnel, for example, should feel free (without any negative consequences) to inform a direct supervisor or an IACUC member of any excessive pain/distress witnessed during the course of a toxicology project. Such information may be kept confidential, but careful documentation must be made. It then becomes the responsibility of site management and/or the IACUC to investigate and take appropriate corrective action as warranted. This process is discussed by Silverman (1994).

There must be a reasonable and specific plan, with an outlined schedule, for the correction of significant deficiencies. Resolution of any major deviation should occur through a joint effort of the researcher, facility management officials of the area of concern (animal care supervisor, environmental or maintenance staff, purchasing agent, etc.), veterinary staff, along with the IACUC. If major deficiencies are not resolved according to the plan and schedule, the remaining deficiencies 'shall be reported in writing by the IACUC, through the Institutional official, to APHIS and any Federal agency funding that activity' (the 'Act'). The IACUC will also withhold approval and is now postponed until at least June 2000 for the specific studies affected until concerns are resolved. Of course, conflicts may occur among staff, researchers, management and study sponsors. These will have to be resolved in the light of the principles presented in the Introduction – ethics, science, social pressures and legal requirements.

Secondary responsibilities of the IACUC are training staff members, providing information on animal use, interacting with other personnel (health and safety officers, managers, legal staff, etc.), developing a liaison with outside organizations promoting humane research and engaging in public education (Dell, 1994). Any complaints or charges of animal misuse made by the public or the media should be investigated by the IACUC. This should aid facility management in arriving at an appropriate response.

Part 2 of 9 CFR, Chapter 1 (Regulations) states:

> Each institution which falls under the authority of the AWA and/or received PHS support for research or teaching involving laboratory animals must operate a program with clear lines of authority and responsibility for self monitoring the care and welfare of such laboratory animals.

The mechanism for such monitoring was given to the IACUC.

In order to monitor the care and use of animals, the AWA requires that an appropriate administrative official at each facility appoint the members of the IACUC ('the Committee"). The Committee serves as the watchdog for the welfare of animals in the same manner as does the Institutional Review Board in the hospital or medical center.

Membership of this Committee must include a scientist from the institution experienced in research involving animals; a doctor of veterinary medicine who either is certified by the American College of Laboratory Animal Medicine or has had experience in laboratory animal medicine; a person who is not affiliated with the facility or institutions; and other members as required by federal, state, or local regulations.

Humane care and treatment of animals used in research, testing, and education requires a certain amount of scientific and professional judgment. The husbandry needs of each species and the special requirements for the species, including humane handling, must be considered in all programs where these animals are to be used. Each facility or institution must establish a written policy for the humane care and use of a given species. This program must be in compliance with applicable federal, state and local laws and/or regulations. Depending on the fund or funding source used to provide for these animals or to provide the basic research, other regulations may be enforced. The established program must follow *Guide for the Care and Use of Laboratory Animals*, published by the US Department of Health and Human Services and the regulations provided by the USDA found in 9 CFR Part 3 entitled "Animal Welfare Standards." The responsibility for directing such a program is usually given to a veterinarian or another qualified professional. At least one veterinarian must be associated with the program.

The Committee is responsible for evaluating the animal care and use program. The duties of the Committee must include:

- Meeting at regular intervals to ensure compliance with the *Guide* (the regular intervals can be no less than annually).

- Ensuring that a mechanism exists to review the humane care and use of animals in research, testing, and education.
- Providing a written report at least annually to the responsible administrative official on the status of the animal care and use program.
- Reviewing and/or investigating concerns involving the care and use of animals at the research facility resulting from public complaints received and from reports of noncompliance received from laboratory or research facility personnel.
- Reviewing, approving, and requiring modifications or withholding approval of those components of proposed or ongoing activities related to the care and use of animals.
- Other duties as assigned.

While the duties of the Committee outlined above fall into a very broad and general category, other duties and responsibilities have historically been assigned to this IACUC (IACUC, 2000). In many institutions the Committee assures that adequate veterinary care is provided. This care can either be provided by an on-site veterinarian or by a contract veterinarian who makes periodic visits to the facility. There must be a *written program* for animal use and care. While it is the responsibility of the institution to ensure that people caring for and using laboratory animals are qualified to do so, it often falls upon the Committee to make sure that these individuals are qualified and bring to the research are the proper qualifications and expertise to conduct the various phases of research that are being considered. While animal care and use programs require professional, technical and husbandry support, the Committee must assure that the institution or facility employs people trained in laboratory animal science and provides the opportunity for continuing education and properly supervised on-the-job training to ensure effective implementation of the program.

Internationally, animal welfare concerns are greater (Europe) or lesser than in the US, as reflected in their current laws (Howard-Jones, 1985; EEC, 1986; Cooper, 1990a,b; Netherlands Animal Welfare Society, 1991; AAALAC International, 1997b,c; Bayne and Martin, 1998; Miller, 1998; Nevalainen *et al.*, 1999).

In the UK, the equivalent regulations are called the Animal (Scientific Procedures) Act and are administered by the Home Office. Corresponding laws are in force in all major countries, as listed below:

1 Australia

- Australian Council, 1990 (ACCART, 1990)

2 Canada

- Guide to the Care and Use of Experimental Animals (Vol. 1, 2nd ed. 1993; Vol. 2, 1984) Rowsell, 1991

3 China

- Animal Protection Law

4 Germany

- German Animal Welfare Act (May 1998)
- German Animal Welfare Act (May 1998) [English translation]

5 Japan

- Law Concerning the Protection and Control of Animals (October 1973)

6 New Zealand

- Animal Welfare Act of 1999 (October 1999)
- Code of Recommendations and Minimum Standards for the Care and Use of Animals for Scientific Purposes (August 1995)
- Code of Recommendations and Minimum Standards for the Emergency Slaughter of Farm Livestock (December 1996)
- Code of Recommendations and Minimum Standards for the Sale of Companion Animals (April 1994)
- Code of Recommendations and Minimum Standards for the Welfare of Animals in Boarding Establishments (August 1993)
- Code of Recommendations and Minimum Standards for the Welfare of Animals at Stockyards (May 1995)
- Code of Recommendations and Minimum Standards for the Welfare of Animals at the Time of Slaughter and Licensed and Approved Premises (July 1996)
- Code of Recommendations and Minimum Standards for the Welfare of Bobby Calves (August 1993)
- Code of Recommendations and Minimum Standards for the Welfare of Deer During the Removal of Antlers (January 1992)
- Code of Recommendations and Minimum Standards for the Welfare of Dogs (May 1998)
- Code of Recommendations and Minimum Standards for the Welfare of Layer Hens (September 1996)
- Code of Recommendations and Minimum Standards for the Welfare of Livestock from which Blood is Harvested for Commercial and Research Purposes (April 1996)
- Code of Recommendations and Minimum Standards for the Welfare of Ostriches and Emus (August 1993)
- Code of Recommendations and Minimum Standards for the Welfare of Pigs (January 1994)
- Code of Recommendations and Minimum Standards for the Welfare of Sheep (July 1996)

7 Norway

- Animal Welfare Act (December 1974)

8 Sweden

- Swedish Animal Welfare Act (February 1998)

9 United Kingdom

- Animals (Scientific Procedures) Act of 1986 (Heath, 1986; Hollands, 1986)

## Occupational health and safety

An occupational health program is mandatory for personnel working in laboratory animal facilities. This program must include a physical examination and a medical and work history prior to work assignment. Periodic physical examinations are required for people in some job categories, especially those individuals who handle hazardous materials. In addition, an educational program must be provided to teach personnel about zoonoses, *personal* hygiene, and other considerations that may be of a prime safety concern to those individuals involved in the program, i.e., special precautions to be taken by pregnant women (Acha and Szyfrei, 1980). Other occupational hazards including animal bites and allergies to animal dander or hair must be considered.

Of special concern is the area of "B" virus exposure to those individuals handling primates. In some areas there may be a requirement for special clothing or breathing apparatus, which may be determined by the risk of exposure to the individual employee.

## Special areas of responsibility

Recently, two areas of consideration have caused the Committee a great deal of discomfort and disagreement in their deliberations. This issue of pain and distress caused by physical restraint and multiple or invasive surgical procedures has caused many members of the Committee to question the use of these procedures with specific concern for the overall welfare of the animals involved. Brief physical restraint of animals for administration of drugs, collection of samples, or other experimental procedures is certainly considered justifiable. The physical restraint can be accomplished manually or with simple restraint devices such as slings, racks or squeeze cages. Prolonged restraint (longer than 4 h) of any animal including the use of chairs for nonhuman primates, should be avoided unless absolutely essential to the research objectives. Prolonged restraint must be specifically spelled out within the research protocol, along with the justification for prolonged restraint in these specific restraint devices. One of the major concerns of the Committee is that these restraint devices should not be used simply as a convenience to the investigator in handling or managing animals. In many cases the use of such devices must be specifically approved by the Committee. Consideration must also be given to the possibility of lesions or illness associated with restraint, such as decubitus ulcers, dependent edema, and weight loss. If these problems develop, adequate veterinary care must be provided to the animals and in some cases may mandate a temporary or permanent removal of animals from such restraint devices.

Multiple or invasive surgical procedures on a single animal are generally discouraged. In any case, the multiple major surgical procedures must be approved by the Committee. It may be necessary for the researcher to develop alternate procedures that do not require the use of multiple surgical procedures. In no case can cost savings be a consideration or adequate reason for performing multiple surgical procedures.

## Pain and distress

While such areas as lighting, food, water, cage size, ventilation, environmental temperature, humidity, and exercise are mandated by regulation; the minimization of stress, discomfort, and pain are poorly defined or understood. The understanding of stress, discomfort, and pain in animals is based primarily on professional and aesthetic judgment made by the investigator(s) and members of the Committee. Without pre- and postsurgical medication, stress and pain will result. What criteria can be used to assure that pain and distress are minimized?

In most research units, pain is recognized by the action of the species under study. By understanding the "normal" posture, movement and attitude of the species, animal care technicians can recognize "abnormal" posture, movement or lack thereof; attitude (aggressive vs. submissive); grooming, eating or drinking habits; or vocalization. Any of these signs may suggest to the investigator or veterinarian that the animal is experiencing pain or discomfort. These observations are straightforward, and judgments concerning their meaning or significance are fairly easy to make, assuming pain in the experimental animal subject is perceived in the same manner as is pain in the human subject.

Behavioral changes as described above can be short-term, in which case the changes may be adaptive. However, long-term behavioral changes can lead to maladaptive mannerisms which are important signs of distress (not necessarily pain) in laboratory animals. These maladaptive behavioral traits indicate that medical intervention is required. Lack of grooming or excessive hair removal in rodents may suggest organic disease before other clinical symptoms are observed. Non-human primates are wasteful eaters and, as such, may become emaciated even though the animal appears to be consuming the appropriate diet.

The areas of stress and discomfort are much harder to quantify, and there is therefore a great need to understand and determine how best to alleviate the cause(s). It is often impossible to determine whether an animal is undergoing a "normal" process of adapting to a stressing agent (stressor). However, an aggressive attempt to alleviate the stress for this individual animal may be ill-advised and counter-productive. The research objective may be compromised to the extent that valid results are not obtained and may require the study to be repeated with more experimental animals involved.

There is a great deal of disagreement on the meaning of terms such as *comfort, well-being, discomfort, stress, fear, anxiety, pain,* and *distress*. The mechanisms that contribute to the functional and recognizable state of well-being vs. that of distress involve biochemical, physiological, and psychological changes which are poorly understood in animals. However, certain factors have been recognized to cause pain and stress and are outlined below. These factors generally apply to all mammalian species. These are summarized in Table 14.3.

Many of these causes of stress (distress) listed in Table 14.4 may become maladaptive responses and can become permanent parts of animals' activities and seriously threaten their overall well-being. Any behavior that relieves the intensity or duration of stress is likely to become habitual. Such behaviors include coprophagy, hair pulling, self-trauma, and repetitive movements (cage circling).

There is a requirement, mandated in 9 CFR, for the *opportunity to exercise* for dogs kept in individual cages. If these cages provide less than twice the space required for the size of the dog (size of dog is determined using formula found in 9 CFR, Part 3 of Subchapter A), these animals must be provided the opportunity to exercise. The exact exercise procedure is determined by the attending veterinarian and approved by the Committee.

*Table 14.4* Observable signs of severe pain or distress

| | |
|---|---|
| *Local effects*[a] | |
| Skin corrosion | Severe erosion or ulceration penetrating most or all of dermis, large areas of severe necrosis |
| Ocular injury | Severe corneal opacity, corneal ulceration, purulent or bloody discharge, severe periocular necrosis |
| Extremity injury | Severe swelling, ulcerative lesions, apparent fractures, severe cutaneous erosion/sloughing, gangrenous appearance |
| *Systemic effects*[a] | |
| Nervous signs | Severe or persistent tremors, convulsions, narcosis, catalepsy |
| Locomotor/muscular signs | Ataxia, paralysis, prostration |
| Respiratory signs | Gasping or labored breathing, very slow or rapid breathing, audible breathing (rales, wheezing) |
| Cardiovascular signs | Very slow or rapid heart rate, severe pallor, redness or cyanosis of extremities |
| Gastrointestinal signs | Persistent vomiting, severe diarrhea, anorexia |
| *Behavioral effects*[b,c] | |
| Excessive vocalization | Squealing, grunting, growling, whimpering, howling (especially during movement or handling) |
| Atypical actions | Self-mutilation, stereotypic activity, restlessness, head shaking, apparent apathy to stimulants |
| Direct response to pain | Licking, biting or scratching of affected area; unusual posture to relieve pressure on affected area (hunched or stretched appearance); excessive struggling or biting during handling; grimacing or baring teeth |
| Absence of actions | Failure to groom, decreased socialization, poor reflexes, marked decrease in feed/water intake |

[a] Myers and DePass (1993).
[b] Mroczek, (1992); USDA (1995).
[c] See also Baumans *et al.* (1994). NRC, 1992.

In those institutions or facilities housing or using primates, there is a mandated requirement to provide "environmental enrichment to promote psychological well-being" of the species involved. These institutions and facilities must have a written plan which addresses the following:

1 *Social grouping*. Based on the social needs of species as it exists in nature.
2 *Environmental enrichment*. Primary enclosures must be enriched to provide species-typical activities, i.e., perches, swings, and foraging or task-oriented feeding methods.

Interaction with the primary care individual is an important element of this enrichment program. The attending veterinarian may exempt an individual primate from participation in the enrichment program based on medical considerations.

The Committee may also exempt an individual from participation in the program based on scientific reasons set forth in the protocol. Regardless of the reason for exemption, the exemption must be reviewed at least every 30 days by the veterinarian and at least annually by the Committee.

### *Code of federal regulations, Title 9, Chapter I, Subchapter A – animal welfare*

Available from: USDA, APHIS/Animal Care, 4700 River Road, Unit 45, Riverdale, MD 20737-1234, USA.

The current version of the regulations developed by the USDA specify how to comply with the AWA and its amendments. The section is divided into four sections: Definitions, Regulations, Standards, and Rules of Practice Governing Proceedings Under the Animal Welfare Act. The Definitions section describes exactly what is meant by terms used in the legislation. "Animal," for example, specifically excludes rats of the genus *Rattus* and mice of the genus *Mus*, as well as birds used in research. The Regulations section includes subparts for licensing, registration, research facilities, attending veterinarians and adequate veterinary care, stolen animals, records, compliance with standards and holding periods, and miscellaneous topics such as confiscation and destruction of animals and access and inspection of records and property. The bulk of the subchapter is the third section which provides standards for specific species or groups of species. Included are sections for cats and dogs, guinea pigs and hamsters, rabbits, nonhuman primates, marine mammals, and the general category of "other warm-blooded animals". Standards include those for facilities and operations, health and husbandry systems, and transportation. The final section sets forth the Rules of Practice applicable to adjudicating administrative proceedings under Section 19 of the AWA. [Full text at APHIS/AC website].

### *Animal care policies*

The policy manual gives policies issued by APHIS/Animal Care that clarify the AWA regulations. Among the topics covered are "Written Narrative for Alternatives to Painful Procedures," "Space and Exercise Requirements for Traveling Exhibitors," and "Annual Report for Research Facilities." Originally issued in April 1997, new policies may be added at any time and included in the manual. [Full text and APHIS/AC website].

## ***Professional organizations***

There are several professional groups which provide special information to the institution and the Committee in areas of education, training, or certification. A brief description of some of these organizations follows.

### *AAALAC*

The AAALAC promotes high-quality animal care and use through a voluntary accreditation program. Institutions using, maintaining, or breeding laboratory animals for biological research may apply for accreditation. The NIH accept full accreditation by AAALAC as assurance that the animal facilities are in compliance with Public Health Service policy. AAALAC is accordingly an NGO (Non-Governmental Organization).

### *American Association for Laboratory Animal Science (AALAS)*

AALAS is made up of individuals and institutions professionally concerned with the production, care, and use of laboratory animals. The organization collects and exchanges information on all phases of laboratory animal care and management through the publication of a journal (*Laboratory Animal Science*), bulletins, and other documents. The AALAS animal technician certification is an important method to develop uniform requirements and training programs which lead to certification of technicians based on their experience and knowledge.

### *American Veterinary Medical Association (AVMA)*

AVMA is a major international organization of veterinarians. Its objective is to advance the science and art of veterinary medicine, including its relationship to public health and agriculture. It sponsors specialization in veterinary medicine through the formal recognition of specialty certifying organizations, including the American College of Laboratory Animal Medicine (ACLAM) and The American College of Veterinary Pathologists (ACVP).

### *ACLAM*

ACLAM is a specialty board founded in 1957 to encourage education, training, and research in laboratory animal medicine (ACLAM, 1996).

Several professional societies such as the Society of Toxicology (SOT), the ACVP and the Federation of American Societies for Experimental Biology (FASEB) have developed and published "position" papers on the use of animals in experimentation (i.e., *Guiding Principles in the Use of Animals in Toxicology*, adopted by the SOT in July 1991).

## Additional resources

American Association for Accreditation of Laboratory Animal Care, 11300 Rockville Pike, Suite 121, Rockville MD 20852-3035, USA

American Association of Laboratory Animal Science, 70 Timber Creek Drive, Suite 5, Cordova, TN 38018, USA

American College of Toxicology, 9650 Rockville Pike, Bethesda, MD 20814-3998, USA

American Society of Primatologists, Regional Primate Center, University of Washington, Seattle, WA 98195, USA

American Veterinary Medical Association, 1931 North Meachum Road, Suite100, Schaumburg, IL 60173-4360, USA

Animal and Plant Health Inspection Service, US Department of Agriculture, 4700 River Road, Unit 84, Riverdale, MD 20737-1234, USA

Animal Welfare Information Center, National Agricultural Library, US Department of Agriculture, 10301 Baltimore Boulevard, Room 205, Beltsville, MD 20705, USA

Australian and New Zealand Council for the Care of Animals in Research and Teaching, Limited, P.O. Box 19, Glen Osmond, SA 5064, Australia

Canadian Association for Laboratory Animal Science, CALAS National Office, Biosciences Animal Service, University of Alberta, Edmonton, Alberta T6G 2E9, Canada
Canadian Council on Animal Care, Constitution Square, 350 Albert Street, Suite 315, Ottawa, Ontario K1R 1B1, Canada
European Center for Validation of Alternative Methods, TP 580, JRC Environmental Institute, 21020 Ispra (VA), Italy
Foundation for Biomedical Research, 818 Connecticut Avenue NW, Suite 303, Washington, DC 20006, USA
Institute for Laboratory Animal Resources, National Academy of Sciences, 2101 Constitution Avenue NW, Washington, DC 20418, USA
Institute of Laboratory Animal Science
University of Zürich, Winterhurerstrasse 190, 8057 Zürich, Switzerland
International Council for Laboratory Animal Science, University of Kuopio, SF-70211 Kuopio 10, Finland
Public Responsibility in Medicine and Research, 132 Boylston Street, Fourth Floor, Boston, MA 02116, USA
Scientists Center for Animal Welfare, Golden Triangle Building One, 7833 Walker Drive, Suite 340, Greenbelt, MD 20770, USA
Society of Toxicology, 1767 Business Center Drive, Suite 302, Reston, VA 20190-5332, USA
The Johns Hopkins Center for Alternatives to Animal Testing, 111 Market Place, Suite 840, Baltimore, MD 21202-6709, USA
Universities Federation for Animal Welfare, 8 Hamilton Close, South Mimms, Potters Bar, Hertfordshire EN6 3QD, UK

## References


AAALAC International. US and European guidelines more alike than different: a side-by-side comparison. *AAALAC Int Connect* 1997a; August 11.

AAALAC International. Snapshots of animal research regulations across Europe. *AAALAC Int Connect* 1997b; August 11.

AAALAC International. US and European guidelines more alike than different: a side-by-side comparison. *Connection*; August 1997c.

AAALC International. *History Abstract, Accreditation Program, and Rules of Accreditation.* Rockville, MD: AAALAC International, 1997d.

ACCART. Australian code of practice for the care and use of animals for scientific purposes. *ACCART News* 1990;3(2):1–5.

Acha PM, Szyfrei B. *Zoonoses and Communicable Diseases Common to Man and Animals.* Scientific Publication No. 354. Washington, DC: World Health Organization, 1980.

ACLAM. *Report of the American College of Laboratory Animal on Adequate Veterinary Care in Research, Testing, and Teaching*,1996.

Acred P, *et al.* Guidelines for the welfare of animals in rodent protections tests: a report from the Rodent Protection Test Working Party. *Lab Anim* 1994;28:13-18.

APHIS. United States Department of Agriculture. (August 31). *Fed Reg* 1989;54(168): 36112–36163.

AVMA. *Report of the AVMA Panel on Euthanasia*, 1993.

Baumans V, *et al.* Pain and distress in laboratory rodents and lagomorphs: report of the Federation of European Laboratory Animal Science Association (FELASA) Working Group on Pain and Distress.

*Lab Anim* 1994;28:97–112,
Bayne K, Martin D. AAALAC International: using performance standards to evaluate an animal care and use program. *Lab Anim* 1998;27(4):32–35.
Bennett BT. Alternative methodologies. In: Bennett BT, Brown MJ, Schofield JC. *Essentials for Animal Research – A Primer for Research Personnel*, 2nd ed. Beltsville, MD: National Agricultural Library, 1994. pp. 9–17.
Cooper JE. Management, health and welfare of laboratory animals in developing countries. *SCAW Newslett* 1990a;13(1):14–15.
Cooper JE. Management, health and welfare of laboratory animals in developing countries (Part Two). *SCAW Newslett* 1990b;13(2):12–14.
DeWoskin RS. *Quality Assurance SOPs for GLP Compliance*. Buffalo Grove, IL: Interpharm Press, 1995.
DIA. *Computerized Data Systems for Non-Clinical Safety Assessments: Current Concepts and Quality Assurance*. Maple Glen, PA: DIA, September 1988.
Eascio J, Woodbridge G, Mitchell P. *ISO 14000 Guide: the New International Environmental Management Standards*, New York: McGraw-Hill, 1996
EEC. Council directive on the approximation of laws, regulations, and administrative provisions of the member states regarding the protection of animals used for experimental and other scientific purposes, 1986.
EEC. Council Directive of 18th December, harmonisation of laws, regulations and administrative visions relating to the application of the principals of good laboratory practice and the verification of their validity for tests on chemical substances. *Off J Eur Commun* 1987;LO15:29–30.
EEC. Council Directive of the 9th June, inspection and verification of good laboratory practices (GLP). *Off J Eur Commun* 1988;L145:35–37.
EPA. Federal Insecticide, Fungicide, and Rodenticide Act. (FIFRA). 7 U.S.C., 136 et. seq., 1970.
EPA. Toxic Substances Control Act (TSCA) 15. U.S.C., 160.1 et. seq., 1976a.
EPA. Proposed health effects test standards for Toxic Substances Control Act: Test rules and proposed good laboratory practice standards for health effects. *Fed Reg* 1976b:44:44054–44059, 44066–44067.
EPA. Federal Insecticide, Fungicide and Rodenticide Act (FIFRA) Title 40. Code of Federal Regulations Part 160. *Federal Register*, 29 November. Washington, DC: US Government Printing Office, 1983.
EPA. Toxic Substances Control Act (TSCA); good laboratory practice standards. *Fed Reg* 1989a;54:34034–34050. (40 CFR 792).
EPA. Federal Insecticide, Fungicide and Rodenticide Act (FIFRA); good laboratory practice standards. *Fed Reg* 1989b;54:34067–34074. (40 CFR 160).
EPA. FIFRA good laboratory practice standards, final rule. *Fed Reg* 1989c;54:34052–34074.
EPA. Toxic Substances Control Act (TOSCA) Title 40. Code of Federal Regulations Part 160. *Federal Register*, 17 August. Washington, DC: US Government Printing Office, 1989d.
EPA. *Good Automated Laboratory Practices*. Research Triangle Park, NC: Office of Informational Resources, August 21, 1995.
FDA. Title 21. Code of Federal Regulations Part 58. *Federal Register*, 22 December. Washington, DC: US Government Printing Office, 1978.
FDA. Good laboratory practice regulations, final rule. *Fed Reg* 1987a;52:33768–33782.
FDA. Department of Health and Human Services. *Fed Reg* 1987b;2(172):33768.
FDA. *Computerised data systems for nonclinical safety assessment. Current concepts and quality assurance*. Washington, DC: US Food and Drug Administration, 1988.
FDA. Guide for detecting fraud in bioresearch monitoring inspections. Office of Regulatory Affairs, FDA, April 1993.
FDA. Electronic signatures; electronic records; proposed rule. *Fed Reg* 1994;59:13200.
FDA. Electronic Records: Electronic Signatures; Final Rule, FDA 21. Code of Federal Regulations Part 2. *Federal Register*, 20 March. Washington, DC: US Government Printing Office, 1997.

French RD. *Antivivisection and Medical Science in Victorian Society*. Princeton, NJ: Princeton University Press, 1975.

Freudinger U. Public initiative for the abolition of vivisection, called the "Weber Initiative." (English translation). *Schweiz Arch Tierheilkd* 1985;127:635–649. (In German with English summary.)

Gad SC and Taulbee SM. *Handbook of Data Recording, Maintenance and Management of the Biomedical Sciences*. Boca Raton, FL: CRC Press, 1996.

Hampson J. Animal welfare – a century of conflict. *New Sci* 1979;84:280–282.

Hearn JP. Introduction. *ILAR J* 1995;37(2):55–56.

Heath M. British law relating to experimental animals – its provisions and restriction. *Anim Technol* 1986;37:131–136.

Hollands C. The animals (Scientific Procedures Act, 1986). *Lancet* 1986;2:32–33.

Howard-Jones N. A CIOMS ethical code for animal experimentation. *WHO Chron* 1985;39:51–56.

IACUC. Institutional Animal Care and Use Committee Guidebook, NIH Pub. No. 92-3415. US Department of Health and Human Services, 2000. pp. E33–E37.

LAWA. Laboratory Animal Welfare Act (1996). United States P.L. 89-544, amended by the Animal Welfare Act of 1970 (P.L. 91-597), 1976 (P.L. 94-279), 1985 (P.L. 99-198) and 1990 (P.L. 101-624).

MAFF. 59 Nohsan, Notification No. 3850, Agricultural Production Bureau, 10 August. Japan: Ministry of Agriculture, Forestry and Fisheries, 1984.

Myers, RC and DePass LR. Acute toxicity testing by the dermal route. In: Wang RGM, Knaak JB and Maibach HI, editors. *Health Risk Assessment – Dermal and Inhilation Exposure and Absorption of Toxicants*. CRC Press, Boca Ratan, FL, 1993, pp. 167–199.

MHW. Notification No. Yakahatsu 313. Pharmaceutical Affairs Bureau, 31 March 1982 and subsequent amendment. Notification No. Yakahatsu 870, Pharmaceutical Affairs Bureau, 5 October 1988 and subsequently revised. Ministry of Health and Welfare Ordinance No. 21, March 1987.

Miller J. International harmonization of animal care and use: the proof is in the practice. *Lab Anim* 1998;27(5):28–31.

MITI. Directive, (Kanpogyo No. 39 Environmental Agency, Kikyoku No. 85), Japan: Ministry of International Trade and Industry, 31 March 1984.

Mroczek, NG (1992) Recognizing animal suffering and pain. *Lab Animal*, 21:27–31.

Netherlands Animal Welfare Society. Dutch animal welfare office makes progress. *ATLA News Views* 1991;19:389.

Nevalainen T *et al.* FELASA guidelines for education of specialists in laboratory animal science (Category D). *Lab Anim* 1999;33:1–15.

NIH. *NIH Almanac*. NIH Publication NO. 92-5. Bethesda, MD: National Institutes of Heath, 1992.

NIH. Department of Health and Human Services, Public Health Service, National Institutes of Health. *Guide for the Care and Use of Laboratory Animals.* NIH Publication no. 85-23. Bethesda, MD: National Institutes of Heath, 1989.

NRC. *Recognition and Alleviation of Pain and Distress in Laboratory Animals*. Washington, DC: National Academy Press, 1992.

NRC. *Guide for the Care and Use of Laboratory Animals.* Washington, DC: National Academy Press, 1996.

OECD. *OECD Principles of Good Laboratory Practice*. OECD Document C(81)30, Annex 2, 1981.

OECD. *Good Laboratory Practice in the Testing of Chemicals*. Paris: Organization for Economic Cooperation and Development, 1982.

Public Health Service, Department of Health and Human Services. (November 20) Health Research Extension Act. *Public Law* 1985;99–158.

Rowsell HC. The Canadian Council on Animal Care – its guidelines and policy directives: the veterinarian's responsibility. *Can J. Vet Res* 1991;55:205.

Russell WMS, Burch RL. *The Principles of Humane Experimental Technique*. London: Methuen, 1959.

Silverman J. IACUC handling of mistreatment or non-compliance. *Lab Anim* 1994;23:30–32.

Taulbee SM, DeWoskin RS. *Taulbee's Pocket Companion: USFDA and EPA GLPs in Parallel*. Buffalo Grove, IL: Interpharm Press, 1993.

UK GLP Monitoring Authority. *The Application of GLP Principles to Field Studies*. Advisory Leaflet. London: UK GLP Monitoring Authority, 1990,

UK GLP Monitoring Authority. *Good Laboratory Practice and the Role of Quality Assurance*. Advisory Leaflet. London: UK GLP Monitoring Authority, 1991.

UK GLP Monitoring Authority. *Good Laboratory Practice and the Role of the Study Director*. Advisory Leaflet. London: UK GLP Monitoring Authority, 1992.

UK GLP Monitoring Authority. *The Application of GLP Principles to Computer Systems*. Advisory Leaflet. London: UK GLP Monitoring Authority, 1995.

UK GLP Monitoring Authority. *Good Laboratory Practice Regulations 1997*. Statutory Instrument No. 654. London: HM Stationary Office, 1997.

USDA. *Subchapter A – Animal Welfare*. 9 CFR Chapters 1, 1.1–4.11. Washington, DC: United States Department of Agriculture, 1995.

Appendix A

# Common regulatory and toxicological acronyms

| | |
|---|---|
| 510(k) | Premarket notification for change in a device |
| AALAS | American Association Laboratory Animal Science |
| AAMI | Association for the Advancement of Medical Instrumentation |
| ABT | American Board of Toxicology |
| ACGIH | American Conference of Governmental Industrial Hygienists |
| ACT | American College of Toxicology |
| ADI | Allowable Daily Intake |
| AIDS | Acquired Immune Deficiency Syndrome |
| AIMD | Active Implantable Medical Device |
| ANSI | American National Standards Institute |
| APHIS | Animal and Plant Health Inspection Service |
| ASTM | American Society for Testing and Materials |
| CAS | Chemical Abstract Service |
| CBER | Center for Biologic Evaluation and Research (FDA) |
| CDER | Center for Drug Evaluation and Research (FDA) |
| CDRH | Center for Devices and Radiological Health (FDA) |
| CFR | Code of Federal Regulations |
| CFAN | Center for Food and Distribution (FDA) |
| CIIT | Chemical Industries Institute of Toxicology |
| CPMP | Committee on Proprietary Medicinal Products (UK) |
| CPSC | Consumer Product Safety Commission |
| CRF | Code of Federal Regulations |
| CSE | Control Standard Endotoxin |
| CSM | Committee on Safety of Medicines (UK) |
| CTC | Clinical Trial Certificate (UK) |
| CTX | Clinical Trial Certificate Exemption (UK) |
| CVM | Center for Veterinary Medicine (FDA) |
| DART | Development and Reproduction Toxicology |
| DHHS | Department of Health and Human Services |
| DIA | Drug Information Associates |
| DMF | Device (or Drug) Master File |
| DOE | Department of Energy |
| DOT | Department of Transportation |
| DSHEA | Dietary Supplement Health and Education Act |
| EEC | European Economic Community |

EM Electron Microscopy
EPA Environmental Protection Agency
EU European Union
FCA Freund's Complete Adjuvant
FDA Food and Drug Administration
FDC Food Drug and Cosmetic Act
FDCA Food, Drug and Cosmetic Act
FDLI Food and Drug Law Institute
FHSA Federal Hazardous Substances Act
FIFRA Federal Insecticides, Fungicides and Rodenticides Act
GCP Good Clinical Practices
GLP Good Laboratory Practices
GMP Good Manufacturing Practices
GPMT Guinea Pig Maximization Test
HEW Department of Health, Education and Welfare (no longer existent)
HIMA Health Industry Manufacturer's Association
HSDB Hazardous Substances Data Bank
IARC International Agency for Research on Cancer
ICH International Conference on Harmonization
id Intradermal
IDE Investigational Device Exemption
IND(A) Investigational New Drug Application
ip Intraperitoneal
IRAG Interagency Regulatory Alternatives Group
IRB Institutional Review Board
IRLG Interagency Regulatory Liaison Group
ISO International Standards Organization
IUD Intrauterine Device
iv Intravenous
JECFA Joint Expert Committee for Food Additives
JMAFF Japanese Ministry of Agriculture, Forestry, and Fishery
LA Licensing Authority (UK)
LAL *Limulus* amebocyte lysate
$LD_{50}$ Lethal Dose 50: the dose calculated to kill 50 per cent of a subject population, median lethal dose
LOEL Lowest Observed Effect Level
MAA Marketing Authorization Application (EEC)
MD Medical Device
MHW Ministry of Health & Welfare (Japan)
MID Maximum Implantable Dose
MOE Margin of Exposure
MRL Maximum Residue Limits
MSDS Material Safety Data Sheet
MTD Maximum Tolerated Dose
NAS National Academy of Science
NCTR National Center for Toxicological Research
NDA New Drug Application

| | |
|---|---|
| NIH | National Institutes of Health |
| NIOSH | National Institute Occupational Safety and Health |
| NK | Natural Killer |
| NLM | National Library of Medicine |
| NOEL | No-Observable-Effect Level |
| NTP | National Toxicology Program |
| ODE | Office of Device Evaluation |
| OECD | Organization for Economic Cooperation and Development |
| PDI | Primary Dermal Irritancy |
| PDN | Product Development Notification |
| PEL | Permissible Exposure Limit |
| PhRMA | Pharmaceutical Research and Manufacturers Association |
| PL | Produce License (UK) |
| PLA | Produce License Application |
| PMA | Premarket Approval Applications |
| PMN | Premanufacturing Notice |
| po | Per os (orally) |
| QAU | Quality Assurance Unit |
| RAC | Recombinant DNA Advisory Committee |
| RCRA | Resources Conservation and Recovery Act |
| RTECS | Registry of Toxic Effects of Chemical Substances |
| SARA | Superfund/Amendments and Reauthorization Act |
| sc | Subcutaneous |
| SCE | Sister Chromatic Exchange |
| SNUR | Significant New Use Regulations |
| SOP | Standard Operating Procedure |
| SOT | Society of Toxicology |
| STEL | Short Term Exposure Limit |
| TLV | Threshold Limit Value |
| TSCA | Toxic Substances Control Act |
| USAN | United States Adopted Name Council |
| USDA | United States Department of Agriculture |
| USEPA | United States Environmental Protection Agency |
| USP | United States Pharmacopoeia |
| WHO | World Health Organization |

Appendix B

# Notable regulatory Internet addresses

| *Organization or publication* | *Web address (URL)* | *Sample main topics* |
|---|---|---|
| Agency for Toxic Substances and Disease Registry | www.atsdr.cdc.gov | |
| Australian Therapeutic Goods Administration | http://www.health.gov.au/tga | Medical devices, GMP codes, parliamentary secretary's; working, status document, party on complementary medicines, medical releases, publications, site map, related sites |
| Canadian Health Protection Branch | http://www.hc-sc.gc.ca/hpb | Medical devices, chemical hazards, food, product safety, Science Advisory Board, diseases, radiation protection, drugs, HPB transition policy, planning and coordination |
| ChemInfo | www.indiana.edu/~cheminfo/ca_csti.html | SirCH: chemical safety or toxicology information |
| Code of Federal Regulations | http://www.access.gpo.gov/ nara/cfr/cfr-table-search.html | NARA code sections |
| Cornell Legal Library | http://www.law.cornell.edu | Code of federal regulations, supreme court decisions, US code, circuit courts of appeal |
| EPA | www.epa.gov | |
| European Agency for the Evaluation of Medicinal Products | http://www.eudra.org/en_home.htm | What's new, documents forum, other sites |
| European sites | http://www.eucomed.be/ eucomed/links/links.htm | European institutions, related sites |
| Food and Drug Administration (FDA) | www.fda.gov | Foods, human drugs, biologics, animal drugs, cosmetics, medical devices/radiological health |
| FDA – CDRH | www.fda.gov/cdrh/index.html | Home page |
| Search site | www.fda.gov/cdrh/search.html | Search CDRH site |
| Comment | www.fda.gov/cdrh/comment4.html | Comment on CDRH site |
| Device advice | www.fda.gov/cdrh/devadvice/32.html | |
| PDF reader | www.fda.gov/cdrh/acrobat.html | |

| *Organization or publication* | *Web address (URL)* | *Sample main topics* |
|---|---|---|
| General Principles of Software Validation (draft) Guidance for Industry | www.fda.gov/cdrh/comp/swareval.html | Text |
| Division of Small Manufacturers Assistance | www.fda.gov/cdrh/dsma/dsmamain.html<br>www.fda.gov/cdrh/dsma/cgmphome.htm | |
| | www.fda.gov/cdrh/humfac/frqar.html<br>www.fda.gov/cdrh/fr1007ap.pdf<br>www.fda.gov/cdrh/gmpasci.zip | Good Manufacturing Practice (GMP) (also known as the quality system regulation) final rule text as published in the Federal Register |
| | www.fda.gov/cdrh/comp/designgd.pdf | PDF version of design control report and guidance |
| Design Control Guidance for Medical Device Manufacturers | www.fda.gov/cdrh/comp/designgd.html | Text |
| Design Control Guidance for Medical Device Manufacturers | www.fda.gov/cdrh/comp/designgd.pdf | PDF version |
| Design Control Inspection Results | www.fda.gov/cdrh/dsma/dcisresults.html | First year roll out final design control inspection results and presentation text |
| Do It By Design: An Introduction to Human Factors in Medical Devices | www.fda.gov/cdrh/humfac/doit.html | By Dick Sawyer, Office Of Health and Industry Programs |
| | www.fda.gov/cdrh/humfac/doitpdf.pdf | PDF version of Do It By Design: An Introduction to Human Factors in Medical Devices by Dick Sawyer |
| Human Factors Implications of the New GMP Rule | www.fda.gov/cdrh/humfac/hufacimp.html | Text |
| Medical Device Quality Systems Manual: A Small Entity Compliance Guide | www.fda.gov/cdrh/dsmzza/gmp_man.html | Text |
| (Document withdrawn) | www.fda.gov/cdrh/comp/ghtfproc.html | Draft Global Harmonization Task Force (GHTF) process validation guidance text |
| | www.fda.gov/cdrh/comp/ghtfproc.pdf | PDF version |
| Guidance on Information Disclosure by Manufacturers to Assemblers for Diagnostic X-ray Systems | www.fda.gov/cdrh/comp/2619.html | |
| | www.fda.gov/cdrh/comp/2619.pdf | PDF version |

| *Organization or publication* | *Web address (URL)* | *Sample main topics* |
|---|---|---|
| Guidance on Quality System Regulation Information for Various PreMarket Submissions text | www.fda.gov/cdrh/comp/qsrpma.html | |
| CDRH Letter to Manufacturers | www.fda.gov/cdrh/yr2000/cdrh/letters/980921/y2kcompltr.html | Letter to Manufacturers – Y2K issue for production processes and quality system software text |
| CDRH Letter to Manufacturers | www.fda.gov/cdrh/yr2000/cdrh/letters/980921/y2kcompltr.pdf | PDF version of Letter to Manufacturers |
| GMP/QS Workshops with CDRH Participation | www.fda.gov/cdrh/dsma/workshop.html | |
| FDA/Industry Exchange Workshops on Medical Device Quality Systems Inspection Techniques | www.fda.gov/cdrh/meetings/qsitmeet.html | |
| Quality System Inspections Reengineering | www.fda.gov/cdrh//gmp/gmp.html | |
| FDA – Field Operations | www.fda.gov/ora/ | What's new, import program, inspectional, science and compliance references, federal/state relations |
| Design Controls | www.fda.gov/ora/inspect_ref/qsreq/dcrpgd.html<br>www.fda.gov/ora/inspect_ref/igs/elec_med_dev/emcl.html | Design control report and guidance text<br>Guide to inspections of electromagnetic compatibility aspects of medical device quality systems text |
| Guide to Inspections of Quality Systems | www.fda.gov/ora/inspect_ref/igs/qsit/qsitguide.htm | QSIT inspection handbook text |
| Guide to Inspections of Quality Systems | www.fda.gov/ora/inspect_ref/igs/qsit/QSITGUIDE.PDF | PDF version of QSIT inspection handbook text |
| Food and Drug Law Institute | http://www.fdli.org | Special interest, publications, multimedia, order products, academic programs, directory of lawyers and consultants, contact us |
| Health Industry and Manufacturers Association (HIMA) | http://www.himanet.com | About HIMA, newsletter, HIMA calendar, industry resources, business opportunities, FDA/EPA/OSHA, reimbursement/payment, global year 2000, government relations, public relations, small company, diagnostics |

| *Organization or publication* | *Web address (URL)* | *Sample main topics* |
|---|---|---|
| International Regulatory Monitor (Monitor) | http://www.go-nsi.com/pubs | Editorial portion of newsletter |
| Internet Grateful Med | www.igm.nlm.nih.gov | |
| Japanese Ministry of Health and Welfare | http://www.mhw.go.jp/english/index.html | Organization, Y2K problem, statistics, white paper, related sites |
| Library of Congress | http://thomas.loc.gov | Searchable database of federal legislation, congressional record and committee information |
| Medical Device Link | http://www.devicelink. com | News, consultants, bookstore, links, discussion, magazines (MDDI, MPMN, IVD technology) |
| National Archives and Public Records Administration | http://www.access.gpo.gov/su_docs/aces/aces140.html | Code of federal regulations, federal register, laws, US Congress information |
| National Library Network | www.toxnet.nlm.nih.gov | TOXNET: toxicology data network, a cluster of databases on toxicology, hazardous chemicals, and related areas |
| National Toxicology Program | http://ntp-server.niehs.nih.gov/ | |
| New Quality System (QS) Regulation | www.fda.gov/bbs/topics/ANSWERS/ANS00763.html | FDA talk paper announcing the GMP final rule text |
| Regulatory Affairs Professionals Society (RAPS) | http://www.raps.org | Certificates, resource center, publications, chapters, related links, contacting RAPS |
| US Department of Commerce | http://204.193.246.62 | Bureau of export administration, international trade association, patent and trademark, national institute of standards and technology |
| University of Pittsburgh | www.pitt.edu | |
| World Health Organization | http://www.who.int | Governance, health topics, information sources, reports, director-general, about who, international digest of health, legislation (http://www.who.int/pub/dig.html) |

# Index